Technology of Paper Recycling

Technology of Paper Recycling

Edited by

R.W.J. McKINNEY
Director
Fibre Research Consultants Ltd
Bookham
Surrey

SPRINGER-SCIENCE+BUSINESS MEDIA, B.V.

First edition 1995
Reprinted 1997

Originally published by Chapman & Hall in 1995
Softcover reprint of the hardcover 1st edition 1995

Typeset in 10/12 pt Times by Colset Private Ltd, Singapore

ISBN 978-94-010-4578-0   ISBN 978-94-011-1328-1 (eBook)
DOI 10.1007/978-94-011-1328-1

A Catalogue record for this book is available from the British Library

Library of Congress Cataloging-in-Publication Data available

Preface

Paper-making has a long tradition of recycling wastes as raw materials. In the Middle Ages, as paper-making was being established across Europe, paper was made from discarded materials - worn out clothes, old sails and hemp ropes. One of the earliest accounts of paper-making in the West was given by Emir Mu'izzi Badis (1007–1061) when he described the production of white paper from hemp rope. As the demand for paper products increased, the limited supply of suitable raw materials created problems, so that in England in the 1860s one of the first pleas for source separation was made when it was suggested that:

Every housekeeper ought to have three bags; a white one for the white rags, a green one for the coloured, and a black one for the waste paper.

It is only recently that the disposal of waste materials has become a matter of wide public debate, and their reuse through recycling has subsequently been enforced through legislation and other indirect measures. The debate has resulted in a high level of interest in paper recycling, but some of the complexities involved in the technology are not adequately covered in existing publications.

Other reference books that describe the technology of paper recycling are not based on the end use of the fibre being produced and it was to fill this gap that this text was conceived. However, it was necessary to place different types of recycled fibre in the overall contexts of wastepaper recovery and processing. The intent was to produce a general review to complement existing texts, and to include information about wastepaper recovery and techniques used in various geographic areas as well as on deinking different printing inks.

This text will be of benefit to newcomers to paper and board recycling, whether attending university, at work in a mill or involved in the supply of chemicals, equipment, services or wastepaper to paper-manufacturing units. Those who have some experience in wastepaper recycling should also find that the width and depth of coverage of some areas allows new insights to be gained. A comprehensive index is provided to enable readers to find topics in which they are especially interested.

Inevitably, some of the views expressed in the book will have to be changed in the light of new information generated by the increased level of R&D within the industry which has followed the considerable expansion of wastepaper use. Existing errors caused by omission or commission are my sole responsibility.

Acknowledgements

Authors who have provided chapters are experts in their fields, and they have had to devote considerable time out of a busy schedule in writing their chapter. My sincere thanks to them.

Finally, I am grateful to many friends and acquaintances within the paper industry who have made my contribution to this book possible by sharing their knowledge and experiences with me.

R.W.J. McKinney

Contributors

R.C. Howard	Fibre & Paper Physics Section, Pulp & Paper Research Institute of Canada, 570 St. John's Boulevard, Pointe Claire, Quebec, Canada, H9R 3J9
J.K. Huston	Weyerhauser Paper Company, Technology Center, Tacoma, Washington 98477, USA
J.B. Morrison	Miami Mill, Cross Pointe Paper Corporation, PO Box 66, West Carrollton, OH 45449, USA
R.W.J. McKinney	Fibre Research Consultants Ltd, 19 Woodlands Road, Bookham, Surrey KT23 4HG, UK
R. Oye	Oye Paper Tech. Consult. Co., 2–12–16 Setagaya, Setagaya-ku, Tokyo 154, Japan
A. Renders	Solvay Interox, rue de Rensbeek 310, B-1120 Bruxelles, Belgium
B. Thoyer	IFRA, Washingtonplatz, D-64287 Darmstadt, Germany
R.W. Turvey	Eka Nobel, 304 Parkway Worle, Worle, Weston-super-Mare BS22 0WA, UK

Contents

1 Wastepaper recovery and collection

R.W.J. McKINNEY

1.1 Introduction

Paper has a long history of use, with its invention credited to Tsai Lun, in China, in 105. Since then, it has been produced from a wide variety of raw materials, including cotton, linen, bark, hemp, jute, straw and wood. For many centuries, the most important raw material was rags, so that there is a long tradition of paper production from secondary raw materials, such as rags and old paper. Many publications, which today would be considered priceless, have been lost because of recycling. The use of virgin resources, based on wood, in this historical context, is thus relatively new, with no more than a century since its widespread adoption. Its use came about due to the explosion in demand for paper, as education became widespread in Europe and North America, with the resultant expansion in demand for printed products, which was also made possible by the developing printing industry. As literacy increased, so did the demand for printed products. Secondary raw materials could not supply sufficient fibres to meet this demand for printed products, such as newspapers, books, journals, etc. The introduction of wood as a raw material for paper allowed these demands to be met in the industrial world. This was followed by the use of paper in packaging since its use had many advantages over more traditional packaging materials, including wood.

During the First and Second World Wars, paper was a strategic product; the transmittal of information was essential and the most useful medium was paper. In many countries, wastepaper was collected and reused, following the historical tradition of recycling. However, paper quality was poor. After the Second World War ended, the pulp industry developed, in order to improve pulping and to meet the continuing expansion in demand. The superior quality which resulted from virgin based products meant that the high recycling rates achieved during the war years fell sharply. However, in countries with limited wood resources, the only way a domestic industry could develop was to import basic virgin raw materials and to recycle the products produced from these after they had been used. Hence, wastepaper became an important raw material in Western Europe and Japan and recovery and collection systems developed to meet this demand.

As the paper industry was developing, fundamental changes were taking place in society. Life styles changed, especially as the woman's traditional role in the home altered to reflect increased employment opportunities outside the home, which limited time available for food preparation, cooking, etc. Semi-processed foods became important, fuelling the retail revolution, thus leading to increased demands for protective packaging during transport.

As the urban sprawl developed, waste disposal became an urgent issue, which was satisfied by the dumping of wastes into areas which had provided the raw materials to build the urban sprawl, such as clay, sand and gravel pits, quarries, etc. As the consumer society matured, dissatisfaction with a materialistic life style grew and inevitably attention centred on resource consumption, with paper products being the groups that were specifically criticised. Although recycling was practised for some board grades, a higher recycling rate was promoted as a desirable environmental policy in the 1970s [1]. Criticism of consumerism increased through the 1970s and the 1980s, with the strong condemnation of the paper industry from environmental pressure groups gaining widespread media coverage from the mid 1980s onwards. As information became available on the quantity of paper in solid wastes more emphasis was given to recycling, to help reduce these volumes. There are many examples of resource conservation attempted through legislation in the late 1980s and early 1990s [2–4], and so the motivating force for the use of wastepaper ceased to be economic, but instead has become an illustration (some would say illusion) of a socially progressive society. Increased demand for wastepapers has had profound effects on many parts of the paper cycle including methods of wastepaper collection, the structure of the wastepaper industry, and wastepaper use by paper and board industries - especially in countries which were, and still are, virgin fibre rich. In the USA, for example, wastepaper recovery was less than 27% of consumption in 1985, but had increased to more than 38% in 1992, an increase of almost 12 million tonnes [5].

1.2 Wastepaper recovery and reuse - the global picture

In 1992, the total world apparent consumption of paper and board was 245.6 million tonnes, although pulp production was only 164 million tonnes, or 67% of the total production [6]. Much of the other third of the raw material was provided by wastepaper; in 1992, the total world consumption was about 96 million tonnes. Pulp and wastepaper inputs do not add up to total paper and board production, due to losses during production, from both pulp and wastepaper, although the losses from wastepaper are much higher. Table 1.1 illustrates how wastepaper use has increased over the period from 1986 to 1992.

Table 1.1 Global wastepaper use [7]

Year	Pulp and paper production (m tonnes)	Wastepaper consumption (m tonnes)	Apparent utilisation rate (%)
1986	202	63	31
1990	237	85	36
1991	239	91	38
1992	246	96	39
2000[a]	307	138	45

[a] See reference [8]

1.3 Wastepaper grading systems

Wood and non-wood fibres can provide a wide spectrum of properties, depending on the fibre type and the production process, giving, for example:

- the protective strength of packaging
- contrast provided by printed papers
- the absorbency of tissue papers

Fibre selection and blending enables the specific properties required from an individual product to be met.

When papers and boards are made or converted, different fibre types from several processes can be combined, e.g. the production of corrugated cases uses strong, unbleached softwood kraft fibres for liner, possibly with a layer of bleached fibres, combined with relatively weak NSSC hardwood (or recycled) fibres as fluting. It is not possible to separate these during recycling of old corrugated cases. Mixing of fibre types occurs in many other products. When wastepapers are a mixture of many fibre types, in different paper or board products, they have a low value, since mixed papers will not normally provide the properties required for a specific end use, such as the necessary strength requirements in packaging, contrast in printing papers or absorbency in tissue. Hence mixtures of fibre types can only be used to produce low grades of paper or board. When wastepapers are sorted into grades with a limited mixture of fibre types, they have a higher value than mixed papers. The purpose of wastepaper grading systems is to provide a framework for sorting, so that grades which are defined within the system represent acceptable levels of non-homogeneity of fibre types, as well as contaminant types and proportions. This allows recycling to be accomplished, even in products in which specific properties are desirable.

There are many different wastepaper grading systems in use, normally restricted to specific geographic regions, for example, CEPAC [9], UK [10] and German [11] definitions in Europe, Paper Stock of America in

Table 1.2 UK standard wastepaper groups

Group 1 (white woodfree unprinted)
Best white shavings no. 1
Best white shavings no. 2
Cream shavings
White and cream envelope cuttings
White coated
White printers' shavings
White soft tissue
Group 2 (white woodfree printed)
Best one-cuts
Black and white best PAMS
Book quite
Sulphate waste
Tear white shavings
White carbonless copy paper (NCR)
White continuous business forms
White heavy letter
White listings no. 1 (CPO)
White listings no. 2 (CPO)
Group 3 (white and lightly printed mechanical)
Lightly printed scanboard
White mechanical coated
White mechanical listings (CPO)
White scanboard (duplex)
White unprinted news
Woody one-cuts
Group 4 (coloured woodfree)
Coloured best PAMS
Coloured carbonless copy paper (NCR)
Coloured continuous business forms
Coloured heavy letter
Coloured mill broke
Coloured shavings
Coloured tissue
Multigrade
Sulphite bag waste
White and light toned shavings
Group 5 (heavily printed mechanical)
Crushed news
Green mechanical listings
Heavily printed scanboard
Mechanical book quire
News and PAMS
Over-issue news
Over-issue PAMS
Telephone quire
Group 6 (coloured krafts and manilas)
Buff and coloured tab cards
Buff envelope cuttings
Dark and light coloured manilas
Kraft liner
Multiply kraft sacks
New brown krafts
Old brown and coloured krafts

Table 1.2 *(cont)*

Group 7 (new KLS)
Double lined kraft (DLK)
New KLS cuttings
Group 8 (container waste)
Container waste (old KLS)
Group 9 (mixed papers)
Mixed papers
Group 10 (coloured card)
Coloured card
Group 11 (contaminated grades)
Beer mats and beer mat board (wet strength)
Foil laminated boards
Label waste (woodfree mixed papers)
Mechanical bookbinders cuttings (latex tips or bindings)
Photographic papers
Plastic granule/poly-lined kraft sacks
Polycoated carton waste
Polythene sprayed kraft sacks
Silicone coated papers
Telephone directories (soft covers)
Wet strength kraft
Wet strength papers (white or coloured)
Wet strength wallpaper base and printed wallpaper
White wax laminated cup waste
Woodfree bookbinders' cuttings (latex tips or bindings)

North America, etc. These definitions have developed over an extended period of time and are generally quite different. Periodic revisions are made, for example, to incorporate new grades, such as the Paper Stock of America definition of mixed office papers in 1993 [12]. All grading systems provide basic definitions, which are usually refined and detailed by purchasers of specific grades. The UK standard wastepaper groups are listed in Table 1.2, German in Table 1.3 and CEPAC in Table 1.4, while the Japanese groups are described in chapter 2. Various grades from the Paper Stock of America system are given in chapter 9.

Each one of the UK groups contains a number of different grades. Considerable emphasis is given in this system to the fibre types in the paper, so that separation of woodfree and woodcontaining paper grades is encouraged. There is much less emphasis on this separation in the German and the CEPAC systems. As a consequence, there is much less sorting of wastepapers according to fibre types in Germany, which is restricted to high-quality grades only. This has restricted the availability of woodfree grades in Germany. Although Germany is one of the largest importers of bleached chemical pulp, the low availability of woodfree wastepapers has inhibited the development of recycling to replace bleached chemical pulp imports. This illustrates the importance of wastepaper grading systems,

Table 1.3 Wastepaper classification in Germany

Grade	Description
Group I: lower grades	
A00	Original mixed wastepaper including domestic wastepaper with no guarantee against foreign bodies or grades of paper which may be detrimental to production
B10	Sorted mixed wastepaper, a mixture of different paper and board grades of which a maximum of 1% contains foreign bodies or grades which may be detrimental to production
B12	Sorted mixed wastepaper, a mixture of different paper and board grades containing less than 40% old news and PAMS, and of which a maximum of 1% contains foreign bodies or grades which may be detrimental to production
B19	Wastepaper from retail outlets, used packaging and wrapping, but containing at least 70% corrugated material, solid board and packaging papers, and of which a maximum of 11% contains foreign bodies or grades which may be detrimental to production
B42	Grey and mixed board grades including imitation leatherboard, but excluding strawboard
C02	Sorted mixed printers' and publishers' wastepaper
D11	Heavily printed materials, brochures, periodicals, subscription magazines, paperbacks, magazines, directories and catalogues
D21	Periodicals and magazines, etc., *with* pins and staples
D29	Periodicals and magazines, etc., with pins and staples but without glue bonding
D31	Old news and PAMS, at least 60% newspaper
D39	Old news and PAMS, at least 60% newspaper and without glue bonding
Group II: medium grades	
E12	Sorted, crushed and over-issue newspapers
F12	Continuous stationery sorted by coloured, mechanical
G12	Carbonless copy paper
H12	White cartonboard excluding B42 grey and mixed grades but *with* pins and staples
H22	Coated cartonboard waste from liquid packaging board manufacturers
J11	Shredded coloured files
J19	Sorted coloured files, without file separators and carbon paper
Group III: upper grades	
K02	Printed papers containing woodfree, coated and non-waterproof wastepaper grades, but free of dyed paper
K12	White files, mixed woodfree and mechanical, without receipt pads and ticket books
K22	Sorted white woodfree files
K51	White woodfree continuous stationery with a maximum of 3% carbonless copy paper and carbon paper
K59	White woodfree continuous stationery, free of carbonless copy paper and carbon paper
L11	Unsorted coloured shavings
O14	Mechanical, lightly printed shavings
P22	Best white newsprint without reel core material
P23	Best white mechanical magazine paper without reel core
P32	Best white mechanical paper shavings free of reel paper
Q14	White uncoated woodfree shavings, lightly printed
R12	Best white uncoated woodfree shavings
S12	Best white coated woodfree shavings
T14	White, coloured or lightly printed substitute chromo board

Table 1.3 *(cont)*

Grade	Description
U31	Coloured woodfree punched cards
U33	Manila woodfree punched cards
Group IV: strength retaining grades	
V11	Used kraft paper sacks (waterproof and non-waterproof)
W12	Used pure kraft paper
W13	New pure kraft paper
W41	Corrugating material from production and converting without cores
W52	Used corrugating material II, double lined with kraft and/or test liner
W62	Used corrugating material I, with kraft liner and corrugating medium of semi-chemical or chemical pulp.

which can have a significant effect on the development of the wastepaper recycling industry.

The German paper and board industry classifies wastepaper grades into four categories:

- Group I - letters A–D, lower grades
- Group II - letters E–J, medium grades
- Group III - letters K–U, upper grades
- Group IV - letters V–W, strength retaining grades

The grades are further classified with a two figure number. The first figure has no meaning, while the second figure identifies the following:

- 0 - unsorted waste
- 1 - mixed waste
- 2 - sorted waste
- 3 - waste of one single colour
- 4 - lightly printed waste
- 9 - special grades

The CEPAC system was developed to provide a common wastepaper grading system in Europe, but has not been widely adopted.

1.4 Definitions used in recycling and wastepaper collection

1.4.1 Pre-consumer and post-consumer

There have been numerous attempts to distinguish between pre- and post-consumer grades of wastepaper, although these have the objective of trying to distinguish between grades of wastepaper which have 'traditionally' been recycled and those which have been present in trash going for disposal by

Table 1.4 CEPAC wastepaper grading structure (Confederation of the European Paper Industries)

Grade	Description
Group A - ordinary qualities	
A0	Unsorted mixed wastepaper shavings
A1	Mixed papers and boards (unsorted)
A2	Mixed papers and boards (sorted)
A3	Board cuttings
A4	Supermarket waste
A5	Corrugated container waste
A6	New shavings of corrugated board
A7	Over-issue pamphlets and magazines free from adhesive bindings
A8	Over-issue pamphlets and magazines free from adhesive bindings
A9	Mixed news and pamphlets
A10	News and pamphlets free from adhesive bindings
A11	Mixed PAMS and magazines
Group B - medium qualities	
B1	Once-read news
B2	Over-issue news
B3	White lined board cuttings
B4	Mixed coloured shavings
B5	Bookbinders' shavings
B6	Bookbinders' shavings without adhesive
B7	Coloured letters
B8	White woodfree books
B9	Bookquire
B10	Coloured best PAMS
B11	White carbonless copy papers
B12	Coloured carbonless copy papers
B13	Coated board
Group C - High qualities	
C1	Mixed light coloured printers
C2	Light coloured woodfree shavings
C3	Coloured tabulating cards
C5	Buff tabulating cards
C6	Mixed white letters
C7	White woodfree letters
C8	White woodfree continuous stationery
C9	White woodfree continuous stationery free from colouring
C10	Printed white multiply board
C11	Unprinted white multiply board
C12	White newsprint
C13	White magazine paper
C14	White woody coated paper
C15	White woodfree coated paper
C16	White woody shavings
C17	Mixed white shavings
C18	Mixed woodfree shavings
C19	White woodfree shavings, uncoated
Group D - kraft qualities	
D0	Brown corrugated
D1	Corrugated kraft I
D2	Corrugated kraft II
D3	Used kraft sacks
D4	Clean used kraft sacks
D5	Used kraft
D6	New kraft

landfill. There are few technical merits in this distinction. There are two fundamental differences between virgin fibres and wastepaper, and these are considered below with reference to this distinction.

The first is that fibres are damaged during the first papermaking operation - fibres collapse during drying on the paper machine and there is hornification at the surface, so that on repulping the properties of the fibres have changed (see chapter 6), with more fines produced. Hence all recycled fibres, whether from mill broke or a contaminated post-consumer source, behave very differently from their equivalent virgin fibres.

Secondly, wastepaper carries with it non-fibrous materials, some of which create difficulties during paper making; for example, the presence of adhesives in wastepaper creates stickies, which cause many problems during paper making. Even paper mill broke can contain stickies, for example, white pitch from the use of coated broke. A very general case is that post-consumer wastes have greater processing needs, but all wastepapers have a need for some contaminants removal.

In paper-making terms, the distinction between pre- and post-consumer wastepapers is irrelevant, which is why traditional wastepaper grading systems have not made this distinction. It creates many anomalies; for example, a printed carbonless copy paper form, supplied by the firm responsible for printing, is a pre-consumer waste. An identical form, once completed by the addition of a relatively small amount of ink by the user, is a post-consumer wastepaper, although both have exactly the same processing requirements. Many types of pre-consumer wastepaper are landfilled, since they are very difficult to recycle, for example, laminated papers. These distinctions have been an area of considerable controversy, with many adjustments and modifications proposed to basic definitions. Typical basic definitions are:

Pre-consumer wastepapers: wastepapers which are collected from converters and printers. The quality ranges from new KLS cuttings, DLK, and unprinted whites, to printed forms. In fact, all of these grades are post-industrial consumer, since the converter or printer is a consumer. High quality pre-consumer grades have long been used by the industry, but there is a limited supply of these so that wastepaper utilisation rates higher than 10–20% in a specific product sector inevitably mean that post-consumer grades are being used.

Post-consumer wastepapers: this usually refers to papers which are in a waste stream usually disposed of by landfill. They can arise from a domestic environment, for example, news and magazines, or from an industrial environment, for example, office wastepaper.

Since the terms pre- and post-consumer have little bearing on the contaminant content or quality of the fibres, they are of no technical value, and where possible should be avoided. If there is undue emphasis on the use of post-consumer wastepaper, grade switching will occur, but the overall utilisation rate may not increase.

1.4.2 Yield

In the production of all paper and board products other than soft tissue, significant quantities of non-fibrous additives are used; for example, many packaging boards or printing papers contain up to 5% by weight of starch. Printing papers contain mineral fillers, about 20–30% by weight, while coated papers have added coating fillers so that they contain up to 40% by weight of mineral fillers and coating binders. Hence paper and board production makes extensive use of non-fibrous materials.

When wastepaper is reprocessed, most non-fibrous materials, as well as some fibres, are discarded or lost during reprocessing, with the fibre yield from wastepaper calculated from the following:

$$\text{yield (\%)} = \frac{\text{bone-dry weight of recycled materials after processing}}{\text{bone-dry weight of wastepaper input}} \times 100$$

Actual yields are dependent on the wastepaper type and the product being manufactured.

1.4.3 Utilisation rate

This is a measure of the amount of wastepaper used in the production of a specific product, or product sector, or by the industry. When utilisation rates are estimated, no corrections are normally made to compensate either for losses during recycling or paper and board making, so for specific product groups the apparent utilisation rates can be overestimated and can be higher than 100%.

The apparent utilisation rate (%) is given by:

$$\frac{\text{wastepaper used (tonnes)}}{\text{paper or board production (tonnes)}} \times 100$$

1.4.4 Recovery rate

This is a measure of the recovery of wastepaper, and is given as a proportion of the total (apparent) paper and board consumption. Recovery rate calculations underestimate the proportion of wastepaper recovered, since no allowance or adjustments are made for products which are not available for recycling, for example, liners used in plasterboard production, tissue, wallpaper, etc. The apparent recovery rate (%) is given by:

$$\frac{\text{wastepaper recovered (tonnes)}}{\text{apparent consumption of paper and board}} \times 100$$

Table 1.5 Apparent recovery and utilisation rates (1992)

Country	Utilisation (%)	Recovery (%)
USA	33.1	38.7
Germany	52.1	50.6
Japan	52.7	51.1
The Netherlands	70.5	54.7
Sweden	14.3	49.7
Australia	45.8	36.8
South Korea	69.6	43.2
UK	60.2	33.9
Denmark	96.6	37.6

Adjustments are not made to compensate for manufactured goods imported or exported as, for example, packaging of consumer goods. The apparent utilisation and recovery rates are given in Table 1.5, which shows the very wide range when countries are compared, i.e. a utilisation rate greater than 70% for The Netherlands, for example, compared to less than 15% in Sweden. There is less variation in the recovery rates, although The Netherlands' recovery rate of almost 55% is a much better performance than that achieved in the UK, i.e. less than 35%.

To a large extent, the structure of the industry determines the utilisation rate. When a high proportion of products in an individual country is board, the recycling rate achieved can be very high, as in Denmark, The Netherlands and the UK. When a high proportion of production is high quality papers, then the utilisation rates tend to be lower, since the wastepaper utilisation rates in high quality printing papers is low. Figure 1.1 illustrates this effect, by showing the relationship between utilisation rates and fine paper production. Countries below the line have relatively low wastepaper utilisation rates and these tend to be fibre-rich countries, with substantial commercial forests and large export volumes of paper and board. To achieve high utilisation rates, these countries would have to import large volumes of wastepaper.

1.5 Recovery of wastepapers

Since papers and boards are used for various specific functions, after use they tend to be concentrated at different locations, for example, packaging at wholesale and retail outlets, news and magazines in domestic homes and white papers in offices. Different systems have evolved, which are continually changing, to recover these wastepapers.

Not all of the recovery systems have concentrated on supplying wastepaper 'in grade'. In the UK, the grading system, and hence recovery systems, concentrate on the production of homogeneous grades of wastepaper, whereas in Germany, emphasis has been on volume recovery, with

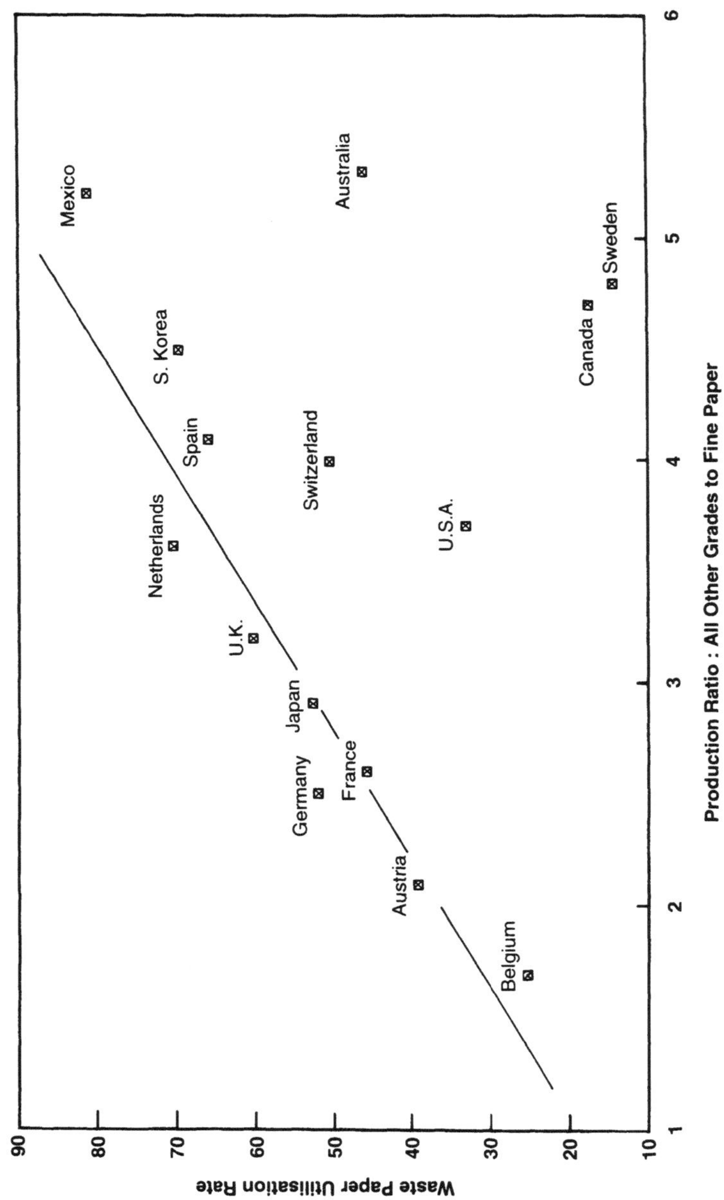

Figure 1.1 Relationship of apparent utilisation rate to industry structure.

less emphasis on 'in grade' recovery. Hence some of the differences between the two grading systems.

In most industrialised countries, the recovery of wastepapers from converters and printers is well developed. Wastes from these sources do not normally exceed 5–15% of the paper or board converted or printed, and so as to enable wastepaper use to grow, it was necessary to develop collection by other recovery systems. Each broad band of wastepapers has recovery systems which overlap, but which are considered separately.

1.5.1 Recovery of packaging grades

High quality grades recovered from converters include double lined kraft (DLK) and different types of cuttings. Although they are relatively free from contaminants, box plant trim and cuttings are a mixture of two very different fibre types, namely kraft liner and NSSC. The properties of a blend of these are different from either one on their own and this is one of the reasons why the strength properties of a recycled (test) liner are not as good as a virgin (kraft) liner, at the same basis weight. Nevertheless, the recovery of these grades from box plants is well organised and recovery rates are high.

A large proportion of fibrous packaging materials is used to protect goods during transit and most of this becomes waste at the point of sale. Since some components of packaging, e.g. fluting and a proportion of liners, do not have demanding specifications, these can be made from mixtures of corrugated and solid boards, with some other fibre types also present. In countries which are fibre deficient, such as Japan and the UK, mills became dependent on waste packaging materials recovered from wholesale and retail outlets and an efficient collection system developed, with high recovery rates achieved, for example, almost 70% of corrugated containers in Japan, in 1991. This was based on the separation at source (the point of sale) of various packaging materials, so that fibrous packaging was separated from plastics, wood, etc. Collection was by wastepaper merchants, either free collection or with a small payment to the retail outlet for the board.

Prices for collected materials are volatile. During periods of weak consumer demand, the need for packaging falls and although typically this is by less than 5–10%, the impact on the collection of used packaging is very marked. Competition increases and so prices for packaging fall. Prices for wastepaper grades also fall, partly to reflect reduced prices for packaging, but also to discourage collection. This adjustment is frequently too severe, with a drastic reduction in supplies, and so as consumer expenditure recovers, demand for board increases, leading to an increased requirement for waste board, resulting in high prices for available supplies. The net result has been a roller-coaster ride in prices; Figure 1.2 illustrates this,

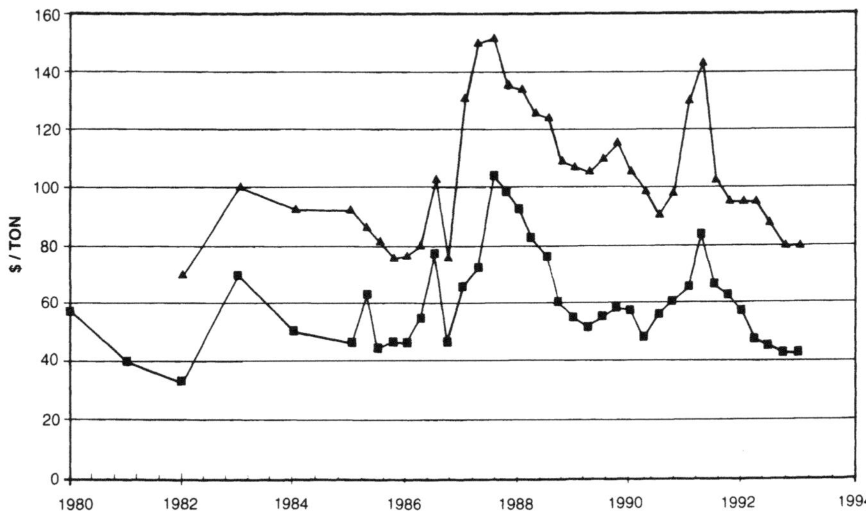

Figure 1.2 US wastepaper prices (west coast): (■) old corrugated containers; (▲) new double lined kraft.

using old corrugated containers (OCC) prices in the USA. Price instability is clear, even though the USA used exports as a safety valve.

In the UK, wastepaper was the most important raw material for board production and so producers became involved in wastepaper collection in the late 1950s and early 1960s, resulting in the concentration of wastepaper trading in the hands of mill-owned wastepaper merchants - about 75% of wastepaper used in the UK was handled by the top five mill-owned wastepaper merchants in 1992. This did not protect wastepaper merchants from wide swings in prices. If low prices are sustained, then inevitably collection falls and some merchants go out of business.

A novel approach to protecting wastepaper collection channels and to minimise price fluctuations was adopted in The Netherlands. Under this scheme, the national government provided subsidised warehousing, so that during periods of weak demand collected wastepaper could be stockpiled, and released during periods of high demand. Under the EXVOPA agreement, Dutch board mills agreed to buy up to 6% of their annual consumption and store it until market conditions allowed its release [13]. Recycling is encouraged nationally and subsidised by local governments, and a large proportion of total collection is carried out by voluntary organisations. Although these must have a permit they may receive subsidy payments during periods of low market prices. A high population density and the industry structure assist The Netherlands to achieve high apparent recovery and consumption rates. Other paper-stocking schemes, for example, in Japan, were unsuccessful.

Sweden adopted a legislative method to control the effect of price swings on wastepaper collection. In 1975, local authorities were given a monopoly for household waste collection, which prevented others moving in and out of the wastepaper collection industry, depending on the market conditions. A single company controls a high proportion, about 70%, of the wastepaper recovered. This focused approach has enabled Sweden to achieve high recovery rates for packaging materials, as well as newspaper, and to avoid excessive price fluctuations [14].

In most countries, stability was established only by normal market forces, but in the late 1980s and early 1990s governments began to introduce legislation which had a particularly destabilising effect on packaging grades recovery by traditional routes. In Europe, this started in Germany with the introduction of 'die Grüne Tonne' in 1985. Under this scheme, householders were provided with a second bin, other than the normal domestic rubbish bin, in which to place their recyclables. After collection, these were sorted in a central sorting operation. In 1986, demand for wastepaper fell, but the Grüne Tonne system did not reduce collection to match demand, since it was subsidised. A resulting surplus of low quality grades pushed down prices in Germany, as well as in other neighbouring European countries, leading to the failure of some parts of the traditional collection systems. In some areas, charges were introduced for collection, although these charges were lower than those placed on other wastes for disposal by landfill. Some generators of wastepaper refused to pay for collection. In other areas, where generators had previously been paid for the wastepaper collected, payment was stopped, although collection was free.

One of the most profound effects of this legislation was that in Germany many wastepaper companies had to become involved in solid waste management, or lose part or all of their wastepaper business. This move towards general waste management was given greater impetus by the second piece of legislation, the Packaging Ordinance, in 1991, which was the first of a series of ordinances designed to fundamentally change the throw-away society, with the first stage being an attack on throw-away packaging. Under the Ordinance, targets for collection and recycling of packaging were established, and in the case of fibrous packaging, recovery targets were set at 80% by 1995, with an 80% recycling target also established [3]. Other European countries followed suit with a series of regulations designed to increase the recovery of packaging materials.

However, wastepaper demand did not increase at the same rate as recovery, so that prices fell in Germany and its neighbouring countries. Surplus wastepaper was exported from Germany, with subsidised collection under the 'Dual System Deutschland' (DSD) system, paid for by the 'Green Dot' Licence fees. Packaging board recovery in neighbouring countries by traditional methods could not match these low prices and so in countries such as France, Italy and the UK, wastepaper merchants were forced out of operation during 1992 and 1993.

Criticism of the German system was intense, since Germany was seen to be solving its waste disposal problem by exporting it. Within Germany itself, criticism was also strong, as much of the recovered packaging waste materials did not have a market and the operating costs of the DSD system were much higher than expected, which resulted in some changes to the system. Partly as a consequence of the introduction of the system, the use of packaging materials fell, by several per cent. However, it is difficult to justify the system in terms of environmental improvements, since the billions of DMs it has cost may have given a greater benefit if spent in a different way. No cost/environmental benefit analysis has been made, so this is difficult to judge.

Given the rash of packaging waste recovery agreements in other countries, it appears probable that a pattern of total waste management for retail outlets will emerge, with waste management companies favoured over wastepaper merchants, since the latter provide a solution to the disposal of only one component of packaging waste streams. Many changes in systems for the collection and disposal of fibrous packaging can be expected up to the end of the 1990s.

1.5.2 Recovery of news and magazines

High quality unprinted news is available from printers - edge trim, end of reels and other wastes. Unprinted wastes command a high price. Newspapers which have not been sold or distributed are returned to the main distributors and sold to recycling mills as 'over-issue' news, which is usually less contaminated and more homogeneous than newspapers from consumers. The supply of news has expanded quite substantially, primarily by collection from domestic sources, which is organised in a variety of ways, with major differences between regions and countries. Many changes are due to government intervention.

When surveys of domestic rubbish have been carried out in industrialised countries, one of the most common components of domestic waste is paper, where it is usually within the range 25-35% by weight [15, 16], or in the range 15-20% of municipal solid waste (combined commercial and domestic wastes) [17]. A substantial proportion of this paper fraction is newspapers and magazines. A solid waste disposal problem has been created, since new regulations have dramatically increased operating costs for landfills and few new licences to operate these are being given by local or national governments. Even though many potential landfill sites are still available (but are not licensed, or are uneconomic to develop) in some areas of North America and Germany a solid waste disposal crisis is imminent. One solution has been to recover as much as possible from domestic wastes and to recycle the material recovered, including newspapers and magazines. Hence the move to mandate recycled fibre content in newsprint, with targets having been set by many states in the USA.

The first country to introduce legislation to assist paper recovery was Sweden, in 1975 [14]. In 1985, Germany mandated the collection of recyclables, as described earlier. Governments in many other countries followed the example of Germany, with for example, the introduction of 'blue box' collections in Toronto and Seattle, and mandatory recycling in New York City, New Jersey, etc. The result was a glut of old newspapers, in many countries, from 1989 onwards. Even though there were large expansions in deinking capacity in the USA, Canada and Europe, supply continued to exceed demand throughout the early 1990s. Low prices, even negative prices, were the result and recovery systems through charities, voluntary organisations, etc. had to stop collection. In other areas, payments which were made for the collection of news and magazines had to stop and in some cases, charges were introduced - a similar pattern to that found in the recovery of packaging wastepaper grades.

Other countries, such as Japan and the UK, have not introduced mandatory recycling. Japan has achieved very high recovery rates for newsprint, greater than 98% in 1991 [18]. This high level was achieved by domestic collection of all grades, pre-sorted by residents, by small wastepaper merchants, and in return householders were given various paper products, such as toilet paper. Instrumental in achieving very high recovery rates were the publicity campaigns by the Paper Recycling Promotion Center, which was established in 1974. Voluntary agreements between industry and the Japanese government have continued, with a target of a 55% utilisation rate by 1995 having been set by the industry.

The development of old-newspaper recovery systems in the UK has also been voluntary. In the late 1970s and early 1980s, many local authorities had collections of wastepapers, separated by householders, collected with the domestic rubbish, but placed in a separate container. When newsprint production capacity in the UK was almost completely eliminated during the recession of the early 1980s, to less than 100 000 tonnes per year, few of these collections survived, and most local authorities were forced to abandon them. When wastepaper based newsprint production expanded in the mid 1980s and later, few local authorities were involved in collections, although in some areas cooperation developed, usually with assistance from wastepaper merchants. However, the 1990 Environmental Protection Act set local authorities a target of recycling 50% of the recyclables in domestic wastes (approximately equivalent to a 25% recycling rate of all domestic rubbish) by the year 2000. Many local authorities have examined systems for wastepaper recovery from domestic houses, in order to assess the costs of different methods. There are basically two systems, which are:

- 'Bring' systems - householders take their already sorted recyclables for deposition in specialised containers at recycling centres; this includes glass, used oil, etc., as well as newspapers. This system was developed extensively by Bridgewater Paper Mill in the UK, and it supplies a high proportion of their wastepaper requirements, and also that of the SCA newsprint mill.
- 'Collect' systems - recyclables are collected from homes, either commingled or separated, for example, a 'blue box' system. The sorting of commingled material can be either carried out at the kerbside, into separate compartments of a specialised vehicle, or in a materials recovery facility (MRF).

There are several variations of these systems, including hybrids, and these are summarised with their respective advantages and disadvantages in Table 1.6 [19, 20, 21].

In the recovery of news and magazines, bring systems can be very effective, as shown, for example, by the results achieved using the 'igloos' installed by Bridgewater Paper Mill. Higher recovery rates are reached by using collect systems, although the costs are usually higher. If collect systems are uncoupled from demand, which is usually the case if operated by local government, supply usually outstrips demand and inevitably, prices fall. In the UK, where the bring system involves cooperation between local governments and wastepaper merchants or newsprint producers, supply and demand are kept in balance. Although surpluses outside the UK can still affect the market for news and magazines, the success of the bring recovery systems, which are demand driven, is a clear illustration that these are preferable to legislation-driven recovery systems.

As more governments establish recycling targets, it is clear that the recovery of old newspapers will be linked to integrated waste management systems, with increased cooperation between local government with either general waste management companies or wastepaper merchants. If recovery targets are set irrespective of demand, surpluses will inevitably result.

1.5.3 Woodfree wastepaper recovery

Woodfree papers are used in a wide variety of applications which require printing after cutting to size and these operations generate wastes. In general, the recovery of these industrial wastepapers is very well organised for high quality print-free grades and to a lesser extent, for printed wastes. There are considerable variations in the way that this material is collected; for example, in the UK and USA printed materials are normally source separated into wood-containing or woodfree grades, whereas in Germany, mixed printed papers are usually collected. With the expansion in use of

Table 1.6 Household waste collection systems in use in the UK

System	Advantages	Disadvantages
Bring, e.g. Bottle banks, paper igloos, etc.	Lowest set-up and operating costs. Flexible and easy to adapt to market trends. Quality of materials generally high	Quality and quantity of material dependent on the householder. Rate of recovery limited and unlikely to be above 10% without a high number of banks. Siting of banks problematical due to noise generated and visual impact
Collect, source segregated (no central processing)	Convenient for householder. High recovery rates of up to 30%. Can be integrated with refuse collection	Expensive both in terms of capital and operating costs. Such a system could well double the costs of collection. With no further sorting the level of contamination may be high
Collect, source segregated (with central processing)	Convenient for householder. High recovery rates of up to 30%. Can be integrated with refuse collection. Recovered materials will have a higher resale value than the equivalent system without sorting	Even higher capital and running costs than above due to the requirement for a sorting plant. A suitable site for the sorting plant needs to be found. Potential health problems with workers in sorting plant, depending on wastes being sorted
Collect, mixed (with central processing)	Minimum effort for householder. High collection rates	The best systems for this type of collection utilise a split wheeled bin or separate collection containers for recyclable elements and remaining refuse. Health problems for workers in sorting plants. Material collected may be contaminated
Hybrid - bring combining collect, with, e.g. paper collection, door to door, etc.	Minimum set-up cost. A paper collection could remove approximately 20% by weight of domestic wastes	Some effort required by the householder. Existing refuse collection vehicles may not be capable of implementing the scheme, therefore either new vehicles or a duplicate collection round would be necessary in the short term

woodfree wastepapers, particularly in tissue production from the late 1970s onwards, the quantities available from these sources were not adequate to meet demand and so wastepaper began to be collected from commercial premises - usually offices with a large number of employees, handling large volumes of paper - banks, government offices, etc. Collection was either of top grades only, for example, computer printout (CPO) or white ledger (heavy letter) and similar grades, or the collection was of all office papers. In the latter case, papers were generally not sorted, giving a low quality mixed office wastepaper, containing white and coloured woodfrees, newsprint, carbon paper, etc., which could be shredded for security reasons.

There have been numerous pilot studies to evaluate the recovery of office wastepapers [22–26] and these show that the office waste composition, yield

Table 1.7 Yields of office wastepaper

Type of office	Geographic area	Yield[a]	
		Total paper	White, woodfree
Finance and insurance	USA	4.9	3.2
Finance and insurance	USA	4.5	3.4
Government	USA	2.7	1.1
Government	Canada	1.3	0.8
General	Canada	2.2	0.6
General	USA	3.4	1.6
General	Japan	1.6	1.3
Civil Service	UK	0.6	0.4

[a] Expressed as kg per week per occupant

Table 1.8 Composition of solid office wastes

	Office		
	Finance/insurance	General	Government
Paper[a]	4.5	3.0	2.7
White woodfrees	3.4	1.8	1.4
Coloured woodfrees	0.5	0.5	0.3
Other papers	0.7	0.9	0.9
Trash	0.5	0.7	0.5
Total waste generated[b]	5.2	3.6	3.2

[a] Expressed as kg per week per occupant
[b] Totals do not add due to rounding

and quality vary, according to the type of office. Yields from some of these studies and others are given in Table 1.7.

A number of studies have also looked at the components of office waste [25, 26], and the results from one of these are given in Table 1.8. The grading systems used do not correspond to wastepaper grade definitions, so the breakdown is given as a guide only.

Since the top quality grades represent only a portion of the wastepapers, recovery systems are becoming based on the mix of papers that a specific wastepaper processing line can accept – this is referred to as 'office pack', and includes all of the white and coloured woodfrees and some other papers. The use of an office pack system instead of the collection of only CPO and white ledger increases yield, in some instances by a factor of two.

Most recovery schemes are based on source separation, so that office occupants separate their wastes into appropriate streams, with separate containers for each stream. Containers can be in each individual office, or located at strategic points throughout the building. These are emptied into one or more exterior containers, usually by building maintenance staff, which are than collected by wastepaper merchants, either on a regular or as-required basis. Frequently, small merchants organise a collection route, designed to match the capacity of a relatively small vehicle, which typically delivers its load direct to a larger merchant, who has sorting and baling facilities. Bales are collected until a full load is available for delivery to a user mill.

It is much more costly to collect office wastepaper than to collect printers' waste; for example, a vehicle could collect four tonnes from a printer in an hour, whereas it could take four hours to collect two tonnes of office paper. High disposal costs for wastes can make office paper collection economic, but in the absence of this, the selling price of the wastepaper has to be high enough to carry the collection, sorting and baling costs.

Government intervention in office wastepaper collection has been relatively small. In the USA, the state of California has mandated a goal of 50% diversion of municipal solid wastes from landfill by the year 2000 and office wastepaper collection is recognised as playing an important part in achieving this goal. Other cities and states in the USA have also targeted office wastepaper collection. Due to the contamination problems encountered when office waste was collected without source separation, for sorting in an MRF, New York City introduced a rule to make sorting mandatory in 1993.

Since considerable pressure to increase recycling rates is being brought to bear on manufacturers of printing and writing papers, and as this is the sector with the lowest recovery rate, more government pressure to increase recovery rates can be expected, with waste haulers becoming more involved as the emphasis switches from only office paper recovery to total waste management, as has happened in the recovery of other grades.

1.6 Government policies regarding wastepaper collection

As discussed in preceding sections, government involvement in wastepaper recovery is inevitable, since most governments in industrialised nations are keen to reduce the volumes of solid wastes for disposal. Targets to this effect have been set in the UK and Denmark (25% of domestic waste), USA (25% of total solid waste), and Canada (50% of total solid waste). Since paper and board are substantial contributors to both domestic and commercial waste streams, it has been inevitable that the role of the paper industry in recycling would come under scrutiny. In addition, the impact of producer responsibility for waste disposal, introduced into Germany, for example, as the Packaging Ordinance, has led to a very measurable reduction in the volume of waste to be collected and disposed of by local government. This represents a very substantial cost saving for government, so it is probable that this will be imitated by other governments seeking to reduce the costs of public services, although in Germany these cost savings are passed on to wastepaper collectors. A number of countries have set targets for board recovery and recycling as part of packaging waste management, for example, Germany and Japan. An EC Packaging Directive also sets targets, but many EC countries established their own targets before this Directive was introduced, following the German example.

Only a few countries have legal requirements for source separation, for example Denmark, Sweden and some states and cities in the USA. Public authorities have organised wastepaper collection schemes from households and commercial premises. In many areas, financial assistance is provided for the separate collection of recyclables, for example, the 'blue box' programme in Ontario, and the recycling credits paid in the UK, primarily to voluntary organisations, for the diversion of recyclables from the domestic waste stream. Other countries support wastepaper collection indirectly, for example, in Austria by co-financing collection equipment.

In the hierarchy of solid waste management, recycling comes second to waste avoidance, and is followed by incineration. There is a growing recognition that incineration with energy recovery is an essential part of a waste strategy, for example, the packaging waste management system introduced in France in 1993. This has provoked fears among paper makers that wastepapers could be diverted away from their mills, into the easier option of incineration. Denmark and some states in the USA have introduced provisions, which ensures that recycling paper is given priority. In Sweden, which has extensive waste-to-energy and associated district heating systems, wastepaper suitable for recycling *must be used* in recycling, and the national government has requested communities to ensure that the paper industry is given first access to this wastepaper. As the importance of incineration grows, more protection for recycling and wastepaper collection will be required.

Governments have also acted to improve the markets for recycled paper products, although generally this followed moves to stimulate wastepaper collection, which resulted in a glut. In the USA, content targets for papers purchased by the Federal government were first introduced in 1988 and then updated in 1993 [2]. Many states have introduced more ambitious targets, especially in the case of the recycled content of newsprint. A voluntary agreement in the UK is that newsprint used should contain an average of 40% recycled fibre by the year 2000. In many countries, national and municipal governments specify the use of recycled papers where possible. Some governments have been involved, directly or indirectly, in the introduction of environmental labelling schemes, which favour the use of recycled products, for example, the Nordic White Swan, covering Sweden, Finland and Denmark, or the Blue Angel in Germany [27]. Further government intervention favouring the use of recycled products can be expected.

1.7 Wastepaper prices

Inevitably, prices are linked to the balance of supply and demand - when supply is greater than demand, then prices are low. In the early 1990s, some grades of wastepaper, for example, news and magazines in the USA, or mixed papers in Germany, had negative prices, since the increase in their collection outpaced the demand. Wastepaper prices are influenced by other factors, including the costs charged by wastepaper merchants.

The cost components of wastepaper collection and delivery include:

- Collection charges - collection transport, including fuel, vehicle, labour; fee to purchase wastepaper; container rental; costs of providing containers, vehicles, etc.
- Processing charges - sorting; baling; administration; cost of buildings and equipment, waste disposal
- Transport to mill - fuel, vehicle, labour; back transport; etc.

Purchasing mills generally pay for wastepaper ex works, although only for deliveries which are accepted. Other costs must be met by the wastepaper merchant. However, it is rare for wastepaper merchants to be able to set prices for their products (many external factors determine these) and so their efforts have to be on minimising costs and maximising revenues other than from wastepaper sales. When prices fall, many merchants introduce charges for collection, rather than pay a fee. This, and the cash from selling wastepaper, are the only two income streams for wastepaper merchants, unless they are subsidised. Since the collection fees are a cash source other than from wastepaper sales, wastepaper merchants' costs are not a predictor of actual wastepaper prices, since collection fees can go up and down, depending on the state of the market for a specific grade and the costs of other disposal routes, such as landfill.

Correlations between wastepaper prices and other prices, or price indices, have been researched. However, no relationship between mechanical pulp prices and packaging was found in a study which examined various models for forecasting wastepaper consumption [28].

In the case of woodfree grades, careful examination shows a much stronger relationship with pulp prices than with any index of prices, or measure of inflation. Pulp prices inevitably affect demand, since in a period of high prices, pulp substitutes are purchased by mills using virgin pulps, whereas when virgin pulp prices are low, the problems associated with even high-quality-wastepaper use can be avoided by purchasing virgin pulp grades.

To illustrate this relationship, the prices of two woodfree grades of wastepaper in the UK were compared with the price of virgin NBSK pulp. The wastepaper grades used were:

- coloured best pam (CBP), the lowest quality available (in terms of ink density)
- white heavy letter (WHL), a good quality deinking grade, equivalent to white ledger in the USA

A mathematical comparison of NBSK and wastepaper prices, for the period 1978–1992, gave the following correlation coefficients:

- CBP - correlation coefficient of 0.61; significant at 0.001 level
- WHL - correlation coefficient of 0.87; significant at 0.001 level

If the price of a high quality grade such as white woodfree continuous stationery in Germany, is compared with the NBSK price, the correlation coefficient is very high, i.e. 0.94, which is a clear indication that this is a pulp substitute grade in Germany. The correlation coefficient is higher than that of white heavy letter in the UK which indicates that pulp price is also the strongest influence on woodfree wastepaper prices in Germany.

A similar comparison for the same grade in France reveals a strong correlation at 0.84, which again suggests that this grade is used as a substitute for bleached pulp, and that woodfree wastepaper prices are determined by pulp prices.

These analyses show that in the case of woodfree wastepapers, bleached pulp prices have a major effect on prices, although this probably falls as the quality of the woodfree grade becomes more contaminated, as illustrated by the difference between correlation coefficients for the two wastepaper grades in the UK. Over a period of time the best predictor for price is likely to be virgin pulp prices.

Long term, this relationship poses problems for wastepaper merchants. The long-term trend for real pulp prices is downwards, which has been achieved by building larger pulp mills, etc., to benefit from economies of scale. This long-term downward trend is likely to continue for virgin pulp, but will be based on new factors, such as pulp production in countries with very fast rates of wood growth and better yields, resulting from forest genetics. In the case of wastepaper merchants, these benefits cannot be realised, and the trend for the real costs of wastepaper merchants is upwards, i.e. labour, fuel, taxes, etc. When pulp prices are very low (as in 1993) wastepaper prices follow. The payment of collection fees is thus the only method by which wastepaper merchants will be able to follow pulp prices, and inevitably these will apply to the collection of all types and qualities of wastepaper.

1.7.1 Wastepaper value

The value of a wastepaper is set by several considerations, such as the following: the cost of the wastepaper and solids losses; its runnability on the paper machine; the types of fibre present and the cost of the virgin fibre the recycled fibre displaces; the impact of recycled fibre use on product sales value; the costs of associated solids waste disposal, etc. A system for establishing real fibre costs is very complex, but could be established by laboratory evaluation coupled with actual performance, with performance tracking on paper machines, etc. Commercial needs or availability of wastepaper or legislative pressures may dictate that some grades are purchased which are not 'good value', but nevertheless, value for money should be part of a wastepaper purchasing policy.

A basic equation for estimating the cost of fibre is given by the following:

$$\text{actual cost of fibre (bone-dry)} = \frac{\text{wastepaper cost (bone-dry tonne)}}{1 - (\text{ash content} + \text{fines loss} + \text{other losses})}$$

When estimating the actual cost of fibre by using this equation, the following applies:

- Use of bone-dry values. This is due to distortions caused by moisture contents, for example, if 100 tonnes of air-dry wastepaper are purchased and converted to 70 tonnes of air-dry pulp, there is a difference of 3 tonnes of moisture, which contributes to 'solids' losses
- Ash determinations must be carried out at a temperature which does not result in ignition losses from calcium carbonate; the Tappi Standard 'Ash in Paper' temperature of 925°C is too high in this respect
- When there is a low ash target, most of the fines will be washed out, with actual losses dependent on plant layout. Some 'good-fibre' losses can be expected, but good plant design will minimise these
- Other losses, which incorporate non-fibrous components, e.g. baling wire, staples, pins, printing ink, and chemicals, such as starch added during paper production, which are lost as dissolved solids in effluent, etc.

Applying the above formula to a typical, predominantly white office grade would give (approximately), using values of £70 per air dry tonne, 10% moisture, 20% ash, 8% fines loss and 5% other losses, the following:

$$\text{actual cost of fibre (bone-dry tonne)} = \frac{77.78}{1 - (0.2 + 0.08 + 0.05)} = £116.10$$

This is only the first step in assessing wastepaper value, as the cost of using specific grades varies, for example, due to different chemical consumption, solids losses producing more or less sludge for disposal, effects on the runnability of the paper machine, product quality variations, etc. The cost of using a specific grade can be estimated, but only after considerable volumes of wastepaper are used and time and effort devoted to looking at variables, and is given by:

Cost of using wastepaper grade = (actual fibre cost × efficiency factor × quality factor) + sludge disposal + effluent load + chemical costs + energy costs

where:

the efficiency factor is the efficiency of the paper machine on an equivalent virgin fibre grade, divided by its efficiency when on a specific wastepaper;
the quality factor reflects its ability to replace virgin pulps, without a loss in sales value;
the sludge disposal costs are the total costs, i.e. chemical processing plus actual disposal;
the effluent load is a measure of the biological oxygen demand (BOD) load on the effluent plant from different grades and the cost of BOD removal;
the chemical costs reflect that different wastepaper grades have different chemical requirements, e.g. white and coloured;
the energy costs reflect different energy costs of different grades, e.g. more dispersion, higher temperatures, etc.

Although a detailed study would be necessary to assess the comparative values of various wastepaper grades, this information would help wastepaper-using mills to spot pricing anomalies, and allow mills to purchase wastepaper grades which give better overall economies.

References

1. Thomas, C. (1977) *The Paper Chain*, Earth Resources Research Ltd, London.
2. Clinton, W. (1993) *Executive Order on Federal Acquisition, Recycling and Waste Prevention*, Federal Register, 20th October 1993.
3. Anon (1990) *Verordnung über die Verordnung der Bundesregierung*, Bundesrat Drucksache 817/90 vom 14.11.1990, Verlag Hans Heger, Bonn.
4. *Japanese Resource Recovery and Recycling Act*, 1992.
5. Anon (1992) *Recovered Paper Statistical Highlights, 1992*. American Forest and Paper Association, Washington.
6. Anon (1993) *Pulp and Paper International - Annual Review*, July 1993.
7. Anon (1987, 1991, 1992, 1993) *Pulp and Paper International - Annual Survey*, July issues.
8. Anon *Worldwide Review of Recycled Fibre*. Jaako Poyry Oy, Helsinki.
9. Anon (1990) *CEPAC Wastepaper Grades*, Confederation of the European Paper Industries, Brussels.
10. Anon (1988) *Standardised Grades of Wastepaper in the UK*, British Paper Board Industry Federation/British Wastepaper Association, Swindon.
11. Anon (1991) *Standard Grades of Wastepaper in Germany*, German Paper Manufacturers Association/German Association for Papermaking Raw Materials/Association of German Refuse and Waste Collecting Companies, Cologne.

12. Anon (1993) *Guidelines for Paper Stock: PS-93*; Institute of Scrap Recycling Industries, Inc., Washington DC.
13. Schaafsma, H. (1989) The impact of mandatory recycling on wastepaper. *Pira Conference - Recent Developments in Wastepaper Processing*, 1989.
14. Luthbom, A. (1989) Swedish wastepaper recovery systems. *Pira Conference - Recent Developments in Wastepaper Processing*, 1989.
15. Anon (1992) Domestic UK waste analyses by Warren Spring Laboratory, in the *Recycling Plan*, Royal Borough of Kingston upon Thames, Kingston upon Thames.
16. Barton, J. (1991) Recycling - can local authorities make it pay?. *NSCAEP Workshop, Beyond the Green Bill*, 1991.
17. Anon (1990) *EPA Characterisation of Municipal Solid Waste in the USA: 1990 Update*, Environmental Protection Agency, Washington DC.
18. Anon (1992) *Statistics of Waste Paper in Japan*, Paper Recycling and Promotion Center, Tokyo.
19. Forrest, P. *et al.* (1991) *Sorting at Source, Separation of Domestic Refuse*, SWAP Report, SWAP Leeds.
20. Anon (1992) *The Adur Project*, Institute of Grocery Distribution and ERRA Report,
21. Gwynn-Jones, R. (1992) *Recycling City and Beyond*, Friends of the Earth Report, Friends of the Earth, London.
22. Anon (1976) *Optimisation of Office Paper Recovery System*, Environmental Protection Agency, Washington DC.
23. Anon (1980) *Project Paper Recycle*, Ministry of the Environment, Ontario, Canada.
24. McKinney, R.W.J. (1989) Recovery of high quality office paper. *EUCEPA Symposium*, Lubjlana.
25. Cesar, M. (1991) *Resource Recycling*, June, 63.
26. Hinshaw, J. and Braun, I. (1991) *Resource Recycling*, November, 27.
27. McKinney, R.W.J. (1993) *Prog. Pap. Recycling*, August.
28. Deadman, D. and Turner, K. (1979) *Forecasting Demand for Secondary Materials, A Case Study of Wastepaper*, Futures.

2 Recovery and utilisation of wastepaper in Japan

R. OYE

2.1 History of wastepaper recovery

The wastepaper recovery rate in individual countries is dependent on the situation regarding its fibre resources and paper industry, as well as the per capita paper and paperboard consumption. In 1991, the world average wastepaper recovery rate was 37% [1], but the recovery rate was below 30% in many cases where the per capita paper consumption was lower than 20 kg.

In 1953, wastepaper recovery in Japan was only 19.6%, with 20.2 kg per capita paper consumption. As consumption of paper and paperboard was at a low level, a considerable portion of paper was reserved as books or documents for long-term use and/or utilised for wrapping and other purposes. Nowadays, paper and board consumption per capita is more than 230 kg, with the recovery rate reaching 50.8% in 1991. A large part of the extra consumption of paper and paperboard is disposed of from houses and offices, as wastepaper in urban garbage, which has contributed to a serious problem in garbage management for local governments.

The wastepaper recovery rate, from 1955 to 1991, is shown in Figure 2.1. It seems that the recovery rate changes periodically, i.e. about every ten years. The period from 1953 to 1964 could be called the first term, during which the recovery rate increased from 25% to 40%. Before 1953, wastepaper was mainly utilised to produce low-grade papers, paperboard and toilet paper. In the 1950s corrugated container made its debut as packaging, and rapidly took a dominant share. For corrugated medium, old news was used, and kraft brown for jute liner. In those days, rice straw had been utilised as a raw material for rope, mats and bags for packing, which were recovered and used for making straw board and low-grade papers. Corrugated board replaced these rice straw products, as well as wooden boxes for packaging, and caused a rapid increase in wastepaper recovery.

Although the quantity of wastepaper that was consumed grew from 1964 to 1973, the recovery rate hovered around 40% in this second term, because the ratio of paper to total production of paper and paperboard increased, and neutral sulphite semi-chemical pulp (NSSCP) was introduced and accepted for corrugated medium, due to its high rigidity and strength properties.

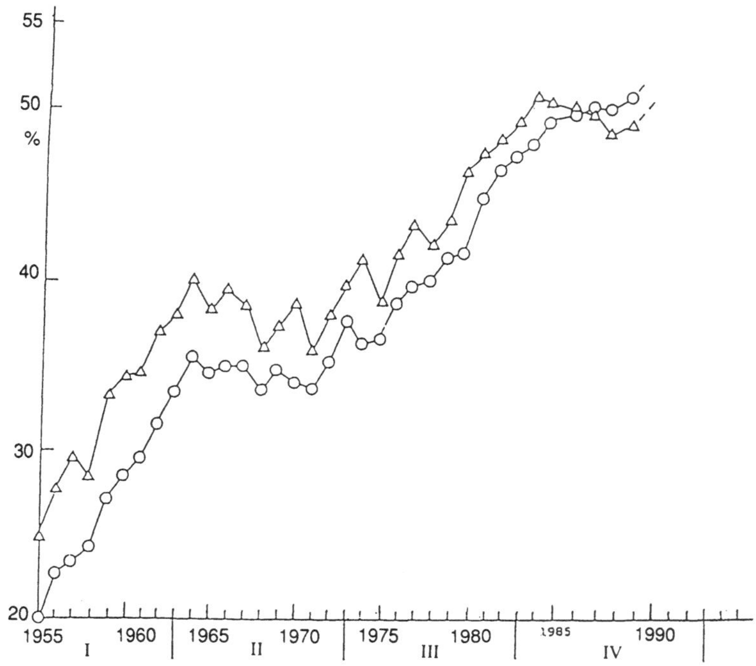

Figure 2.1 Trends of wastepaper recovery (△) and utilisation (○) in Japan.

After 1974, the recovery rate started to increase again and reached 50.4% in 1984. NSSCP production decreased due to technical and financial difficulties arising from the need for waste liquor treatment and for associated water pollution control. In addition, the oil crisis from 1973 very rapidly and substantially increased power costs for pulp production, especially mechanical pulps. This had the most serious impact. Furthermore, at the same time the cost of importing softwood chips from North America, which had started in 1964, increased. The supply of wood chips from sawmill waste is dependent on timber production, which was depressed by the recession in the early 1970s and a subsequent reduction in new home construction. However, when wood chips are prepared from roundwood, the cost must cover felling, handling and transportion of lumber, so these costs are higher than the costs associated with the production of chips from sawmill wastes, hence the increases in chip costs.

Newspaper publishers wanted to maintain a stable supply of newsprint from domestic papermakers. To overcome the increased cost burdens from energy and raw material, the only choice was to utilise deinked old news as a substitute for mechanical pulp, in order to continue the production of newsprint. Therefore, the unit ratio of old news for blending in newsprint furnish increased from 0.10 in 1977 to 0.47 in 1982. (Table 2.1). As the

Table 2.1 Unit consumption of wastepaper for paper and paperboard in Japan (Source: Wastepaper Annual Statistics, Paper Recycling Promotion Centre, Tokyo)

	1977 April–September	1982 April–September	1989 April–September	1990 April–September
Paper	0.156	0.251	0.224	0.240
Newsprint	0.101	0.466	0.423	0.474
Printing, writing paper	0.102	0.142	0.132	0.150
Wrapping paper	0.087	0.034	0.011	0.050
Household tissue	0.905	0.819	0.703	0.616
Miscellaneous paper	0.085	0.037	0.023	0.013
Paperboard	0.693	0.829	0.889	0.907
Kraft liner	0.243	0.446	0.659	0.666
Jute liner	0.886	0.890	0.983	1.031
Interior liner	1.117	1.059	1.095	1.091
Pulp medium	0.477	0.817	0.998	1.021
Low-grade medium	1.103	1.033	1.120	1.120
Manila board	0.331	0.641	0.515	0.514
White paperboard	0.916	1.028	0.919	0.953
Chipboard	1.201	1.127	1.131	1.134
Coloured board	1.097	1.099	1.121	1.091
Total paper and paperboard	0.396	0.485	0.498	0.516

Table 2.2 Trend in pulpwood consumption in Japan

	1955	1977	1982	1988
Domestic softwood	684	684	783	902
Domestic hardwood	89	1125	932	962
Imported softwood	–	907	626	797
Imported hardwood	–	481	515	849
Total pulpwood	773	3196	2856	3510
Increase	1	4.1	3.7	4.5
Imported pulp	–	97	158	283
Paper and board production	220	1570	1745	2464
Increase	1	7.1	7.9	11.2
Wastepaper recovery (%)	24	43	48	48
Wastepaper unit ratio of newsprint (%)	–	10	47	42

Note: Units are $\times 10^4 m^3$ for wood, and $\times 10^4 t$ for pulp, paper and paperboard.

result of this increase in wastepaper utilisation, there was a drastic decrease in imported softwood chips, as shown in Table 2.2.

Although the recovery rate reached 50.4% in 1984, it became rather stagnant in the fourth term until quite recently, in spite of a social trend to promote recycling. After the unit consumption of old news for blending in newsprint furnish reached 0.47 in 1984, it dropped to 0.42 in 1988 and then increased to 0.48 in 1991, as shown in Figure 2.2. Possible reasons why it was difficult to blend more deinked old news could be the decrease in grammage of newsprint from 49 to 46 g m^2, as well as the fall in the market price of pulp.

2.2 Classification of wastepaper

Wastepapers are classified into 49 grades within 4 groups in Europe by CEPAC [2], 51 grades in USA [3] and 26 grades in the 9 groups in Japan [4, 5]. Details of the latter are given in the Appendix. There are two types related to woodfree office wastepaper. They are the so-called manifold white ledger or continuous business forms, and old PPC (plain paper copying) or xerography paper. The former is classified into 4 grades, such as A (printed by a monochromatic impact printer on a white base), B (impact printed on a coloured base), C (printed by a non-impact electrophotographic printer on a white base), and D (non-impact printed on a coloured base).

Standard quality purchase specifications criteria for wastepaper, such as the content of objectionable materials, the condition of baling and the method for labelling, is specified for each grade in the 9 groups by the Paper Recycling Promotion Centre, in Tokyo.

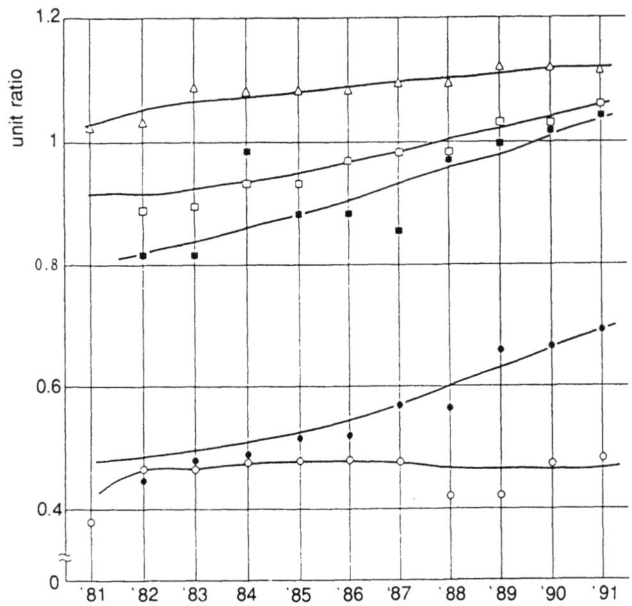

Figure 2.2 Trend of unit ratio of wastepaper (weight of raw material/weight of product) for paper and paperboard in Japan from 1981–1991 (estimated values): (○) newsprint; (●) kraft liner; (□) jute liner; (■) pulp medium; (△) low-grade medium. Source: Koshi Handbook 1992, Paper Recycling Promotion Centre, Tokyo.

2.3 Collection routes

The collection routes of wastepaper are illustrated in Figure 2.3, with estimates of numerical values. Twenty per cent of the total recovered wastepaper is collected from large-volume sources, such as converters, publishers and printing houses, by the buyers for industrial wastepaper. This is known as industrial wastepaper.

Recently, collection by dealers of industrial wastes from large-scale sources, such as department stores, supermarkets and other industries, has increased. Dealers sometimes pay for wastepaper and also sometimes receive payment for managing wastes, depending upon the case in question.

Chirigami-kokan, or collectors, visit houses, on a door-to-door basis, to collect old news, magazines and other wastepaper, and in exchange for this they give toilet rolls or pay cash. Usually a collector drives a small truck and calls out to houses with a microphone. Once the high recovery rate of old news was attributed to these chirigami-kokan, but their share in total wastepaper collection is now diminishing, as a result of increased labour costs.

On the other hand, group collection for recycling by block associations

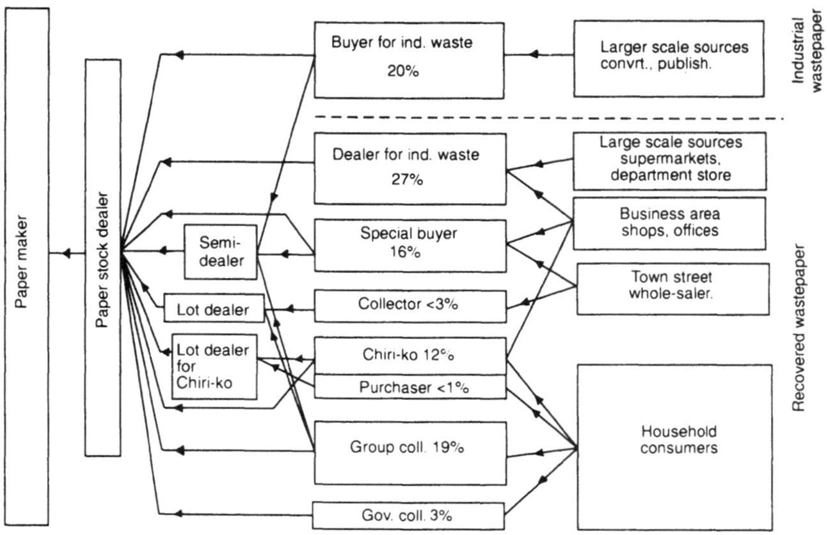

Figure 2.3 Wastepaper generation and collection routes; percentage figures given are estimated (by courtesy of Y. Takimoto, Tomizawa & Co.)

in residential districts, pupils and PTAs of schools, and other volunteer groups, is increasing. As group collection is based on a voluntary movement for environmental protection, through resources and energy preservation, minimal payment is usually made for wastepapers. One interesting instance is the collection of milk cartons. It was reported that about 10 000 t of waste milk cartons were collected, out of a total of 230 000 t, in 1991 [6]. After being used, these cartons are cut out, washed, dried and brought to a Cooperative store or supermarket by consumers. There are Cooperative stores which recover 80% of the milk cartons sold by them. The recovered cartons are bought by a paper stock dealer and sent to paper mills, where they are used in the production of toilet rolls. The proceeds are donated to UNICEF through a regional branch of the Japan Cooperative Society.

Most of the wastepapers collected through the many channels illustrated in Figure 2.3 are gathered together at the paper stock dealers and trading concerns. Their functions are as follows: cleaning up, sorting and baling, holding stock to control supply and demand, supplying wastepaper to the contract or affiliate paper makers according to their request, and functioning in some cases as financial support to collectors.

Most local governments are worried to some extent about garbage management. Some of them are encouraging inhabitants to recover wastepaper, in order to reduce the volume of garbage, by paying for this wastepaper, although the payment is small. It is estimated that it costs about ¥25 (15 cents) to treat 1 kg of garbage by the Tokyo Metropolitan

Government. According to the Ministry of Health and Welfare, 46 940 000 t out of a total of 48 390 000 t of garbage in 1988 were treated by local governments. Its volume was reduced by incineration, or by other means, to 16 900 000 t, i.e. 36% of the original, in 1988. Industrial wastes in 1987 were 253 million t, of which 43% was recycled and the final disposal was 36 million t, which represented 14% of the initial discharge. However, it is very important to maintain an essential final disposal site. In the Tokyo Metropolitan area, for example, the capacity of the final disposal site fell from $23 \times 10^6\,m^3$ in 1986 to $10 \times 10^6\,m^3$ in 1989. Hence the strong desire to recover wastepaper from garbage, or other wastes, for recycling, in order to mitigate the burden of waste management on local governments as much as possible.

As paper consumption increased, the wastepaper content in urban garbage became as high as 49% in the Tokyo area in 1987, compared with 27% in 1967. Most important has been the introduction of computer and copying machines in offices and the subsequent generation of so-called OA (office automation) wastepaper - this was rather unexpected following predictions of the paper-less society. However, this type of wastepaper has been collected rather like ordinary wastepaper (at a low level) by collectors. Many movements to recover OA wastepapers at a number of offices in various districts have been reported. One of these was the study carried out by the Paper Recycling Promotion Centre, supported by the Ministry of International Trade and Industry since 1990. They organised a committee to study a practical way to collect OA wastepaper, classified into several categories. According to their report, 20 146 people in 10 major offices participated in the project and recovered 130 150 kg of classified wastepaper in one month [7]. This means that an average of 6.5 kg per month was collected by each person. The amount of copying and computer papers consumed in this period was 85 828 and 17 218 kg, respectively. Wastepapers were classified into 2 or 3 grades, depending on the office type, which were then re-sorted into 3–6 grades by collectors before shipment to paper makers.

A survey on the participants' attitude to recycling was carried out at the same time. This showed that every person who was involved in this project recognised the significance of recycling, and also that they felt no reservations about recycling, or to classifying wastepaper according to given instructions, on disposal. In addition, they considered the importance of recycling to be in the following order:

(1) recycling of resources;
(2) saving forest resources;
(3) preservation of global environments;
(4) reduction of garbage.

One difficulty in recovering OA wastepaper is the treatment of

confidential documents. Most offices use shredders to cut paper into 3–6 mm width strips. It was proved that the effects of shredding, in this size range, were minor compared to that of recycling [8]. Another solution to this problem is where some contractors receive confidential wastepapers and transport them to a recycling mill to be defibered in a pulper, in the presence of members of the company's staff.

2.4 Recovery and utilisation rates

The trend of the total wastepaper recovery rate is shown in Figure 2.1, with details of the estimated recovery rates (by grade) given in Table 2.3.

The recovery rate of old news was as high as 98.2% in 1991, increasing from a value of 88.7% in 1982. Newspapers delivered to houses and offices usually contained advertising inserts weighing (on average) about 35% of the newspaper. Inserts are printed on woodfree and coated papers. In addition, the moisture content of old news is allowed to be 12%, which is 5% higher than newsprint shipped from paper mills, so the actual recovery rate for newsprint is somewhat lower than that recorded. The recovery rate of old corrugated containers was 73.5% in 1982, 79.5% in 1984 and 69.2% in 1991.

Wastepaper from printing papers has been recovered at low levels, when compared to the total recovery rate of 50.8%. It has remained at ~30% for the last decade. In order to raise this recovery rate, the collection of more wastepaper from printing and information grades is necessary. As a whole, the utilisation rate for printing and information papers still remains at ~16%, which would have to increase to 30% to realise a recovery rate of 55%, which is a paper and board industry target. An upper limit for the recovery of paper and board is widely believed to be 66.1%, i.e. 50.8% is already recovered, with an additional 15.3% recoverable, while 19.8% is unrecoverable and 14.1% is unknown.

Table 2.4 gives an estimate of wastepaper utilisation by products in 1991. It seems that there is very little room to utilise more wastepaper for paperboard, as the unit ratio of wastepaper for paperboard has exceeded 0.92 (average value). Even kraft liner has 30% on average fibres blended from wastepaper, so some liner grades are called giji-K or pseudo-kraftliner. The trend of the unit ratios for paperboards and newsprint is shown in Figure 2.2. As can be seen in this figure, the utilisation rate in newsprint is fairly constant. It is rather unrealistic to expect that more old news can be utilised in newsprint, because of the decreasing grammage, to 43 g m^2.

The unit ratio of wastepaper to household tissue was 0.90 in 1977, but this value gradually decreased to 0.62 in 1991. This means that products made from new pulp by major manufacturers are more accepted by

Table 2.3 Estimation of the recovery rate of wastepaper by grade for 1990 and 1991 (Source: Koshi Handbook 1992, Paper Recycling Promotion Centre, Tokyo)

Grade	Recovery (t)	Consumption of paper and paperboard (t)	Recovery rate in 1991 (%)	Recovery rate in 1990 (%)
Hard white shavings, white card White woody shavings, white manila Fine paper printed (including coated paper) Quires woody paper, printed Old magazines	3 947 239	12 403 176	31.8	30.8
Old news	3 649 743	3 718 129	98.2	96.0
Kraft browns	508 147	752 514	67.5	56.9
Old corrugated containers	5 973 414	8 636 586	69.2	68.3
Box board cuttings	588 710	3 357 665	17.5	18.5
Total	14 667 253	28 868 070	50.8	49.7

Table 2.4 Estimation of unit ratio of wastepaper by grades in 1991 (April–September) (Source: Koshi Handbook 1992, Paper Recycling Promotion Centre, Tokyo)

Material	Wastepaper (total)	White shavings, white cards	White woody shavings, manila	Fine paper, printed	Kraft browns	Quires, woody paper, printed	Old news	Old magazines	Old corrugated containers	Box board cuttings
Newsprint	0.4828			0.0005			0.4820	0.0003		
Printing and information paper	0.1629	0.0008	0.0004	0.0154	0.0037	0.0219	0.1154	0.0053		
Wrapping, kraft paper	0.0063				0.0004		0.0059			
Sanitary tissue	0.6253	0.0161		0.6092						
Miscellaneous	0.0251	0.0095		0.0014		0.0021	0.0014	0.0031	0.0069	0.0007
Total paper	0.2474	0.0025	0.0002	0.0596	0.0021	0.0126	0.1666	0.0033	0.0005	0.0000
Kraft liner	0.6935	0.0006		0.0058	0.0596		0.0047	0.0274	0.5811	0.0143
Jute liner	1.0605	0.0184	0.0093	0.0032	0.0768	0.0008	0.0736	0.1979	0.5965	0.0840
Interior liner (all wastepaper)	1.1045	0.0362	0.0412	0.0259	0.0135	0.0104	0.3659	0.2424	0.2845	0.0845
Pulp medium	1.0419				0.0597		0.0189	0.1413	0.8045	0.0175
Low-grade medium (all wastepaper)	1.1164	0.0025	0.0004	0.0317	0.0401		0.0045	0.1172	0.8855	0.0345
Manila board	0.5137	0.0039	0.0147	0.0489		0.0511	0.1979	0.0931	0.0694	0.0347
White paperboard	0.9540	0.0230	0.0080	0.0978		0.0270	0.2073	0.5390	0.0068	0.0451
Chipboard	1.1168	0.0530	0.0021			0.0044	0.1350	0.5034	0.1344	0.2845
Coloured board	1.1040	0.0037	0.0029	0.0096	0.0253	0.0117	0.0885	0.5242	0.1217	0.3164
Construction board	1.1110	0.0165	0.0308			0.0005	0.3812	0.3868	0.1254	0.1698
Other boards	0.9073	0.0139	0.0132	0.0400	0.0319	0.0059	0.0676	0.1540	0.5214	0.0594
Total paperboard	0.9200	0.0092	0.0060	0.0223	0.0441	0.0067	0.0754	0.1734	0.5325	0.0504
Total	0.5256	0.0053	0.0026	0.0442	0.0195	0.0102	0.1289	0.0736	0.2206	0.0209

Note: unit ratio = wastepaper consumed/unit of product.

consumers, perhaps because of effective sales promotion campaigns and high productivity. It is to be regretted that people who are concerned about recycling nevertheless often choose products from new pulp.

2.5 Legislation

Not only paper and paperboard, but other kinds of waste generated throughout the country are seriously affecting environments. In this connection, the *Law for Promotion of Utilisation of Recyclable Resources* was promulgated on 26th April 1991, and became effective on 25th October 1991 [9]. According to Article 1, the purpose of the law is described as: "Considering that Japan relies on importing many important resources, but a large part of these resources are now being discarded without being used, the purpose of this law is to provide the basic mechanism for promoting the use of recyclable resources, and thereby promote the healthy development of the nation's economy." The paper industry was specified as being one of the specially designated industries, and required to make an effort to increase the utilisation rate of wastepaper according to Ministerial Ordinance No. 53, Ministry of International Trade and Industry. Article 1 of this states that "taking into consideration that the low ratio of wastepaper used in printing paper, information-related paper, and wrapping paper, and the different uses of wastepaper depending on the kind of paper, a business shall, in cooperation with consumers as well as central and local governments, strive to increase to 55% the amount of wastepaper that is used in domestic paper production by 1994."

In addition paper manufacturers are required to execute the following:

1. Provide the necessary stock yard for wastepaper and the installation of facilities or equipment necessary for wastepaper use.
2. Improve technology for wastepaper treatment, such as pulping, screening and cleaning, deinking and strength improvement, in cooperation with machinery and chemical suppliers.
3. Preparation of a plan for wastepaper utilisation for each business year, which should include concrete utilisation conditions.
4. Give information on the percentage of wastepaper used.

Reaching the target of 55% was once considered to be possible without any difficulty, as inferred from an extrapolation of the increase in utilisation rate in previous years. However, it might now be delayed due to the recession of 1992, when production of paper and paperboard had fallen when compared to the previous year.

Over the last 40 years, wastepaper has been utilised merely because of fibre cost reductions, and as much as was necessary to meet demand was collected. However, increased recovery is now generating more wastepaper than economically required, for example, by environmental and political movements.

2.6 Quality of recycled paper

Following a memorandum issued by Government authority, Governmental offices over the country have now been using recycled printing and copying papers for several years. In addition, a number of corporations are using recycled paper in offices, even though the price of recycled paper is never lower when compared with that of regular paper. This can only be possible as a result of the public attitude to resources and environments.

The definition of Saisei-shi or recycled paper is still not clear. The literal translation of Saisei-shi is 'regenerated paper'. Hundreds of brands of paper and paperboard have been designated as environmentally friendly (Eco-marked) or as saving forest resources (Green-marked) by semi-Governmental organisations, but the recycled fibres content in these products is not clearly defined. Paperboards and newsprint are blended with recycled fibres to such a high extent that they are not usually called recycled paper. As a rule, no restrictions have been given to quality specifications regarding these grades of paper and paperboard.

There is no common consensus of opinion about which quality attributes of recycled paper could be accepted as lower than that of regular paper. Sometimes it seems that there is a trend to accept the lower brightness of recycled paper as being evidence of recycling. However, no paper maker accepts that insufficient quality in practical use can be accepted as a result of recycling.

A survey on the quality of so-called recycled papers was carried out in 1992 by Shigyo Times Co. Ltd [10]. Questionnaires were sent to 253 industrial companies, which were selected as representing the 5 top-ranking companies from each of the major industries. According to the results obtained, 56% of the respondents replied that recycled paper was comparatively inferior to regular paper in quality, but 42% admitted to observing no differences between them. Various shortcomings of recycled paper were pointed out by 99 companies (plural answers were included) as follows:

- generation of paper dust - 16
- lower brightness - 12
- lower stiffness - 11
- curl - 8
- lower gloss - 7
- easy picking - 7
- less ink transfer - 6

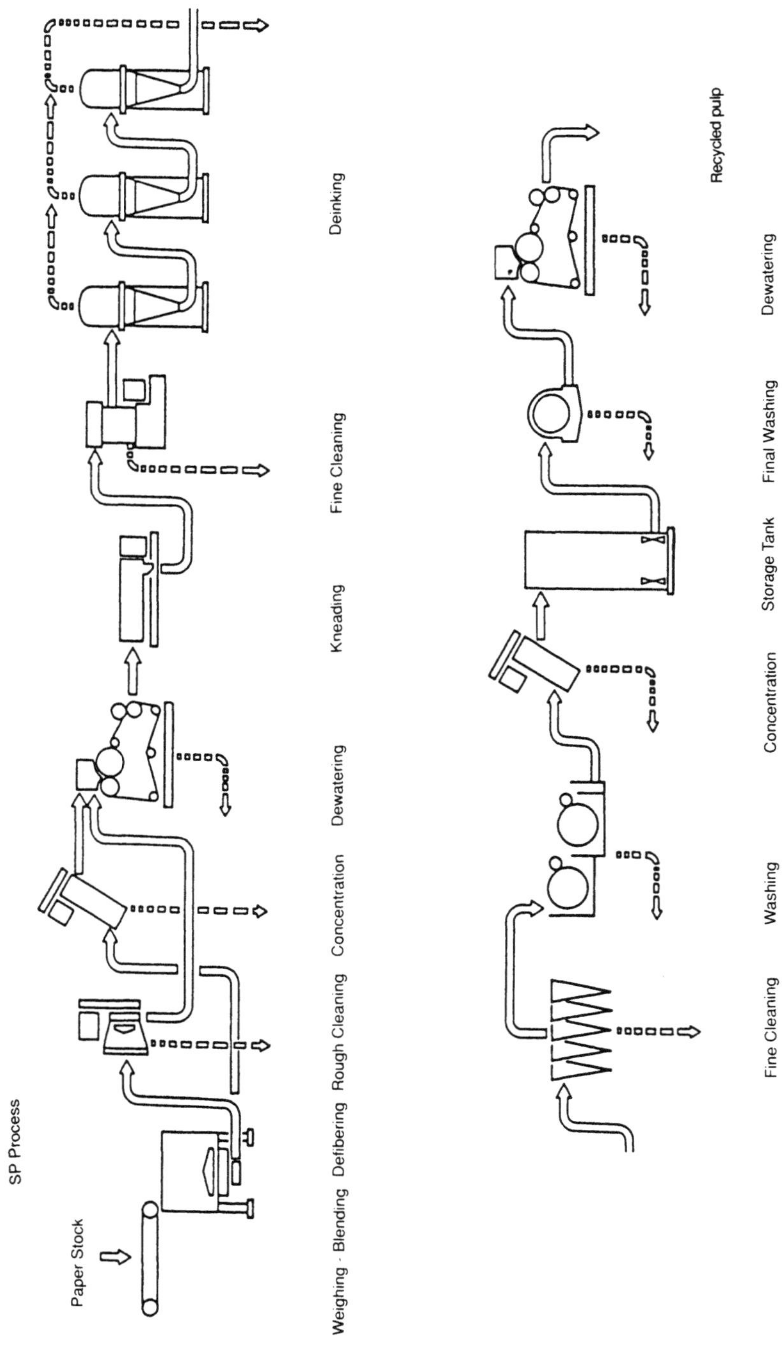

Figure 2.4 Schematic representation of a wastepaper treatment process for producing household tissue from printed woodfree (by courtesy of Ehime Pulp Cooperative Association).

- inferior runnability - 5
- poorer register - 4
- lower strength - 4
- poorer ink absorption - 3
- lower smoothness - 3

In addition 85% of those questioned replied that they purchased recycled paper at a higher price than regular paper.

2.7 Outline of wastepaper treatment

The utilisation of wastepaper has a long history, especially in the case of toilet paper. Several decades ago, printed woodfree was cooked in a special digester, with sodium sulphite and sodium carbonate for defibration, in many small paper mills. Nowadays, almost all tissue paper mills use pulpers. Figure 2.4 is an example of a wastepaper treatment process which is used to produce recycled fibre, for toilet tissue, from printed woodfree.

Treatment of wastepaper, particularly deinking techniques, has been much improved over the last two decades. Processes for wastepaper treatment may be conveniently classified into three main types as follows:

(1) paperboard from old corrugated containers;
(2) newsprint from old news;
(3) printing and information paper from OA wastepaper.

Processes generally include pulping, screening, cleaning, and deinking - by a combination of kneading, soaking, flotation and washing. Most equipment is in world-wide use. However, some specialised equipment items, which were originally designed and developed by domestic Japanese machinery manufacturers, are also popular. The process adopted in a specific mill uses a combination of these equipment items, according to the product grade, depending on the quality of wastepaper and the individual mill's condition. So each paper mill has its own process. It is not an exaggeration to say that there is no tailor-made process for wastepaper treatment.

However, the most important technical problems have changed with time. In the early periods (see Figure 2.1), removal of objectionable materials, such as hot melt and stickies or adhesives, was the most important, whereas during the third term (1973–1983) much of the effort was on deinking of old news that had been printed by coloured offset. Currently, deinking of toner ink from xerography or laser printing is the greatest concern.

2.8 Specified facilities and chemicals

Equipment and facilities for wastepaper treatment in the Japanese paper industry have been reviewed in several publications [11–13].

2.8.1 Pulping

A conventional pulper is still the most popular for old corrugated container systems, with a maximum capacity of 1000 tonnes per day. For pulping in deinking systems, there is some controversy as to whether low or high consistency pulping is the most effective. The latter is based on a concept that a high consistency is more favourable for ink particle dispersion into minute particles. However, this increases the number of ink particles and lowers the brightness after pulping. It is said that 55% of ink can be separated from fibres by low consistency pulping, which can be easily removed by washing before kneading and flotation [14]. The remaining ink can be removed effectively after high density kneading. The choice of pulping consistency should be judged by considering the complete deinking system.

For continuous pulping of old news, a horizontal high-density drum-type pulper has been used, by which large-sized foreign non-fibrous materials can be removed without disintegration. This is beneficial in controlling the number of fine contaminant particles that are generated.

2.8.2 Screening and cleaning

As a coarse screen, a plate with hole perforations is widely used. However, a plate screen, having both holes and slots, has been developed.

One of the most effective means for the removal of objectionable materials, such as hot melt or stickies, is the fine slot screen. Screen plates, with 8–10 cut slots, are commonly used and even a 6-cut (0.15 mm in width) version is available. This size is quite fine, considering (i) the diameter of pulp fibres is in the range 0.02–0.04 mm, but with an even larger hydrodynamic diameter, and (ii) the effective width of a slot to an inclined suspended pulp fibre in a pressure screen.

It is a very popular technique to apply a reverse-type centrifugal cleaner, or similar, to remove light-weight contaminants, such as pieces of polyethylene film, polypropylene string and polystyrene foam, as well as stickies. A gyro-type horizontal cleaner has also been developed, from which rejects contain less fibres.

2.8.3 Kneading

In the process illustrated in Figure 2.4, a combination of kneading and flotation is shown. By this process, toilet papers of 84% brightness are being produced, without bleaching, from printed woodfree.

Many types of kneading machines are used to detach ink particles from fibres under high shearing force, by using chemicals, at ~30% consistency. Some of these are known as processors, dispersers or deflakers. The introduction of kneading has developed deinking efficiency to a high extent, and even two-stage kneading is sometimes recommended.

2.8.4 *Soaking*

After kneading, a so-called soaking stage is introduced in many cases. The function of soaking is considered as increasing the effects of deinking agents, such as surface active agents, caustic soda, and sometimes hydrogen peroxide (with sodium silicate) for bleaching [15]. So the soaking stage can therefore have a bleaching function.

At the beginning, soaking was introduced as a result of mill experience. According to a laboratory test, it was known that the effects of soaking were as follows. In the case of deinking old news, soaking promotes paper strength, except zero-span tensile, as well as brightness, but yellowness increases slightly with conditions of 50°C for 8–10 h with the addition of 1–2% caustic soda at 20% pulp consistency [16]. Conditions involving the use of 0.6% NaOH, at 25% consistency, were recommended for old corrugated containers in order to improve defibration degree, and burst and compressive strengths [17]. However, there is an opinion that soaking for a long period of time is not always necessary if homogeneous mixing of the chemicals is carried out during the kneading stage [18]. To avoid yellowing, soaking was omitted in the process shown in Figure 2.4.

2.8.5 *Flotation*

About 11 types of flotation unit are available from domestic machinery suppliers, and many of these were originally designed and developed in Japan. A characteristic of these units is a high air-to-liquid ratio, for the effective separation of ink particles. Recently, a unit was developed to produce minute air bubbles by the use of a turbine blade, which can remove ink particles smaller than several micrometers, which were once considered as being impossible to remove by flotation, with removal originally achieved only by washing.

2.8.6 *Washing*

Minute ink and filler particles detached from fibres are removed by washing. As is well known, resolution by the naked eye is only possible for particles larger than ~35 μm. Even though smaller ink particles are not visible, the brightness of pulp is reduced if these small particles remain, so their removal is necessary. Conventional decker and screw-press type washers are used. However, wire type washers are becoming popular because they are more effective in removing fine particles.

2.8.7 *Deinking chemicals*

Caustic soda is a basic deinking agent and also promotes defibration in a pulper. Pulping is usually carried out at ~50°C. However, a cold pulping system is gaining attention, because the lower temperature is favourable for controlling the disintegration of stickies, thus easing their removal by screening.

The use of a surface active agent is essential for promoting defibration by improving alkali penetration, especially during soaking, wetting ink during kneading and generating foam to capture aggregated ink particles. There are a number of so-called deinking agents available on the market, such as non-ionic detergents, fatty acids (and fatty acid emulsions), fatty acid derivatives, higher alcohol derivatives and fatty oil derivatives. Generally, the last of these is reported as being more effective in the removal of ink and stickies and in improving brightness, plus greater ease in handling this chemical [19].

The application of kneading has influenced surfactant use, and the type used depends on the site of addition. A surfactant with high penetrability is favoured for pulping. To remove undetached and small ink particles remaining on the fibres after kneading, a surfactant capable of both ink dispersion and ink particle flocculation is suitable.

As flotation units with a high air/liquid ratio have been introduced, low-foaming deinking surfactants with short-lived foam were developed. In addition, it is reported that a deinking agent which has a strong affinity for toner ink has been developed by a major domestic chemical manufacturer [20]. However, no exact information on the chemical composition is available.

Hydrogen peroxide is widely used as a bleaching agent, in the presence of sodium silicate. It is added before the soaking stage, sometimes to the pulper. Usually, no sequestering agent is used with hydrogen peroxide.

Some small tissue manufacturers are using hypochlorite for bleaching. There was some initial concern that this produced dioxins, but this was disproved by a precise analysis of the deinking pulp both before and after bleaching [21]. Most of the dioxins contained in wastepaper go into the sludge and only very low concentrations remain in the product and effluent. Dioxins in sludge can be safely decomposed by incineration at 1000°C, followed by rapid cooling in a desulphurisation unit.

Several Japanese patents have been published describing the utilisation of enzymes, such as cellulase or lipase, for deinking [22]. However, there is no report concerning this type of deinking agent, which is different from the enzymes used in slime or resin control, in practical use.

Appendix Classification of Japanese standard qualities of wastepaper[a]

Statistical group	No.	Grade[b]	Contents	Objectionable materials (%)
Hard white shavings, white cards	1	Jouhaku (white shavings)	Shavings and sheets of white unprinted woodfree paper, from bookbinders, printers and sheeting converters	0
	2	Kuriim Jouhaku (cream shavings)	Shavings and sheets of cream coloured unprinted woodfree paper, from bookbinders, printers and sheeting converters	0
	3	Keihaku (ruled lines shavings)	Shavings and sheets of white or cream coloured woodfree paper, having red or blue ruling or register mark, from bookbinders, printers and sheeting converters	0
	4	Kaado (cards)	Used tabulating cards, domestic or imported	0
White woody shavings, white manila	5	Tokuhaku (high-grade white wood-containing shavings)	Shavings and sheets of white unprinted high-grade wood-containing paper, from bookbinders, printers and sheeting converters	0
	6	Chuuhaku (white wood-containing shavings)	Shavings and sheets of white unprinted wood-containing paper, from bookbinders, printers and newspaper printing factories	0
	7	Shiro Manira (white manila)	Cuttings and sheets of coloured and uncoloured manila boards, from carton makers	0
Fine paper, printed	8	Mozou (printed woodfree)	White woodfree paper, black printed; computer printout	⩽1%
	9	Irojou (coloured printed woodfree)	White woodfree paper, printed with various colours, including coated papers	⩽1%
	10	Kento (woodfree shavings including some colour printed)	Shavings of white uncoated and coated woodfree paper, including some colour printed, from bookbinders and printers	⩽1%
	11	Shiroaato (white coated shavings)	Shavings and sheets of unprinted coated woodfree paper, from bookbinders and printers	0
Quires, woody paper, printed	12	Tokujougiri (high-grade colour printed wood-containing shavings)	Shavings of high-grade wood-containing white paper, printed with various colours, from bookbinders and printers	⩽1%

Appendix *(cont)*

Statistical group	No.	Grade[b]	Contents	Objectionable materials (%)
	13	Betsu Jougiri (colour printed wood-containing shavings)	Shavings of white-containing paper, printed with various colours, from bookbinders and printers	≤1%
	14	Chuushitsu Hogo (high-grade wood-containing waste)	Sheets of high-grade wood-containing paper, from bookbinders and printers	≤2%
	15	Kento Manira (coloured manila)	Cuttings of manila board, printed with various colours, from carbon makers	≤1%
Old news	16	Shinbun (news)	Old news	–
Old magazines	17	Zasshi (magazines)	Old magazines	–
Kraft browns	18	Kiricha (new brown kraft cuttings)	Cuttings of unprinted brown kraft paper, from kraft paper sack factories	0
	19	Mujicha (unprinted brown kraft)	Waste sheets of unprinted brown kraft paper, from kraft paper sack factories	0
	20	Zattai (used brown kraft sacks)	Brown kraft sacks, used for cements, chemicals, fertilizers, foods and others	0
	21	Krafuto Dambouru (kraft liner corrugated waste)	New kraft corrugated cuttings and old kraft corrugated containers, mainly imported	≤2%
Old corrugated containers	22	Danbouru (corrugated container waste)	Old corrugated containers	≤3%
Box board cuttings	23	Wanpu (mill wrapper)	Wrapping paper, used for newsprint rolls and other rolls	≤2%
	24	Joudaishi (white paperboard cuttings)	Cuttings of white paperboard, from carton box makers	0
	25	Daishi (chipboard cuttings)	Cuttings of chipboard and coloured chipboard, from carton box makers	≤3%
	26	Bouru (carton box waste)	Cuttings of straw board, from carton box makers, and used carton boxes of white paperboard, chipboard, coloured chipboard	≤3%

[a] Source: Koshi Handbook 1992, Paper Recycling Promotion Centre, Tokyo.
[b] Japanese name is given first, with English translation in parenthesis.

References

1. Anon (1992) *Pulp Pap. Int.*, **32**(10), 32.
2. Anon (1988) *List of Standard European Qualities of Wastepaper*, CEPAC, Brussels.
3. Anon (1993) *Paper Stock Standards*, Paper Stock Institute of America, New York.
4. Anon (1993) *Statistics of Wastepaper in Japan*, Paper Recycling Promotion Centre, Tokyo.
5. *Koshi Wastepaper Handbook 1992*, ed. Oye, R., Paper Recycling Promotion Centre, Tokyo (1992) pp. 3–17.
6. Nihon Keizai Shinbun (Nikkei newspaper) June 4th (1992).
7. Report on Recovery System of Office Wastepaper (in Japanese) Paper Recycling Promotion Centre, Tokyo (1991).
8. Oye, R. and Toyoda, Y. (1991) *Jpn TAPPI J.*, **45**(11), 1261.
9. *Law for Promotion of Utilisation of Recyclable Resources* and *Cabinet Order and Ministerial Ordinances Under the Law for Promotion of Utilisation of Recyclable Resources* (in English), Clean Japan Centre, Tokyo (1991).
10. *Recycling of Paper* and *Recycled Paper* (in Japanese), Shigyo Times, Tokyo (1992) pp. 171–182.
11. *Proceedings of Japanese Tappi Annual Meeting* (1990), pp. 111–150 (in Japanese).
12. *Proceedings of Japanese Tappi Annual Meeting* (1991), pp. 107–127 (in Japanese).
13. *Wastepaper Handbook 1992*, Paper Recycling Promotion Centre, Tokyo, 1992 (in Japanese).
14. Eguchi, M. (1992) *Jpn Tappi J.*, **46**(11), 1390.
15. Yamamoto, T. (1987) *Kami Pulp no Gijutsu*, **38**(2), 7.
16. Yamamoto, T. (1987) *Kami Pulp no Gijutsu*, **38**(2), 1.
17. Yamamoto, T. (1988) *Kami Pulp no Gijutsu*, **39**(1), 19.
18. *"AIPA" Ehime Pulp Cooper. Assoc.* (415-1, Kewanoe-cho, Kawanoe, Ehime). (1992).
19. Okada, Y. (1991) *Jpn J. Pap. Technol.* (*Kami Pulp Gijutu Times*) Tech. Ann. (Special Issue), 96.
20. Okada, E. (1992) Deinking technology and chemicals system in Pan Pacific countries. *Proc. Pan-Pacific Pulp and Paper Techn. Conf.*, Tokyo, pp. 121–129.
21. Yamakoshi, A. Interim report of Japanese Consumers' Cooperative Union, Tokyo (1992).
22. Sugi, T., Numakura, T. and Tyuu, K. (1992) *Jpn TAPPI J.*, **46**(9), 1102.

3 Wastepaper preparation and contamination removal

R.W.J. McKINNEY

3.1 Introduction

Equipment used in processing wastepaper was initially developed from pulp production and stock preparation equipment, with the addition of specialist stages, for example, flotation deinking. Changes have been by a process of evolution, rather than revolution. The development of equipment has been strongly related to the increased use of wastepaper and the variety of grades being made. In western Europe, the growth of wastepaper use has been remarkable, illustrated by the statistics in Table 3.1 and Figure 3.1 for 9 countries of the EC (Belgium, Denmark, France, Germany, Italy, Ireland, Luxembourg, The Netherlands and the UK).

As Figure 3.1 illustrates, the astonishing increase in wastepaper use has been almost linear, starting as Europe recovered from the Second World War. It also shows very clearly that although there has recently been considerable emphasis on wastepaper use for environmental reasons, there has not yet been a strong move away from the trend line, which suggests that by the year 2000, the wastepaper utilisation rate average would be about 56%. It should also be noted that the growth in wastepaper use has been not only to replace virgin wood pulp - other secondary raw materials have also been replaced. In these countries straw, other vegetable fibres and rags provided considerable quantities of fibre - in 1957 16% of raw materials were from these sources - but this fell to 5% in 1973, to 3% in 1980 [1] and to <1% in 1990. Over the same period the proportion of virgin fibre used has stayed fairly constant, but wastepaper has grown - during

Table 3.1 Wastepaper use and utilisation rate (in nine EC countries)

Year	Total paper and board production (Mt)	Total wastepaper consumption (Mt)	Utilisation rate (%)
1950	7.0	1.9	27.1
1960	13.3	4.2	31.6
1970	20.7	7.7	37.2
1980	24.2	10.5	43.4
1990	33.7	16.8	49.9
1992	35.5	18.5	52.0

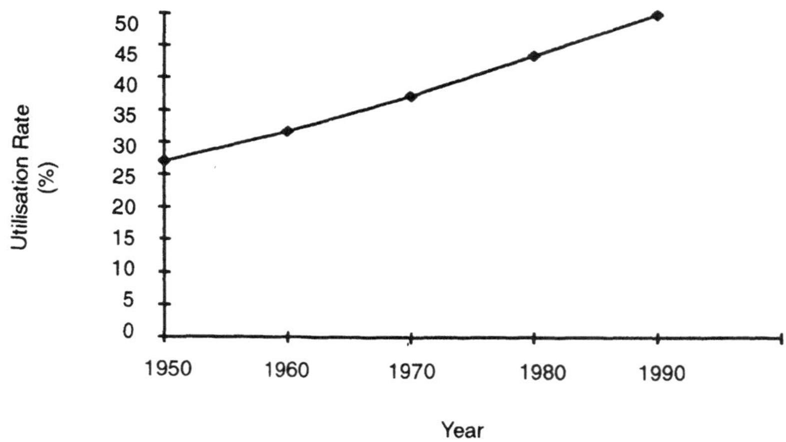

Figure 3.1 Wastepaper utilisation rate.

the period 1957 to 1980 the wastepaper utilisation rate increased from 30% to 43% - the increase during this period matching almost exactly the fall in the use of straw, rags, etc.

Reasons behind the growth in wastepaper use were primarily economic. The EC was (and is) fibre deficient and has to import pulp or paper, so the only way domestic producers could develop was by the use of wastepaper. Its use was initially in packaging, then tissue, and a little later, in newsprint. In the late 1980s and early 1990s it began to be used in graphics papers other than newsprint.

During this period, there have been many changes in wastepaper collection systems. Achieving good wastepaper quality is dependent on sorting by hand, either at source, or in a sorting plant, after collection, or both. At the same time as inflation pushed up the costs of human involvement, the real cost of paper and board was falling, due largely to the economies of scale achieved in pulp and paper making. There are no similar economies of scale in wastepaper sorting, so the quality of wastepaper has fallen, due to cost pressures that limit sorting in a wastepaper plant.

Wastepaper preparation systems have had to respond to the challenges posed by these changes - increasing use of wastepaper in a wider range of products, more contaminated wastepapers and lower product prices. In addition, the diversity of contaminants increased, as paper and board became substrates for a wider range of processes and new printing techniques were adopted. Pressure to use less fresh water has also increased the complexity of processing lines. Nevertheless, a wastepaper line of the 1950s, using packaging grades, would have the basic components of a 'state of the art' plant today - pulper, screening and cleaning, and perhaps dispersion. What a 1950s recycling plant operator would fail to recognise is the control system. This has allowed processes to come under control, leading to more stable operation and higher and more consistent quality. Continual improvements in equipment design have also contributed to improved recycled-fibre quality, allowing the expansion in wastepaper use illustrated in Figure 3.1.

3.2 Wastepaper contaminants

Contaminant removal, essential to convert wastepaper into a reusable fibre, is one of the most important factors influencing the economics of the recycling operation, since this has a direct bearing on the yield from wastepaper and total costs. Unsuccessful contaminant removal reduces paper or board machine efficiency and lowers product quality. The level of contaminants expected in wastepaper grades to be used has a major effect on the complexity of a processing line and hence its capital costs.

3.2.1 Sources of contaminants

As paper and board products are prepared for sale to consumers, by converting and finishing, many types of processes and materials are used to produce finished goods; printing inks, polyethylene or foil liners, adhesives etc. As these products are used and enter the wastepaper collection system, many opportunities are presented for additional non-fibrous materials to enter the wastepaper stream, such as staples, rubber bands, food debris etc. Consequently, wastepapers contain materials ranging from gross contamination by car engine components to microcontamination from ink capsules in carbonless copy papers. All contaminants reduce wastepaper value.

No matter what product is made from wastepaper, its quality, in terms of visual appearance and strength properties, is improved by contaminant removal. This is the objective of all wastepaper stock preparation systems. However, the end product to be made from specific wastepapers determines which non-fibrous materials are regarded as serious contaminants. In fluting medium, for example, ink is not regarded as a serious contaminant. In tissue, a clean, free draining stock is required, consequently coating and filler clays are regarded as contaminants. When printing and writing papers are made from wastepaper, fillers are not regarded as contaminants. Thus, it would be incorrect to define all non-fibrous materials as contaminants, though strictly it is true.

Most non-fibrous materials reduce fibre web strength. As particle size increases, this effect increases, so large particles cause weak spots in the fibre web. Removal is essential to achieve maximum and consistent web strength, but is difficult due to the diversity of materials, organic and inorganic, natural and synthetic, etc. Some reviews of contaminants have

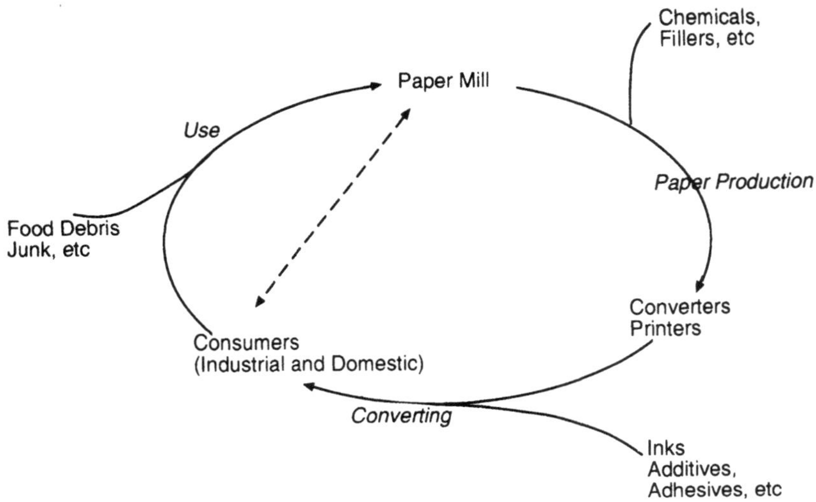

Figure 3.2 Contaminant entry into the paper cycle.

focused on their chemistry or solubility [2, 3], but they can be classified into three overlapping groups, according to their point of entry to the paper cycle, illustrated in Figure 3.2. This is discussed below with respect to the classification into two grades of wastepaper, pre- and post-consumer grades. Groups of contaminants are:

1. Paper mill additives - enter the paper cycle during the manufacture of paper and board - clays, chalk, dyes, and many other additives.
2. Converting additives - enter the paper cycle during conversion and finishing of basic paper and board products - printing inks, foil, polyethylene, adhesives, staples, pins, starch and other chemical additives, etc.
3. Consumer debris - enters the paper cycle during use and subsequent collection of used paper and board products, for example, food debris and junk. Although these can enter the paper cycle at any point, in a paper mill or printing plant, opportunities for their introduction are much greater after use.

The effects of contaminants from these groups on paper or board making processes and product quality are discussed below.

3.2.1.1 Paper mill additives. Clays, chalk and chemical additives, added during the manufacture of paper and board are typical. Wastepapers containing only these additives are high quality and are mainly pre-consumer wastes from converting operations or mill broke. Provided equivalent grades are being made, their re-use should cause few problems. However, some chemical additives can create problems; binders in coating mixes can cause deposition problems if they agglomerate into larger, unsightly, stickie particles – 'white pitch'. Other additives which create problems during their re-use as mill broke include dyes and microcapsules used in carbonless copy papers, wet strength resins, deep colours, etc.

Where contamination is low, processing needs are minimal, though in the case of tissue, the presence of fillers reduces drainage and thus machine speeds, reduces creping blade and possibly Yankee cylinder life, and gives tissue products a coarse feel. Consequently, inorganic fillers and coating materials are significant contaminants in tissue machine systems and washing stages are included to remove various clays, chalk and other inorganics added as filler or coatings.

Coloured papers (from dye additions) can be recycled to produce papers of the same colour, or bleached to improve whiteness. In this latter case, the colouring agent is regarded as a contaminant.

Although contamination is low, fibres from these wastepapers are recycled fibres. Fibres which have been pressed and dried on a paper machine have different properties from equivalent virgin fibres and behave as recycled fibres in physical properties (cf. chapter 6).

3.2.1.2 Converting additives. These include printing inks, pigments, foil, polyethylene, adhesives, staples, chemical additives such as wax, etc. added during the converting and finishing of paper and board.

Wastepaper grades are lower quality than those already described, but are classed as pre-consumer grades, derived primarily from printers and converting operations. This group contains contaminants which are difficult to remove and which create major problems during paper and board production. Hence processing needs are significant, and in most cases are the same as is required for equivalent post-consumer wastepapers.

3.2.1.3 Consumer debris. A very wide range of contaminants are introduced following use, as well as at the various stages involved in the collection of used paper and board products. These can range from hazardous materials, for example, residues in containers or gas cylinders inadvertently included when wastepaper is collected, to contaminants such as food debris. A wider variety of paper types is present in these post-consumer grades.

Almost any material can end up as a contaminant via this route. Sorting by wastepaper merchants of bulk quantities of wastepapers, for example, old corrugated containers from supermarkets, can help to reduce the contamination. However, thorough sorting is labour intensive, hence low-priced grades cannot be thoroughly sorted. Source separation of wastes assists in the production of clean wastepaper, but this requires a well educated, environmentally aware and motivated population.

In many ways, a distinction between these latter two groups of contaminants is artificial, since wastepaper processing needs are determined largely by converting additives, such as inks and adhesives. There is little difference, for example, between over-issue and once read news (pre- and post-consumer grades of newsprint). Thus, in the case of processing needs to recycle printed papers, distinctions between pre- and post-consumer grades of wastepaper are arbitrary. In terms of physical fibre properties, there is no difference between pre- and post-consumer wastepapers or mill broke.

3.2.2 Stickies composition and formation

Adhesive particles, generically referred to as 'stickies' are the single greatest problem for many wastepaper users and are thus classed as a special type of contaminant. Their removal by mechanical means is discussed later in this chapter, but other means of control are outlined below. Stickies control has been the subject of a great deal of research but it is clear there is not a single answer to the many problems they cause on paper and board machines using recycled fibre. A difficulty faced by researchers is that not only is the subject very complex, but individual mills have different names for various types of stickies, sometimes dependent on their effect, softening temperature, or tack. In addition, test methods used vary enormously so that data from different sources cannot be compared. Literature reviews of test methods suggest there are almost as many stickie test methods as there are types of stickies [4, 5]. Stickies have been a problem for many years and despite much research and advances in equipment, remain a problem, for a variety of reasons, which include:

- increased use of adhesives in paper converting and more end uses, for example, in automatic mailing systems, some of which use pressure sensitive adhesive labels. Increased levels of use result in increased contamination of wastepaper;
- the lack of one single best way to remove or control stickies;
- the inability to control precisely the quality of the raw material, wastepaper;
- increased use of chemicals by the paper industry, which increases the complexity of the chemical 'soup' created when wastepaper is pulped;
- not all mills are able to install new equipment to remove or control stickies;
- the wide size range of stickie particles created during wastepaper processing.

The chemistry of adhesives is complex; various formulations are used for different purposes, frequently as a blend of different materials. Hot melts, for example, used in magazine binding, etc. are frequently blends of ethylene vinyl acetate (EVA) and wax. The composition of a typical

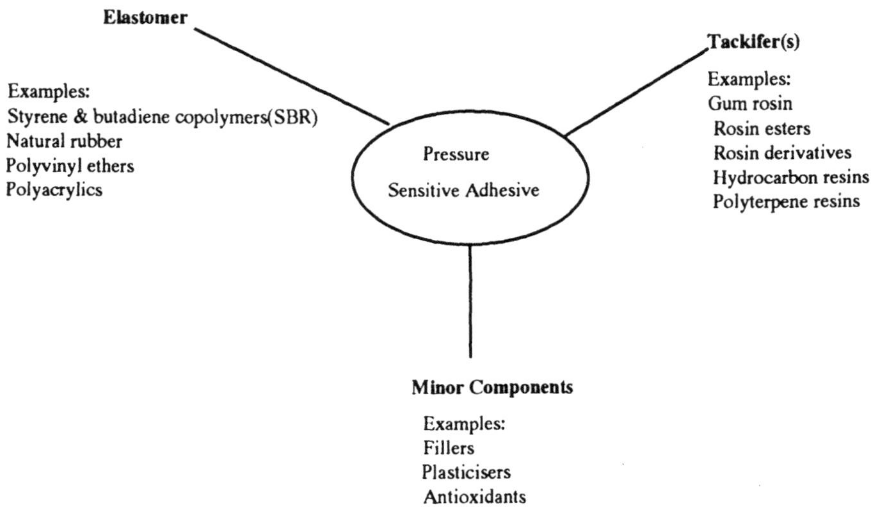

Figure 3.3 Typical components in pressure sensitive adhesive.

pressure sensitive adhesive is given in Figure 3.3. Some of these components are also used in hot melt pressure sensitive adhesives.

Stickies are complex, both in terms of composition and surface properties. It is not usually possible to determine if the components of stickies, when analysed, are derived from the original adhesive, or if they or their tackiness are a result of interactions between chemicals and/or colloidal particles in the system. Many of the tackifiers listed in Figure 3.3 occur naturally in pulp or are added to paper as sizing agents. An analysis of stickies causing a severe deposition problem in a tissue mill, which was not using wastepaper, showed that stickies were formed by an interaction between naturally occurring wood resins (pitch) and a creping additive [6]. In kraft pulping, the addition of oil-based defoamers exacerbates pitch problems [7] and it is probable that similar interactions occur in wastepaper systems. Laboratory prepared polyisoprene, an adhesive used in self-seal envelopes, when pulped with wastepaper and clean water, gave hard, non-tacky particles. However, samples of stickies from a tissue mill causing a deposition problem were polyisoprene, and were soft and tacky [8]. When stickies from a newsprint mill were analysed, the type which caused most problems was found to be predominantly EVA, a component of hot melt adhesives, which are normally hard and not tacky at mill system temperatures. These had become sticky, soft and flexible in the mill system by the absorption of low molecular mass materials, which were found to include mineral oils, fatty acids and phenol sulphonates. Deposition of these was primarily by filling in the voids in the bottom wire of the twin wire machine and in a thickening stage in the deinking plant. Mineral oils alone were also found to fill the wires, sometimes in combination with fatty acids [9]. In the same study, deposits on the wire (rather than in) were also analysed and were styrene diene copolymer, shown to be from coated papers, plus fatty acids, resin acids and hydrocarbons from mineral oils, resins or asphalt [9].

Another study examined stickies being created by the use of about 20% coated broke in the production of lightweight coated paper, with no wastepaper use. These proved to be a styrene–butadiene copolymer, but with some unidentified wood extractives included. Another stickie, depositing on foil blades in a liner board mill, using some coated board wastepapers, was shown to be a mixture of a styrene–butadiene copolymer and polyethylene [10].

Stickies removed from other areas of a machine have also been shown to be mixtures of chemicals. Scrapings removed from a calendar in a newsprint mill included a vinyl acetate polymer, pitch, fatty and resin acids, wax and styrene–butadiene. Fatty acids were from the wastepaper, not additives. In the same mill, deposits from piping were a mixture of resin and fatty acids – 'pitch'. No microbiological details were given [11]. Inorganic materials were also present. Clay, talc, etc. are frequently found in deposits.

In laboratory studies, chemicals added to known adhesives have been shown to increase stickie tackiness. In one case, phenol sulphonates increased the tack of an EVA based hot melt, when phenol sulphonate was 5% or more of the mass of the complex [9]. In another laboratory study, a paraffin (kerosene)-based solvent cleaner increased deposition of acrylic copolymer stickies on a polyester wire by 30%, at an addition level of 0.008% [12], which indicated tackiness had increased.

These studies reveal not only the complexity of the composition of stickies, but also routes for their formation. There appear to be two distinct classes, those that are formed from the direct breakdown of adhesives in and on wastepapers and those that develop from chemical interactions. These are sometimes referred to as primary and secondary stickies, respectively. However, the formation of primary stickies is not restricted to the production of tacky particles during, for example, pulping, when adhesives on wastepaper are broken up into discrete particles. In at least some cases, there appears to be an interaction between hard, non-tacky adhesive particles produced from the breakdown of adhesives by pulping with other chemicals, to create a soft, tacky stickie, which causes problems. Some natural components of wood, or additives such as oil-based defoamers and fatty acids probably contribute to this interaction, as well as to the agglomerate which causes secondary stickie problems. These 'stickie activators' were frequently found in stickie analyses [7–11].

A knowledge of the chemical composition of stickies is useful in some circumstances, for example, when resolving a persistent problem which has no obvious explanation, such as the use of a contaminated wastepaper, or malfunction of some part of the cleaning system. Stickie analysis may reveal the source, or show that agglomeration of some components is responsible for stickie formation. If this is the case, then simple changes, such as of defoamer type, may eliminate the problem. Analysis is also useful in confirming that the source of problems is a specific wastepaper, for example, confirming hot melt bound magazines are responsible for a problem, enabling steps to be taken to eliminate these from the wastepaper furnish.

In the case of deposition, probably the most common stickies problem, there has to be an interaction between the surface and the stickie particle; adhesion is due to interactions at the molecular level. Wetting of a surface, for example, a polyester wire, is required. The stickie particle must have a lower surface energy than the surface to which it adheres and the greater the difference, the more likely the stickie is to adhere to the surface. Adhesives are designed to have low surface energies, in the range 20–40 $mN\,m^{-1}$.

Wetting of a surface of a solid (s) by a liquid (l) is illustrated in Figure 3.4, in which γ_s and γ_l are the surface tensions of the solid and liquid respectively and γ_{sl} is the interfacial tension between the liquid and the solid surface. In Figure 3.4(a) the contact angle, θ, is large, and the liquid does not spread over the surface of the solid, that is, there is no wetting. This would be the result of, for example, a water drop on a waxed board, or an adhesive particle with a surface energy of 40 $mN\,m^{-1}$ on a smooth teflon coated surface. Surface energies for smooth surfaces of named solids are given in Table 3.2.

In Figure 3.4(b), the contact angle, θ, is much lower and there is wetting of the surface, for example, a water drop on a medium sized paper. Instantaneous wetting is illustrated in Figure 3.4(c), corresponding to a water drop on filter paper. The greater the difference between γ_s and γ_l, the greater the adhesion tension, defined by:

$$\gamma_s - \gamma_l = \gamma_{sl} \cos \theta$$

However, surface free energies are not the only determinant of adhesion. Surface roughness is also important, as is cleanliness. Anecdotal information from some mills suggests that in some cases chemical treatment to inhibit deposition on wires or felts is unnecessary with worn wires and felts. This may reflect wear on the surface, producing a smooth surface, or there may be chemical interactions changing surface properties. In laboratory studies it was shown that when a polyester fabric was cleaned by a paraffin (kerosene)-based cleaner, subsequent deposition of acrylic copolymers increased. If the polyester wire was exposed to virgin pulp for a short period (5 min) before running the deposition test, deposition fell, which suggests both treatments affected the surface free energy of the polyester. The complexity of interactions was illustrated by different responses of acrylic

Table 3.2 Critical surface energies of named solids (mN m^{-1})

Teflon	18
Polyethylene	30
Polyester	43
Water	72
Cellulose	200
Copper	2700

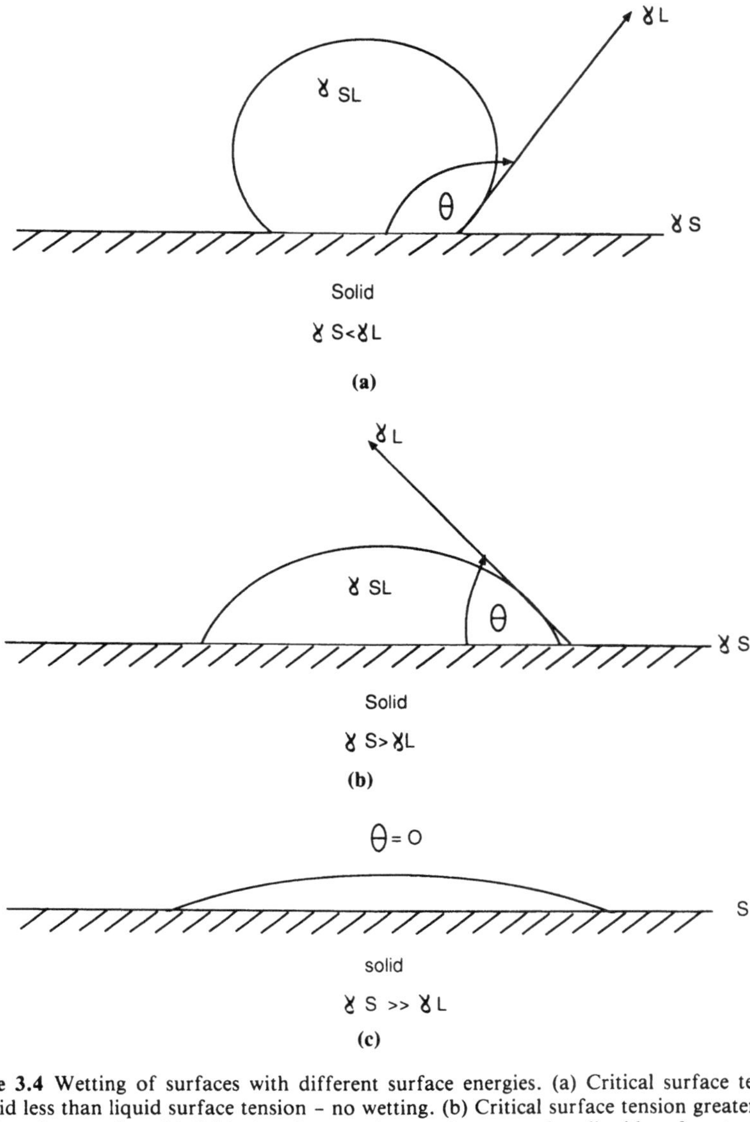

Figure 3.4 Wetting of surfaces with different surface energies. (a) Critical surface tension of solid less than liquid surface tension - no wetting. (b) Critical surface tension greater than liquid surface tension. (c) Critical surface tension much greater than liquid surface tension - spontaneous wetting.

copolymer stickies in different systems. When printed wastepapers, both woodfree and wood-containing, were pulped with acrylic copolymer and not virgin kraft pulps, deposition did not increase when the polyester wire was cleaned. This may have been due to a component of ink coating the surface of the wire, reducing the surface free energy or to changes in pH [12] or the inability of the cleaner to remove a deposit which had effectively reduced surface free energy.

Fatty acids and other extractives or additives have been shown to reduce the surface free energy of various paper grades, thus reducing their coefficient of friction (angle of slip). Additives which increased the angle of slip were absorptive (porous) fillers such as a synthetic silicate and calcined clay. Their effectiveness in increasing the coefficient of friction (increasing the surface free energy) was thought to be due to their absorption of fatty acids and other components which reduce the coefficient of friction. Even a monolayer of organic material at the surface of paper affects its coefficient of friction and surface free energy. Fatty acids and other extractives could affect stickie deposition in a similar way, by coating the surface of an adhesive particle, reducing its surface free energy and making it more likely to deposit, or they may increase the surface free energy of the deposition area.

Fatty acids have been shown to concentrate at the surface of paper, thought to be due to migration as the sheet was dried [13]. Another explanation for deposition could be that as water is evaporated off drying cylinders, a residue coats the drying cylinders and is adsorbed onto the fibre web, similar to the buildup of a natural coating on a Yankee dryer. If non-tacky adhesive particles were at the surface of the fibre web, these could be softened by activators in the surface layer on dryer cans, for example, fatty acids, and possibly removed from the web, leading to deposition problems in felts, on doctor blades and dryer cans, etc.

Although there is a great deal not known about factors controlling surface energies and deposition in mill systems, this provides a mechanism to prevent deposition by the use of additives. Chemicals can be added to increase the contact angle of treated stickies. When chemicals were used, reductions in deposition problems were claimed [14].

The mechanism behind fillers reducing the coefficient of friction of paper was believed to be by adsorption [13] but adsorption by talc of pitch materials was shown to be low - about 10% of softwood pitch and up to 20% of aspen pitch. Since talc controls pitch deposition, the mechanism was postulated to be due to talc coating pitch particles, resulting in detackification [15]. In the study examining the effect of fillers on the coefficient of friction, talc was shown to have minimal effect. This information suggests that two classes of fillers may be used to control stickie deposition in two different ways:

Table 3.3 Relationship between electrostatic climate and felt wash frequency [16]

Time period	Average felt wash frequency (number per week)	Measurements inside conditions (%)
1/84–6/84	5.2	46.3
7/84–6/85	4.1	62.8
7/85–6/86	2.1	73.3
1/83–6/86	0.6	93.3

1. Adsorptive fillers, such as synthetic silica and calcined kaolin may adsorb extractives and other materials responsible for a reduction in surface energy, so helping to prevent deposition. They may be better if used at the beginning of the system.
2. Talc may be used to detackify stickies, by coating stickie particles, and would be more effective if used towards the end of a deinking system or in a machine stock approach system.

The importance of the electrostatic climate was demonstrated by frequent measurements in a Dutch wastepaper-based tissue mill, which showed the relationship of the system zeta potential and conductivity to downtime for felt washing. A combination of high conductivity and a slightly negative zeta potential resulted in a reduced incidence of felt washing. Results are summarised in Table 3.3.

Improvements were believed to be due to increased retention of colloidal organics in the web preventing a buildup of colloidal organics in the machine white water. Conditions considered ideal were conductivity $>900\,\mu S\,cm^{-1}$, and zeta potential in the range 0 to $-10\,mV$. Paper machine pH was about 7 [16, 17]. Under these conditions repulsion between colloidal particles should be reduced, which could be expected to result in agglomeration. However, the results suggest that colloidal particles were carried out of the system, and a hypothesis for a mechanism which would achieve this is given below.

Potential energy curves for two types of colloidal particles are illustrated in Figure 3.5, which shows the idealised behaviour of two differently and two similarly charged hydrophobic particles.

The two lines given are the total potential energy curves, which are the sums of attractive and repulsive forces, as a function of the distance between the particles. As two particles with like charges approach, the total energy is repulsive, apart from a weak attractive interaction at a relatively large distance apart, which may not exist for small particles. Other than this, the curve represents a stable dispersion until the two particles are very close together and Van der Waals attractive forces are effective. When particles have opposite charges the sum of potential energy provides no barrier to agglomeration, since there is an electrostatic interaction as well

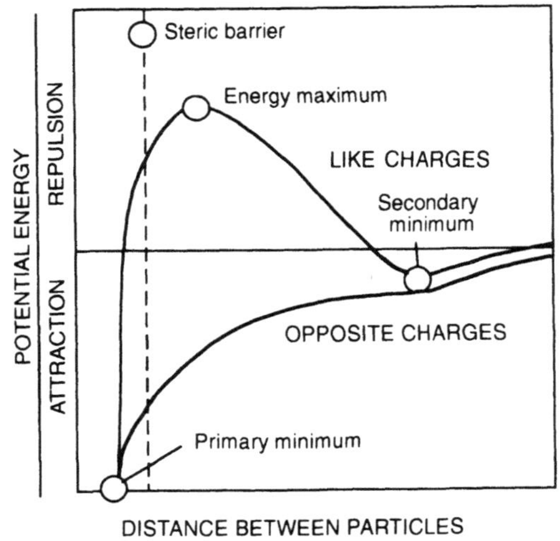

Figure 3.5 Total potential energy curves between particles of like and unlike charges.

as Van der Waals attractive forces. When these particles collide, the total potential energy reaches a deep minimum and a stable, essentially irreversible agglomerate forms. However, if there is a steric barrier, for example, due to surface roughness or adsorbed material, particles cannot get close enough to reach the minimum energy and particles can be separated, for example, by kinetic energy from mixing.

Positively charged particles can be made to deposit on fibres; for example, polystyrene latex, which has a positive charge below pH 8.0, was shown to deposit on fibres in the pH range 3-6. Deposition onto fibres occurred only when there was repulsion between latex particles, which increased the stability of the latex dispersion. Deposition was enhanced by the opposite charges of fibre and the latex particles. Attachment was shown to be weak in some circumstances, and was dependent on the size of the particle and the ability to make good contact, so that the effective distance between fibres and latex particles was critical in determining the behaviour of depositing particles. When particles do not make good contact, for example, in the case of large particles, they respond to higher salt concentrations, or anionic compounds, so that even gentle agitation removed the latex particles. Irreversible deposition was not achieved, believed to be due to the steric effect illustrated in Figure 3.5 [18].

A similar mechanism may be the reason for deposition control being achieved by the conditions in the Dutch tissue mill - pH 7.0, process water zeta potential 0 to $-10\,mV$ and conductivity $>900\,\mu S\,cm^{-1}$. In a recycle mill, which has many types of colloidal particles that differ in size, surface charge, charge density and Van der Waals attractive forces, coagulation may occur between like particles (homocoagulation) as well as between unlike particles (heterocoagulation). The colloidal system can be destabilised by reducing repulsive forces, for example, by adsorption of counterions (to reduce surface charge), salt addition (compresses the electrical double layer, reducing the magnitude of the double layer potential (cf. Figures 7.13 and 7.14)) and pH changes (alters surface potential). In wastepaper processing the best known destabilising mechanism is the acid shock, used to agglomerate colloidal material for removal by screening or cleaning. (cf. Figure 9.2). In the case of the Dutch tissue mill, the pH of 7 and a process water zeta potential of 0 to −10 mV must mean that some of the colloidal particles have a slight positive charge. This would result in the deposition of some of the positively charged particles onto the negatively charged fibre surface. At pH 7.0, fibre still has a negative zeta potential. Due to the high conductivity, the magnitude of the double layer potential is reduced for all particles. If large agglomerates form (which could cause stickie problems) they do not deposit on fibres due to steric hindrance and their low surface potential reduces the stability of the agglomerate, so they are redispersed, but they can be removed as small particles by deposition onto fibres. Although it is unlikely that all of the various types of particles present are removed in this way, those that contribute most to deposition problems may be removed. There is no experimental evidence available to support this hypothesis, though a study confirmed that system charge has a major effect on stickies deposition and that machines may have an optimal colloidal charge 'window' for operation with minimal deposition problems. The window varies from machine to machine [19].

Pulping pH can also have a marked impact on primary stickie formation and deposition. One study pulped a board contaminated with sticky material using a pilot scale drum pulper and a hydrapulper. Although a reduction in fibre flakes was achieved, the stickie count (number per 100 g bone dry fibre) increased four fold when wastepaper was pulped with 1.0% NaOH [20]. In a laboratory study, both the number and tack of acrylate stickies (pressure sensitive) increased when pulped at a pH of 10.5, compared to 7.0 and 4.5, using a bleached kraft virgin-fibre furnish. The pulping pH determined deposition, not the test pH [12]. Chemical analysis of the stickies formed may have revealed the reason for increased tack; it may have been due to dissolution of residual extractives in the pulp or sodium hydroxide dissolving some of the components of the adhesive. However, adjustment to pH 10.5 following pulping did not increase either tack or number, suggesting another mechanism. The energy supplied during pulping may be significant, perhaps by providing the kinetic energy required for 'effective' collisions between similarly charged particles, which are present only at the higher pH. Alternatively, after pulping, consistency was reduced from 3.0% to 0.5%, so that the concentration of an unknown tack enhancer produced at the higher pH may have fallen below a threshold value and thus did not affect tack.

When calcium chloride was added to the pulp/deionised water system, deposition was reduced very significantly, to about 70% of the original value, at 30 mg l^{-1} water hardness, as calcium [12]. Since calcium is frequently present in recycling systems, from either naturally hard water or from calcium carbonate coatings, this may be less important in mill situations. In the Dutch mill, process water hardness was very high, more than 700 ppm calcium [17]. Water hardness was shown to affect 'synthetic' pitch deposition, but was dependent on the 'pitch' components. Increased hardness usually increased pitch deposition [21]. These results appear contradictory but probably illustrate the dependency on components.

The evidence for high pH increasing stickie problems is quite strong so a reduction in pulping pH to neutral should reduce stickie problems. It appears that pH after pulping can be increased without creating more stickies or increasing their tack, so that optimal conditions for flotation deinking or peroxide bleaching can be met. When pulping wood-containing wastepapers at a high pH, peroxide is usually added, to prevent alkali darkening. As well as wood extractives dissolved by the high pH, some resin acids are dissolved by peroxide and these may contribute to increased tack or agglomerates. Avoiding the use of a high pulping pH also avoids using hydrogen peroxide. However, not all mills will be able to avoid using a high pH, due to reduced pulper capacity at lower pH and higher specific energy requirements. Advantages of not using a high pulping pH need to balanced against these disadvantages.

3.2.3 Control of stickies and other contaminants

Without doubt, the best controls for problems due to contaminants and stickies are either not to buy them as wastepaper, or to remove them during wastepaper processing. Unfortunately, it is almost impossible to eliminate totally contaminants from wastepaper and the removal of 100% of contaminants during processing is not possible, even when operating at a low yield, so other control strategies may be required.

Options for control of stickies have been extensively reviewed [22–25]. There are many different approaches which can be adopted, including:

- control of wastepaper quality;
- selection of appropriate recycling technology;
- mechanical removal;
- mechanical dispersion;
- chemical control, by the use of additives;
- design of reject and water systems, to ensure no recycling of adhesives;
- selection of machine clothing and cleaning systems;
- use of recyclable adhesives.

None of these are mutually exclusive, so a blend of these approaches can be adopted. Each is briefly described below.

3.2.3.1 Control of wastepaper quality. Quality control of all raw materials is essential and wastepaper is no exception. However, it is very difficult to assess accurately quality from a small sample or visual inspection of a truck or box-car load of wastepaper. Nonetheless, it is necessary to inspect all wastepaper, if only to check moisture content and apparent cleanliness. Individual bales can be labelled, either by the supplier, or the mill, with labels removed as bales are used. This can be useful when tracing problems, but is not possible with loose deliveries of wastepaper, which is becoming more common.

Specifications for individual wastepaper grades should be developed, to detail acceptable moisture content, as well as prohibited and restricted materials. Wastepaper suppliers need to understand if prohibited means totally unacceptable, or the level at which rejects and downgradings will occur; similarly for restricted materials. When revisions to specifications are made, wastepaper merchants should be informed well in advance of the implementation of the change, with samples. Specifications should be reasonable, and focus on exact requirements. In the UK, for example, one grade allows blends of different types of wastepaper (multigrade). This allows wastepaper sorters to concentrate on the removal of contraries, not on sorting different, but acceptable, types of wastepaper into tightly defined paper grades. There has also to be an acceptance by purchasers that if specifications are higher than the industry norm for a specific grade, a higher price will have to be paid. Within limits, this can be a cost effective approach to contaminant control.

Even 'high quality' pre-consumer grades need to be checked - press room wastepaper can include papers or synthetic cloths used to clean presses, leading to problems with ink and synthetic fibres.

In a liner mill, 26 unacceptable materials in manifold white ledger wastepaper were identified and given to suppliers. This grade was used in the production of mottled white liner. The wastepaper inspection process resulted in a reject rate of 3% and pulper dirt count testing in another 2% of rejects, fed to an old corrugated container (OCC) line [26]. A reject rate of 3–5% is fairly typical for woodfree grades of wastepaper, though this depends on rigorous inspection against specifications. Similar levels have also been reported in a newsprint mill [27].

If problems occur, a review of the wastepapers used and an examination of conveyor belts supplying pulpers, or other wastepapers from the same suppliers may reveal the cause of the problem. Suppliers need to be kept aware of contaminants identified, as well as materials which are the cause of a downgrading or rejection of a load. If a 'new' contaminant is identified, suppliers and checkers need to be made aware of the contaminant immediately.

Following pulping, the quality of a wastepaper batch can be examined, for example, visually for ink speck size distribution and count, and/or by screening for stickies. If a problem batch is detected, the wastepaper responsible should be identified. The contaminated batch can be used, for example, via a bad batch system, or by increasing the rejects rate from screening and cleaning stages. More intense mechanical dispersion can be used, for instance, by the use of higher temperatures, or by operating with a narrow clearance between the static and rotating elements in a disc-type dispersion unit. Alternatively, chemicals may be used to try and clean up the bad batch.

3.2.3.2 Selection of appropriate recycling technology. Normally, when a wastepaper processing line is being designed, a great deal of effort is expended on the process sequence, equipment selection and choice of supplier. However, an important choice is how the equipment installed will be operated, in terms of chemistry and physical conditions. With respect to stickies removal, neutral pumping and low temperature appear best, but these conditions may affect subsequent ink removal [4] and so would have to be allowed for in the design since pulping capacity would be considerably reduced by these conditions. Other technology options are discussed in section 3.3, and illustrated in Figures 3.6–3.8 and Figure 10.2. If colloidal stickies were likely to be a major source of problems, for example, in newsprint deinking with high proportions of resinous wastepapers anticipated, then a second screening and cleaning loop, with a shock to destabilise colloidal particles at the beginning of the loop, such as a pH shock, may be appropriate. Under these circumstances the use of talc or an adsorptive filler, as discussed earlier, may be appropriate.

If wastepaper specifications are written to allow some grades of wastepaper not always acceptable in similar recycling systems, this will influence equipment choice. If, for example, hot melt adhesives are permitted, drum pulping, which has the most gentle defibering action may be the most appropriate pulping method, especially in the case of newsprint and magazines.

The selection of a recycling technology is thus a crucial initial design consideration. After a plant is built and equipment installed, operating options are more restricted and efficiencies can be lost if the basic concept for plant operation is changed.

3.2.3.3 Mechanical removal. After wastepaper checking and given a fixed recycling technology, mechanical removal is by far the most preferable control strategy. Whilst chemical additives may help to suppress problems, stickies removal should always be the preferred option and recycling plants should not be built dependent on chemical control, otherwise they will experience severe problems as the nature of stickies changes and a selected chemical programme becomes ineffective.

Published and anecdotal information indicates that centrifugal cleaning and fine screening are the best systems for stickies removal [20, 28, 30]. Of these, fine screening has been found to be the more effective [20, 28–30]. However, this should not prevent considerable attention from being given to removal early in the wastepaper sequence, by coarse screens, including pulper dump screens. Effective removal of large pieces of contaminants reduces the load on later cleaning stages and makes a significant contribution to successful contaminant removal.

More effective removal of stickies and other contaminants could be anticipated from a successful combination of chemical modifications to enhance physical removal. The use of a pH shock, to destabilise a colloidal suspension, is an example of the successful application of this approach. Some analyses of pitch and stickies show they can have a substantial inorganic component; for example, it is considered normal to have talc in pitch deposits, when talc is added for control of pitch deposition [15]. As well as reducing tack, if enough is adsorbed onto stickie particles the density of stickie particles will change, which may allow increased removal by centrifugal cleaners.

Very effective agglomeration and hardening of stickies was shown to be possible, in laboratory studies. Hard, large particles were produced and their removal by screening using laboratory screens was very effective. However, mill trials revealed that agglomeration and polymerisation of stickies did not occur to the same extent in mill systems as in the laboratory and this was believed to be due to the presence of a natural dispersant, probably lignosulphonate [31].

3.2.3.4 Mechanical dispersion. Dispersion and kneading units have been installed to break up residual visual contaminants, which they do very effectively. However, they appear to have different effects on stickies and other contaminants. Relatively slow-speed kneaders are claimed to roll up stickies and polyethylene fragments [32], whereas high-speed disc units reduce stickies in size [33]. High-speed units have been used for many years in Europe as a final wastepaper processing stage, not only to break up ink particles, but also to protect the machine from stickies deposition, at which they have been very successful. A slow-speed kneader in this position would not give protection. Size reduction of stickies appears to be dependent on dispersion temperatures and the gap between the discs. Higher temperatures (115°C) and narrow disc gaps (0.07 mm) were shown to result in less retention of stickies on a 6 mm slotted screen, when compared to dispersion at 90°C and a disc gap of 0.40 mm [33] (cf. section 3.4.3).

It is clear from this that dispersion and kneading do not remove contaminants, or stickies, but merely have an impact on their size. Since particle sizes are reduced, fine screening and centrifugal cleaning after high-speed dispersion is unlikely to be as effective as screening and cleaning before dispersion. If dispersion creates colloidal particles, deposition could still be a problem, via aggregation. Dilution after dispersion, followed by washing and flotation may help to remove colloidal particles and stickies. One study suggested that stickies were easier to remove following high-speed dispersion [34]. Since mills report fewer stickies problems after high-speed dispersion units have been installed, it follows that primary stickies, in general, cause more problems than secondary stickies. In one study, the addition of talc and Fullers earth to a high-speed, high-temperature Asplund unit reduced stickie transfer in a laboratory simulation of a back to top ply transfer problem. Higher temperatures (150°C) were more effective [20]. It was not clear if stickie problems were reduced due to the size reduction by dispersion, by the additives or a combination of these.

Another way to prevent colloidal stickies formed by high-speed dispersion causing machine problems would be to identify the electrostatic potential which allows small particles to deposit onto fibre surfaces, so ensuring that the concentration in machine water systems does not increase to a level which will result in the deposition of large stickies agglomerates. In all cases where this has been studied, protection was only given when process water had an anionic charge [16, 19].

Since low-speed kneading is claimed to increase the size of some stickies and other contaminants [32], cleaning and screening following kneading should result in reduced contaminant levels. There is much less experience in Europe of this approach than in Japan.

3.2.3.5 Chemical control by the use of additives. Many types of additives have been used to try to reduce problems from stickies and there are a large number of reports available in the literature [2, 8, 14–17, 20, 35–44] some of which have been reviewed [45]. Additives include:

- Ethoxylated surfactants [2, 41];
- Synthetic fibres [8, 35];
- Zirconium compounds [36, 47];
- Solvents and dispersants [37, 41, 43, 44];
- Talc and other inorganic additives [20, 38];
- Cationic and anionic additives [14, 16, 17, 39, 40, 42, 44].

The wide variety of control chemicals available, combined with the uncertainties regarding interactions which contribute to stickies and their origin, results in an inability to predict the efficiency of specific chemical treatment. In turn, this has resulted in unsuccessful attempts by mills to control specific problems using chemicals [29, 30] which has discredited this approach. If methods of stickies formation and basic interactions were understood, then this would help to eliminate a substantial proportion of unsuccessful trials and restore some confidence to mill personnel in the use of chemicals to solve stickies problems. Mechanisms believed to be involved in control are reviewed briefly below.

In a laboratory study, followed by a field trial, ethyoxylated surfactants were shown to result in a reduction (not elimination) in deposition of acrylate stickies when added to the machine wire showers, but were not effective when added to stock chests [2]. These polymers produced a reduction in tack. Adding chemicals to showers has become the most common treatment to limit deposition on wires and felts [46].

Another additive which reduced deposition of acrylate stickies was synthetic fibrils [8, 35], shown to coat these stickies, but not others that had not developed a tacky surface in the experimental system used. They created a steric barrier but if new surfaces of stickies were exposed, they were still tacky. An effective coating was not only dependent on the type of stickie, but also on the system; in some cases fibrils did not coat acrylate stickies and thus appeared to be dependent on factors such as pH and the presence of lignin derivatives.

Steric barriers are also created by zirconium salts, usually acetates and nitrates. In aqueous solutions zirconium chemistry is dominated by polymeric species and these create an effective barrier if adsorbed at the surface of the particle. These salts have an affinity for hydroxyl groups on the surface and the presence of the zirconium complex may lead to the development of a hydrous zirconia layer at the surface. Laboratory studies showed that zirconium was most effective with acrylates and an EVA-based hot melt adhesive [36, 47].

The use of solvents which result in the emission of volatile organic compounds is becoming less acceptable and controls on their release are likely to be introduced [43]. Solvents are used not only to clean the contaminated areas of machines, but in combination with surfactants as dispersion aids. This combination, or surfactants alone, is used to produce small stickies and to stabilise these. Non-ionic surfactants form a steric barrier around the stickie, since the hydrophilic ends of these molecules will remain in the bulk solution, but hydrophobic ends are adsorbed onto the surface of the particle. The hydrophilic ends will hold rigid a layer of water molecules, forming an effective barrier. They may also form, or contribute to the formation of, a barrier on surfaces, preventing deposition. Anionic surfactants function by charge repulsion; they keep dispersed material in the colloidal state and prevent agglomeration from occurring. With the exceptions of some ethoxylated derivatives, anionic and non-ionic surfactants were reported as having been unsuccessful in reducing tack [2].

Dispersant addition needs to be considered carefully. If a recycling system has a normal screening and cleaning system, its early use is likely to be counterproductive, since it will reduce contaminants in size and so reduce removal efficiencies through screening and cleaning. In pulping, for example, the use of dispersion chemicals is contrary to the principles behind the use of high-consistency pulpers. Dispersant use before a washing stage is more rational, provided the wash backwater is either not reused, or is effectively clarified before re-use. Dispersants can be counterproductive in a paper machine system, since they may adversely affect retention.

Talc is widely used in pitch control in pulp and newsprint production and its use in the control of stickies was an extension of this role. Initially it was postulated that talc functioned by adsorbing small organic molecules, or by adsorption onto the surface of large organic particles [48]. Adsorption of specific components of pitch was shown to be ineffective, in that only 10–20% the amount present was adsorbed, which suggests the adsorption of talc onto stickies is more important. Talc thus forms a steric barrier, but if new stickie surfaces are exposed, these will be tacky. It is widely used in newsprint recycling mills and can contribute to the bulk of deposits by being incorporated into the bulk of the deposit. Other inorganic particles such as precipitated silicates and silicas have been shown to be more effective in the sorption of organic species which contribute to reductions in the coefficient of friction, such as natural fatty acids and soaps [49]. These may have a role in stickies activation and so their removal by sorption could prove effective in helping to prevent agglomeration or the development of increased tack on primary stickies.

Cationic treatment of wires and felts was introduced during the mid 1980s, as an additive to showers on machine wires. Surface modification helps to reduce or eliminate deposition. The barrier which forms on the wire is believed to be a combination of anionic and cationic compounds; the machine wire changes colour, believed to be from lignin derivatives or from other anionic trash [40]. This is probably the most common chemical control mechanism. Cationic polymers can also function as retention aids; they retain anionic trash, which may contribute to agglomeration, in the sheet, but in recycled systems a high dose is nominally required due to the presence of high concentrations of anionic trash. Since the anionic charge falls as pH falls, the effective dose required is dependent on pH.

3.2.3.6 Design of reject and water systems. When contaminants, including stickies, are removed by various cleaning and screening devices, they have to be removed from the system. Where secondary and subsequent cleaning stages are similar to the primary, for instance, hole screen size is the same in both secondary and primary; no reductions in efficiency are likely, provided operating conditions are the same. However, if a different type of screen is used in the rejects treatment loop, its efficiency will determine the overall efficiency of that cleaning stage. Vibrating screens are sometimes used, as illustrated in Figure 3.6; in this the vibrating screen would have a much lower efficiency that the pressure screen and would recycle a large proportion of contaminants rejected by the pressure screen. As these are recycled, they are eventually broken down by the mechanical forces of pulp impellers, agitators, etc. and so are accepted by the primary or secondary screen. In Figure 3.6, accepts from the secondary stage are

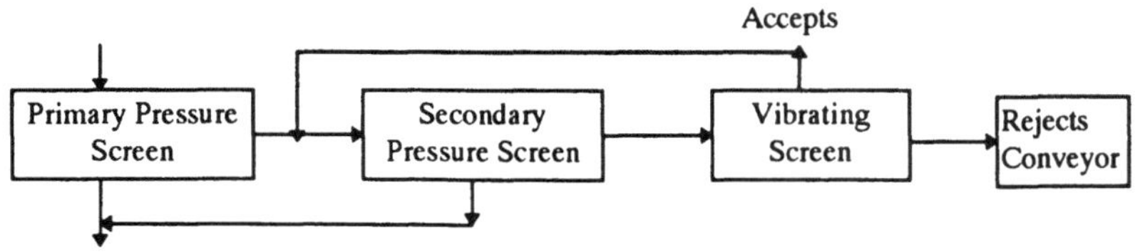

Figure 3.6 Inefficient screening system.

fed forward. Although these will have more contaminants than the primary accepts, this system avoids any cycling of contaminants. However, the alternative of taking secondary screen accepts back to the feed of the primary is also valid but they should not go further back, since this introduces more impellers, etc. (cf. section 3.4.2).

If rejects are discharged to a water system, it is important to ensure there is no possibility of water re-use without treatment, since this would result in the recycling of contaminants. Even after water treatment some contaminants are recycled. In one unpublished study, recycled water following sedimentation, biological treatment and sedimentation, was filtered to check for the presence of stickies. Although the numbers found were low, when the extrapolation was made to account for water consumption, the numbers were significant. This illustrates the difficulties in achieving total contaminant removal by water treatment. If reject flows are dewatered and rejected as solids, the dewatering system has to avoid losing rejects to the water stream, especially if backwater is reused without treatment.

Where possible, rejects should be conveyed to a specific area, rather than be dispersed throughout a plant, especially when volumes of rejects are large. When rejects have to be moved manually to a central area, it is inevitable that some spillage will occur. If these are coarse rejects, such as polythene fragments, wires, etc. and spills are washed to drain, unless a screen protects pumps, pump blockage will occur. If pipelines convey rejects to a central area, the pipework design must reflect the nature of the rejects. High-density rejects from high-density cleaners will very quickly block a pipe if it has sharply angled bends, or if the velocity in the pipe is low.

Equipment used in dewatering and rejects treatment has been reviewed [50–52] and includes:

- belt presses;
- inclined screw thickeners;
- cone presses;
- grit dewatering inclined screw conveyors;
- rejects dewatering mesh screens;
- pressure screen rejects dewatering devices.

3.2.3.7 Selection of machine clothing and cleaning systems. Machine clothing has to be designed to fulfil its primary functions of water removal, fibre support and sheet transport. Secondary considerations are the prevention of contamination which would affect performance or quality. Although anti-deposition coatings are possible, wear reduces the effectiveness of the coating. With this exception, there are no major differences due to the use of recycled fibres. Emphasis on keeping fabrics clean is usually on cleaning systems. Frequently a combined approach to cleaning is used, showers with added chemicals. Different systems are used with forming or press fabrics.

When a polymer treatment is being used, it is usually better to apply via a dedicated shower just before the fabric touches the sheet, for example, a shower just ahead of the breast roll on a fourdrinier, or after the Uhle box on a pick up felt. Best results are achieved by diluting the polymer with fresh water. Nozzles should be positioned to give a 90° spray pattern to the wire, or angled slightly with the run of the wire. Water pressure should be 30–50 psi and spray angles 90° to 100°, with a diameter not less than 1.5 mm, to avoid plugging. In the case of press felts, there is a much greater surface area to be treated and polymer addition rates have to be increased by a factor of 3–5 [53]. In press fabrics, the chemical should be added into the nip of a 90° wrapped roll, to give maximum dwell time before a suction box.

There are three basic types of showers used on forming fabrics, which can be used individually or in combination. These are flooded nip, outside wire return roll and high-pressure needle jet.

Water consumption is highest with the flooded nip shower, so the high-pressure needle jet shower has become common, with typical pressures in the range 2.0–3.5 MPa. A clean water source prevents plugging and allows low diameter orifices to be used, typically 1.0 mm. Use of a return roll as backing helps to prevent fabric damage. Operating stroke is twice the spacing of the nozzles.

The first outside roll after the press nip should be doctored and lubricated to remove loosely attached stickies. Teflon-coated rolls help to prevent stickies from adhering to the roll. Stickies removed have to be collected in a tray and removed from the system. High-pressure oscillating showers can be used on both sides of the fabric. Suction boxes are used to remove water from the fabric [54].

When forming and press fabrics are treated, the result can be that problems are moved from the wet end to the dry end. Cleaning and conditioning systems have been introduced for the dry end, using chemical treatment and a single traversing water and air jet shower. Drying cylinders can be doctored, though this is usually limited to the first one or two drying cylinders. Double doctoring is more efficient than single doctors.

3.2.3.8 Use of recyclable adhesives. Recyclable adhesives are difficult to define, due to the variety of recycling processes, but adhesives themselves are not recyclable; it is their effect on the recycling process which is at issue. There are three approaches to providing an adhesive which causes no problems during the recycling process and these are dispersible adhesives, water-soluble adhesives and adhesives which are easy to remove during processing.

Tests commonly used to assess recyclability [55, 56] are not a measure of recyclability, since they measure dispersion. However, adhesives which disperse may not cause difficulties during recycling, shown by extensive trials carried out with 3M Post-it™ notes [57]. Other adhesives which are dispersible, for example, pressure sensitives used on self-seal envelopes, or adhesive labels, frequently cause stickie problems. The difference must be in the adhesive formulation - the Post-it™ adhesive tested did not contain rubber, waxes or rosin tackifiers, which suggests these play an important role in creating stickies.

Water-soluble adhesives may cause no difficulties, though in recycling systems with low water consumption, as concentrations increase in water systems, a buildup on dryers (or other evaporation points) can result in paper sticking to dryers. If adhesives are designed to be soluble, the use of different conditions such as pH and temperature will lead to varying solubilities. Soluble adhesives have limited water and humidity resistance, so end use applications are restricted.

Some adhesives are easier to remove in some process steps than others; for example, hot melts used on magazines can be removed during drum pulping but would not be removed by low-consistency pulping. Ideally, adhesives would have resistance to softening which would aid their removal by screening, or they would have an appreciable density difference from both water and fibre to increase the efficiency of removal during centrifugal cleaning, or both properties. Since water has a density of 1.0, there is a range on either side of this in which centrifugal cleaning is not very efficient and so adhesives should have a density $<0.95\,\mathrm{g\,cm^{-3}}$ or $>1.05\,\mathrm{g\,cm^{-3}}$, with a high softening point, so that typical temperatures in recycling processes will not soften adhesives.

3.3 Equipment layout

Contaminant removal obeys a law of diminishing returns and follows a classic Pareto curve. Removal of contaminants is relatively easy at the beginning of a wastepaper line - rejects are large and easy to remove, so a large proportion can be removed at a small cost. At the end of a process, contaminants are much smaller and their removal is more difficult and costly - a small proportion is removed at a high cost. Due to this, equipment stages and complexity are determined by the needs of the product and the quality of the wastepaper; for example, wastepaper plants for packaging are less complex than deinking systems for newsprint. Similarly, tissue deinking systems are less complex than deinking systems for high-quality printing and writing grades. Nevertheless there are some similarities in equipment layout, and the basic blocks are illustrated in Figure 3.7.

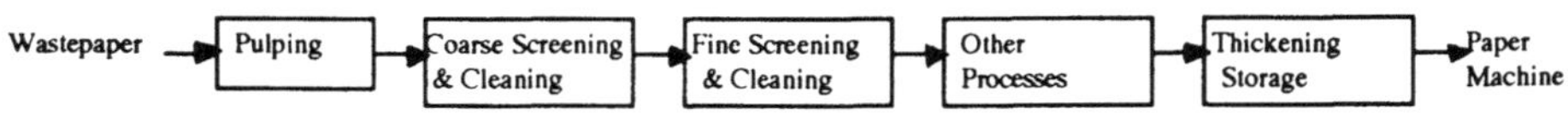

Figure 3.7 Basic wastepaper processing units.

Equipment used in processing waste packaging grades has evolved to become quite different from that used in grades in which deinking is important and so is considered separately, in chapter 8. This chapter is thus concerned primarily with deinking systems. In these, equipment choice and layout are product specific, in that in some grades some non-fibrous materials are undesirable but are acceptable in other grades. An indication of requirements for three deinking grades is given in Table 3.4, to illustrate how quality attributes affect equipment choice.

Table 3.4 Quality attributes for recycled products

Grade	Ash removal	Speck count	Stickies removal	Brightness
Tissue	+++++	+++	+++++	+++
Newsprint	+++	++++	+++++	++++
Printing and writing	++	+++++	+++++	+++++

Key: +, Not very important; +++++, very important.

Due to low-ash requirements for tissue, washing, which is the most effective ash removal stage is included, whereas there is less emphasis on ash removal in printing and writing grades. In the case of speck removal, this is very important in printing and writing grades, but less so in tissue. Stickies removal is important in all grades, since not only do stickies influence finished product quality, but they can have a very marked impact on machine productivity. Although different requirements result in different equipment sequences, almost all deinking plants include pulping, coarse screening, centrifugal cleaning, fine screening, etc. as illustrated in Figure 3.7.

Wastepaper quality and type also influence equipment choice and sequence. Offset printed magazines are easier to deink than electrostatic printed office papers, so a plant designed to cope with electrostatic printed papers would be more complex. Alternatively, a plant designed to allow the use of direct entry grades would be very simple and since direct entry grades are not usually printed, it would not include deinking.

The usage rate of recycled fibre is also significant in determining the complexity of the processing line. If recycled fibres were to be included at a 25% rate, a fairly simple plant would ensure adequate quality, whereas if recycled fibres were to supply 75% of fibre, then high quality is essential to ensure machine efficiency is not sacrificed, which requires a more complex processing system. Similarly, if product quality is low, for example, an industrial paper towel, then the processing needs (or wastepaper quality) will be less than for a premium grade of kitchen towel.

Another factor which determines the equipment sequence is the proposed chemistry of the plant. Although some European deinking mills operate without chemical use, this is rare. Deinking based entirely on chemicals is the other extreme, which has been tried in some mills in North America. Normally there is a balance between these, but the equipment sequence will be partially determined by the chemical philosophy adopted; for example, approaches include two loop systems, with chemistry (or physical conditions) varying between the loops. Possibilities include:

- Alkali first loop, acid second loop
- Acid first loop, alkali second loop
- Cold first loop, hot second loop
- 'Conventional' all alkali system
- Chemical free system
- Chemical deinking

Alkali first loop, acid second loop. Two loop systems were first developed into full scale production in the UK at Bridgewater Paper Mill, a newsprint deinking mill, which uses alkali and acid loops [58, 59]. A typical system is illustrated in Figure 9.2. Acidification is used to destabilise colloidal materials, to form agglomerates, which are removed by additional cleaning and screening. This is most useful when the paper machine operates under acidic conditions, since it prevents agglomeration in the paper machine system, following alkaline processing in the wastepaper processing line.

Acid first loop, alkali second loop. When newsprint contains flexographic inks, an acid first loop, followed by an alkaline loop has been shown (on a pilot plant basis) to result in more efficient removal of flexographic inks. Acid (or neutral) conditions minimise the dispersal of flexographic inks, which are removed by flotation, followed by normal alkaline processing, to remove conventional inks [60]. This is illustrated in Figure 3.8.

The location of fine screening and centrifugal cleaning would be

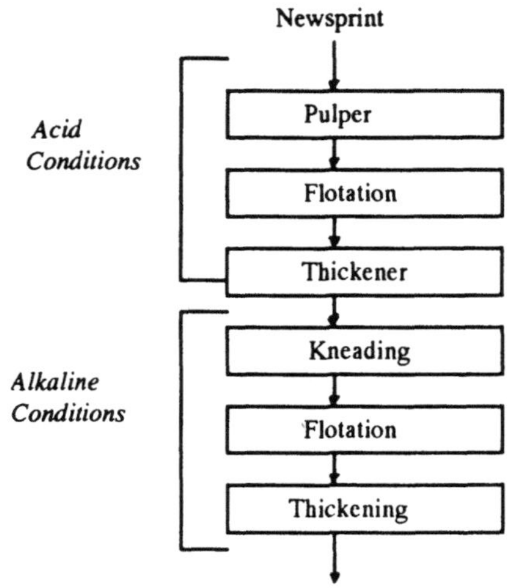

Figure 3.8 Two loop acid–alkali system [60].

determined by the deflaking device; a kneader is illustrated in Figure 3.8. This process has been patented in Europe, North America and Japan [60].

Cold first loop, hot second loop. An early two loop system recommended keeping pulper temperatures low, followed by deinking, cleaning and screening. A second, high-temperature loop included a high-energy dispersion or kneading stage, followed by further deinking, illustrated by Figure 3.9. In the mill illustrated, deinked fibre supplied about 40% of the furnish.

Low temperatures were considered essential to prevent softening of thermosensitive adhesives, such as hot melts, and to permit maximum efficiency of screening and cleaning operations to be achieved. Maximum temperatures of 40–45°C were recommended, with temperatures of 25–30°C considered excellent. Since temperatures inevitably rise in closed water systems, water-to-water heat exchangers have been used to reduce cold loop temperatures, as illustrated in Figure 3.9. Several of these systems are operating in Europe and the concept has been patented in many countries [61].

Other than those described, another advantage of two water-loop systems is that, provided the water loops are treated for solids removal and there is a water bleed from the two systems, a chemically cleaner pulp will be made, which is much less likely to have adverse affects on the paper machine.

Conventional alkali system. A ‘conventional’ process flow is illustrated in Figure 3.10, which is suitable for an addition rate of 25–40% of total fibre, for instance, in newsprint.

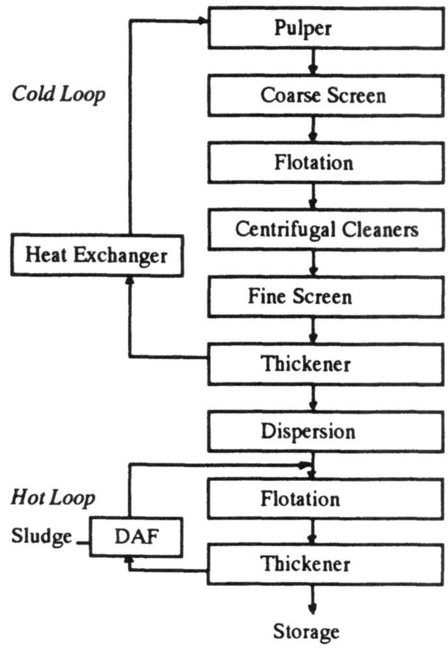

Figure 3.9 Two loop - hot and cold loops [61].

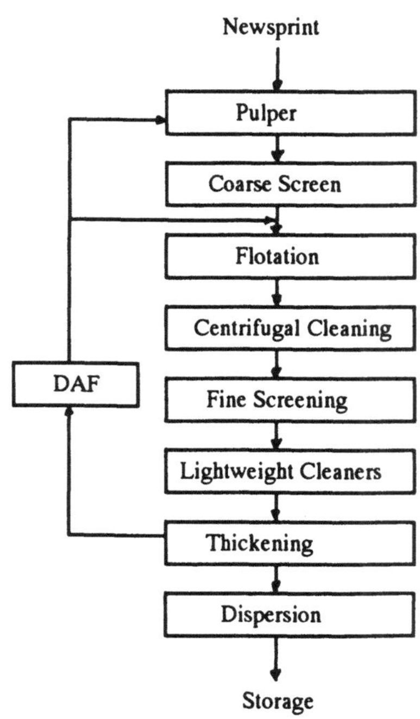

Figure 3.10 Conventional alkali deinking system.

At this addition level of recycled fibre, more sophisticated cleaning using two loops is not necessary.

Chemical free deinking. With increased pressure on all aspects of chemical use, a few mills have operated without any chemicals in their deinking mill, though chemicals are still used for water clarification and sludge treatment. Although these mills were designed as conventional alkaline deinking mills, they have produced acceptable products. Despite lower product quality, there are a number of advantages from the use of no chemicals. The decision on chemical use depends on the balance of advantages and disadvantages from the use of individual chemicals.

Advantages of not using sodium hydroxide in pulpers include:

- reduced dispersion and tack of pressure sensitive adhesives [12, 20];
- reduced dissolution of organics, leading to lower effluent chemical oxygen demand (COD) and biological oxygen demand (BOD) as well as reduced problems from foam;
- lower costs;
- increased operator safety;
- when lignin-containing papers are present, reduced yellowing, and lower costs since hydrogen peroxide is no longer necessary to prevent yellowing.

However, if caustic is not used, the effective pulper capacity falls, since caustic aids wastepaper disintegration, probably by faster wetting of inter-fibre hydrogen bonds. At a neutral pH, ink dispersion and final fibre strength drop. With woodfree wastepapers, treatment with sodium hydroxide helps to reverse some of the effects of drying (during the first cycle) so that, for example, strength properties are improved and water-retention values increase, since fibre swelling increases. In the case of wood containing fibres, hornification from the first cycle drying is much lower (or zero) due to the lignin matrix limiting the collapse of cellulose fibres (cf. chapter 6). In this case, improved bonding is probably due to some lignin extraction, increasing the flexibility of fibres, so that the effect of not adding sodium hydroxide is the same. However, an alkaline stage later in the process may confer these properties, without the same degree of stickies dispersion, since most would have been removed. Better contaminant removal would be achieved by low temperature and neutral pH pulping conditions. Both of these reduce pulper capacity, so pulper size would have to be larger than conventional. If no sodium hydroxide is used at any stage, then some fibre swelling can be restored by dispersion or refining. There is a trend towards lower sodium hydroxide use in pulpers; typical pulper pH in Europe is 9.5–10.0, compared with 10.0–11.0 previously used.

If no colour stripping chemicals are used, then brightness is lower, with more shade variations. In many products this would be unacceptable.

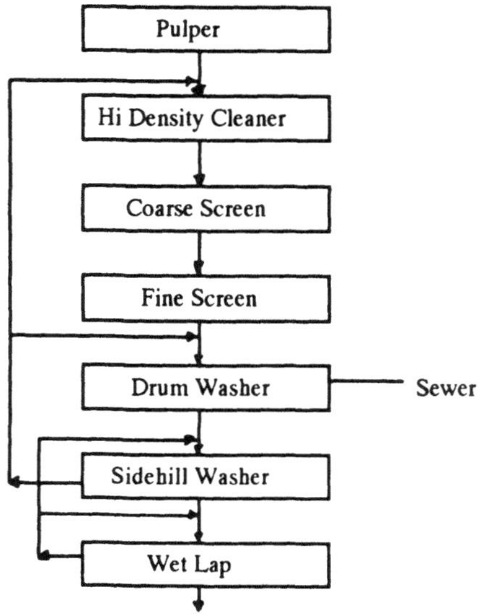

Figure 3.11 Low-cost deinking system.

In flotation deinking, the addition of suitable ink collector chemicals is normally regarded as essential. However, several mills claim successful operation of flotation without chemical addition, though this has not been documented. In one mill, using woodfree furnish, the author observed an increase of brightness across flotation of 3–7% ISO brightness units. However, residual ink levels would not have been acceptable in many products.

Overall, although zero chemical use may give a quality which is acceptable in, for example, some tissue products, quality levels would not be generally acceptable. Environmental advantages are claimed by some producers not using chemicals, though the precise nature and benefits are not usually well documented. Other mechanical systems for the removal of ink would have to be developed before chemical-free deinking became widely adopted.

Chemical deinking. There are essentially two approaches to chemical deinking. In both approaches flotation deinking is not used, but screening and cleaning are included. The first of these uses a dispersant to disperse ink in the pulper, followed by ink removal either during a washing or thickening stage. Provided dispersed ink is removed by the washing stage and from backwater (or it is not reused), this can give acceptable results. A system based on this concept allowed the replacement of groundwood pulp, at 10–15% of a directory paper furnish [62] and is illustrated in Figure 3.11. If there is no ink removal by a thickening or washing stage,

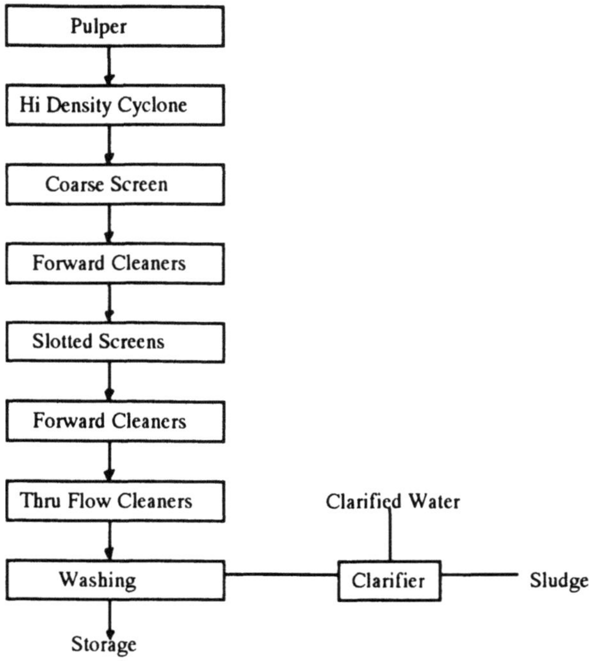

Figure 3.12 Laser deinking process. Temperature 65°C; pH 11.0; time 30–60 mins.

the paper machine will serve as a washing device, so some white water will have to be periodically discharged, to prevent an excessive buildup of ink. If some reduction in product quality is accepted higher proportions of recycled fibre can be used.

The other chemical deinking approach is the use of chemicals to densify and agglomerate electrostatic inks, for removal by cleaning and screening [63, 64]. This system is used, for example, to remove toners used in laser printers and xerographic copiers, and so is most commonly found in wood-free wastepapers. A system which has been recommended for this approach is illustrated in Figure 3.12 [64].

Where this chemistry has been used, laser printed papers have been successfully deinked, though some deposits of stickie-like material were reported [65]. Although the chemicals are effective in the removal of electrostatic inks, they are not with other inks, so that if wastepapers contain a mixture of ink types, achieving good quality is dependent on the efficient removal of other inks in a single washing stage, as illustrated in Figure 3.12.

3.4 Wastepaper processing

The purpose of wastepaper processing is to take a heterogeneous raw material and convert it into a homogeneous stock for paper or board making. This is achieved by the removal of non-fibrous materials. However, the level of homogeneity achieved cannot match that achieved in virgin pulp systems, since the proportions of fibre types, long, short, coarseness, etc. can be controlled through pulp purchasing. Since non-fibrous materials cannot be removed at 100% efficiency, stock from a wastepaper processing system is thus less predictable than virgin fibres, despite intensive cleaning operations. Individual stages typical of deinking lines are reviewed below, whilst those used in the production of brown grades, such as liner and medium are reviewed in chapter 8. Some pulping comparisons are made below.

3.4.1 Pulping

Wastepaper pulping is relatively new, with the installation of the first hydrapulper in 1939 in the USA, though it was not until the 1950s that they became common. Since then pulping has been extensively modified and many different options are available. Initially, the pulper screen plate had large perforations, 8–20 mm diameter. Rotors had high peripheral speeds, about $20\,m\,s^{-1}$ and defibering was a result of hydraulic action, rather than attrition. Cooking after pulping was quite common, though conditions and chemicals used varied [66]. The introduction of more efficient rotors, such as the Vokes rotor, led to the use of lower peripheral rotor speeds, $16\text{–}20\,m\,s^{-1}$ and smaller extraction perforations in the pulper, 3–6 mm diameter. This system led to some problems, as contaminants were left in the pulper, to be broken down and heavyweight materials caused rapid wear of the pulper rotor and tub. Another pulping strategy was adopted, which employed a secondary pulper. The main pulper was used to produce fibre flakes, rather than a defibered stock, and this was dumped via large extraction holes to a secondary pulper, essentially a deflaker, but also incorporating a screen plate. This reduced the buildup of rejects in the pulper and was more energy efficient, and is similar to the system still used in the continuous pulping of OCC and related grades (cf. chapter 8).

A recognition that equipment which deflaked also reduced the size of contaminants, so making their subsequent removal more difficult, led to the rapid adoption of another pulping system which was developed in the mid 1970s, high-consistency pulping. This was widely adopted in Europe, but a different system was developed in Japan, though the objective was the same. In the Japanese system, which is a development of post-pulping cooking, a low-consistency pulper is used to produce flakes, but stock is held in a soaking or reaction tower, with or without chemicals, for an extended period of time, up to 24 hours, to allow fibre flakes to be thoroughly wetted. Deflaking then requires a very low-energy input, so that contaminant size reduction is minimised.

Developments in pulping have resulted in the availability of a range of options including:

- low-consistency, continuous pulping - 3–4% consistency range;
- low-consistency, batch - 5–7% consistency range;
- high-consistency, batch - >12%;
- high-consistency, continuous - >12%;
- explosion pulping systems;
- soaking or reaction tower, to follow coarse (usually low-consistency) pulping.

There are several variants for each pulping option. There is a widespread recognition that the pulper is the heart of a wastepaper processing plant, yet it is frequently the production bottleneck. Many variables affect defibering so if these change, pulper capacity is altered and it is possible to undersize the pulper during the design stage. Some of the different pulping systems have different objectives; for example, as well as defibering the pulper can be regarded as a first cleaning stage [67]. However, this is not always the case; for example, pulping can be in pulpers which do not discharge through a screen plate. The overriding objective of the pulping system (which can include soaking towers and secondary pulpers) is to provide a completely defibered stock to subsequent cleaning stages. If fibre flakes remain, cleaning efficiency falls and fibre losses increase. Steps in achieving defibering include:

- separation of papers contained in wastepaper bales and water penetration into the paper;
- wetting of fibres, to weaken inter-fibre bonds;
- breakdown of inter-fibre bonds, to allow fibre separation;
- separation of fibres from contaminants such as adhesives, inks, laminated materials etc., with minimal degradation of contaminants.

An energy input is required to achieve the defibering goal and it is this that provides the main differences between pulping options. Rapid and high-energy inputs inevitably lead to the breakdown of contaminants. The speed of wetting is largely dependent on pulper conditions, paper type and basis weight. OCC grades have a higher air content and so are more difficult to wet than, for example, newsprint. Energy required to break wetted bonds is low and varies with the type of paper, illustrated by Table 3.5. Although strengths given refer to the dry state, similar differences exist in the wet state.

Of more importance is the degree of sizing of the paper. Sizing or wax treatment slows down water penetration, so that these papers can be very

Table 3.5 Breaking lengths of paper types (dry) [68]

Paper type	Breaking length (m)
Newsprint	3300
Writing paper	5860
Kraft paper	8100

slow to wet; for example, liner grades when hard sized can have a Cobb value of 15 to 20. In the case of a 120 g m^{-2} liner, thorough wetting could require 60–70 min [69]. Wet strength is also very important, since wet strength resins confer much higher bond strengths to paper when wet, so that more energy is required to break inter-fibre bonds. Given these differences in energy requirements for defibering, it is important to match the pulper type and conditions to be used with the required capacity of the wastepaper processing system.

3.4.1.1 Low-consistency pulping. Although many low-consistency pulpers have been installed, recent installations (in systems processing grades other than OCC and related grades) have been predominantly high-consistency pulpers, due to their lower level of contaminant degradation and lower energy requirements. In low-consistency pulping, mechanical defibering forces are high. When unpulped paper (or contaminants) is wrapped around both a static and a rotating part, or contacts a rotating part, the forces are very high. With high speed (16–20 m s^{-1} peripheral speeds), low-consistency rotors and pulp at a relatively low viscosity, mechanical forces tear papers and contaminants such as plastics, etc. High attrition rotors are also common and these will also tear contaminants. Shear forces are created when the two ends of paper are caught in stock flows moving in different directions. The force which holds paper in the suspension is related to the viscosity of the stock in the suspension. Due to the presence of the large amount of water present, a substantial portion of the energy is used in moving the water so that no load power can be 60–80% of total load. The flow pattern creates a vortex around the rotor and baffles are necessary to improve mixing. Mechanical forces between the rotor and the discharge screen plate are high, whilst fibre-to-fibre interactions are low.

Low-consistency, continuous pulping has been used primarily with news and OCC grades and is now primarily used only with OCC grades. Stock is removed continuously from the pulper, through a screen plate with large perforations, typically 18 mm. A secondary pulper provides additional defibering and screening. Some features typical of low-consistency continuous pulping are given in Table 3.6. However, values given are typical of operation and are not to the same flake content, which are compared in Tables 3.10 and 3.11.

Low-consistency batch pulping was used extensively in deinking but is

Table 3.6 Comparison of low- and high-consistency pulping systems

Pulper type	Consistency (%)	Specific power consumption ($kWh\,t^{-1}$)	Retention time (min)	Extraction screen perforation, Φ (mm)
Low consistency, continuous	3–4½	30–45 [a]	5–8	18
Low consistency, batch	5–8	30–45 [a]	10–20	3–16
High consistency continuous	15–20	15–25 [b]	15–20	4–10
High consistency batch	12–18	20–25 [b]	5–10	(10–20) [c]

[a] OCC grade. Includes secondary pulper power consumption. [b] Newsprint. Pulper only. [c] Pulper discharge may not be through a screen plate.

Table 3.7 Comparison of pulper types [70]

	Low consistency	High-consistency batch
Consistency (%)	3–6	12–15
Rotor:tank volume (%)	0.1	8
Specific power (kW inst.)	6	22
No load power (%)	70–80	50–60
Rotor speed $m\,s^{-1}$	16–21	8–15
Shear edge lengths		
rotor	40	33
screen	58	(43) [a]

[a] Screen is not always installed.

now rarely installed, unless followed by a soaking tower. Comparisons with high-consistency pulping are made in Tables 3.6 and 3.7.

Low-consistency pulpers can be converted to medium consistency pulping (8–12%) by the installation of a new rotor. Helical rotors can rarely be used, due to design considerations, illustrated in Table 3.7.

3.4.1.2 High-consistency pulping. High-consistency batch pulping normally uses a helical rotor, illustrated in Figure 3.13, which is very much larger in relation to the pulper tub than a low-consistency rotor, as illustrated by the ratio of the rotor to tank volume, given in Table 3.7. Although the specific power input is much higher, specific energy consumption is lower. Stock above 10–12% loses its fluidity and is quite viscous and the flow circulation patterns are very different from those in a low-consistency pulper; no vortex is created and the circulation pattern is top to bottom. The top of the rotor remains above the stock during pulping. Internal baffles (impact bars) are fitted to direct stock back into the rotor. The relatively low rotor speed reduces the cutting of plastics and other

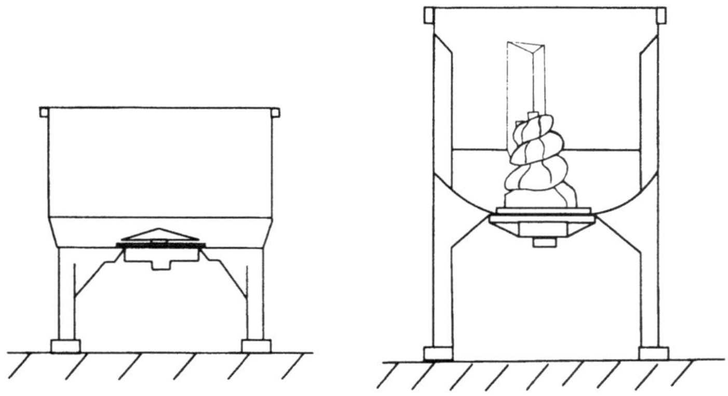

Figure 3.13 Pulping systems. (a) Low-consistency, batch; (b) high-consistency, batch.

contaminants, but shearing forces are high, since the velocity difference between the stock and the rotor is high. Mechanical forces are low and when no screen plate is fitted mechanical forces are very low. As stock crumbles and is pulled into the rotor, an intense defibering action is provided by fibre-to-fibre rubbing.

Discharge from the pulper can be through a screen plate, or direct, with no screening. When discharge is through a screen, stock must be diluted before discharge and so the pulper vat has to be large enough to accommodate diluted stock. Contaminants retained in the pulper are flushed out and dewatered; extracted water can be returned to the pulper. If stock is removed from the pulper without screening, this can be with or without dilution. If there is no dilution, no screening is possible before dilution in a dump chest and in this case, there is a risk that gross contaminants will accumulate in the dump chest, giving rise to problems with pump blockages, etc. If stock is diluted in the pulper, or during discharge, a detrashing screen can be used for gross contaminant removal. Contaminants are retained in the body of the screen, washed and discharged at the end of the pulper dump cycle.

Advantages claimed for high-consistency pulping when compared with low-consistency include:

- reduced contaminant breakdown, leading to better contaminant removal;
- reduced power consumption – estimates range from 25% [71] to 70%. Savings of 70% were seen when pulping magazines to the same flake content [70];
- chemical savings, through higher chemical concentrations as consistency increases;

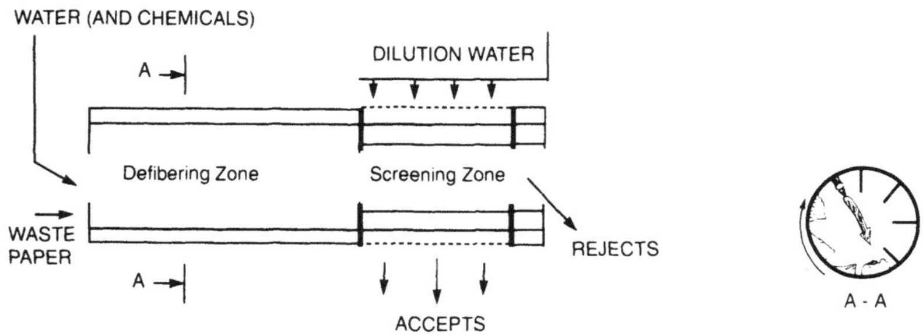

Figure 3.14 High-consistency continuous pulping - drum pulper.

- more complete ink removal [72]. However, the intense fibre interactions result in reduced ink particle size and so the brightness from high-consistency pulping is usually less than for low-consistency pulping. Nevertheless, after washing, when most comparisons are made, brightness of high-consistency pulped stocks is higher, due to better ink removal during the washing which occurs as the test sheets are made.

High-consistency continuous pulping - drum pulpers. Although the discharge from drum pulpers is in the range 3-5%, defibering occurs at 15-20% and so drum pulping is a variant of high-consistency pulping. Pulping drums are divided into two sections, the wetting and defibering section and a screening section. These can be either one complete unit, as illustrated in Figure 3.14, or two separate drums, rotating in the same or opposite directions. The interior of the wetting and defibering zone is ribbed internally so that the ribs pick the wastepaper up and as the drum rotates carry it round until gravity makes the mass of paper fall. As it impacts, the energy transferred wets and defibers the wastepaper. There are no mechanical forces and no high-shear zones; wetting and defibering is only by the tumbling and falling action within the drum. For news and magazines, a suitable retention time in this zone is 15-20 min, but grades which are more difficult to wet, or have higher basis weights, or have higher bond strengths will require a longer retention time. Energy input from tumbling and falling can be increased by increasing the diameter of the drum and the retention time can be altered by changing the length of the drum, rotating speed or angle of incline.

The second stage is screening, and stock consistency is reduced to 3-5% in the perforated screen zone. Fibres and small fibre flakes are washed through the perforations, which are 4-10 mm in diameter. All materials not reduced to below the perforation size in the first zone are rejected out of the end of the drum. The very gentle defibering action is evident by the condition of the rejects; sheets of plastics are rejected whole, without rips.

A cost of using this system is a higher solids loss, if compared to other pulping systems. Increased losses are 1-3%, dependent on grade and subsequent cleaning stages. An advantage is a less contaminated stock. Another disadvantage of this drum is the inflexibility of the pulper in coping with different types of paper; for example, wet strength or hard sized papers will probably be rejected, unless the drum is oversized. Drum pulping is most widely used in recycling news and magazines.

In pilot studies a similar system (but dry) was used to separate fibre by type, mainly into woodfree and wood-containing fractions, though this was not developed into a commercial installation [73].

3.4.1.3 Explosive pulping. Explosive pulping was patented in 1928, to produce a building board from wood wastes [74]. Two variations have been proposed for wastepaper pulping: the Siropulper [75] and the RECOUPE or the StakeTech process [76]. Both systems rely on the rapid penetration of the pulping liquor, due to the high pressure. Wastepaper is defibered by the explosive rapid release of the digester contents. Commercial experience with explosive pulping of wastepaper is limited.

3.4.1.4 Soaking towers. Although 'cooking' after pulping (usually in a high caustic alkalinity environment, up to 5% on fibre, at elevated temperatures, sometimes under pressure) was widely used in the 1950s and 1960s, it is now rare. Amongst the objectives of cooking were defibering flakes, but it was mainly used to cook out contaminants, including lignin, and to develop fibre properties. Soaking towers are a similar concept, and have wetting of fibre bundles and development of fibre properties as objectives, although cooking out contaminants is no longer a target. This system has been widely adopted in Asia, but there are few installations in Europe. Pulping prior to the soaking tower can be either high or low consistency, though low-consistency stock is thickened prior to the soaking tower. Defibering in the pulper is deliberately incomplete to avoid contaminant breakdown, and after soaking it is easily completed by low-energy deflakers, avoiding contaminant degradation. Only very coarse screening is employed, if screening is used as pulpers are discharged. Residence time in soaking towers can be up to 24 hours, but is normally 12-14 hours. Advantages claimed [70] for soaking towers include increased fibre yield compared with other pulping systems, higher brightness or chemical savings of 3-5% and energy savings.

3.4.1.5 Pulping variables. Temperature and pH each have a significant effect on both the speed of wetting and defibering as well as on ink and other contaminant dispersion. Changes in temperature alter water viscosity, leading to changes in the speed of penetration into the fibre web. Higher temperatures increase the speed of penetration, illustrated by the data in Table 3.8. Increased temperatures also reduce stock viscosity and can lead to changes in circulation patterns within a pulper.

Table 3.8 Increased temperature reduces pulping time [66]

Temperature (°C)	Pulping time (min)
21	40
32	30
49	20
66	12

Increased temperature alone can have a significant effect on properties. One study showed than an increase in temperature from 27 to 71°C, using refined bleached kraft, held at the high temperature for only one minute, increased tear by almost 10%, and reduced Williams freeness by 25%, with minor reductions in burst and porosity. Effects were enhanced only slightly by holding the temperature for ten minutes. Above about 70°C effects began to be reversed [77]. Higher pulping temperatures also increase ink and other contaminant dispersion, and the trend is towards lower pulping temperatures to reduce stickies problems. Temperature increases the dissolution of organic material, illustrated by values given in Table 3.9 for yield losses on tab card stock, as measured by losses on a Williams tester, using a 70 mesh screen [77].

Table 3.9 Effect of temperature and pH on yield loss [77]

Temperature (°C)	Distilled water	5% NaOH (on fibre)
38	4.0	6.4
65	4.1	7.3
93	5.9	9.0

It is clear from Table 3.9 that pH also has a significant effect on yield loss - dissolution of fibre components, and probably dispersion of ink increased with a higher pH. Wetting speed is increased; thorough wetting of a 120 gm^{-2} liner with a Cobb value of 15–20 requires 60–70 min at neutral pH, but only 1–2 min at pH 11, and 10 s at pH 12.0 [69].

The effects of temperature and consistency are illustrated in Tables 3.10 and 3.11. These compare pulping to a standard fibre flake content. A Sommerville screen with a 25 mm slotted screen plate was used to collect undefibered flakes. Data reported in Table 3.10 is from pulping a mixture of woodfree and wood-containing papers, to a fibre flake content (as measured by the Sommerville screen) of 1%.

When low- and high-consistency pulping is compared on this basis, it is clear that energy consumption in low-consistency pulping is much higher. However, final defibering in a secondary pulper is more energy efficient than in a low-consistency pulper so the high energy is not typical of a

Table 3.10 Effects of pulping variables [78]

Consistency (%)	Temperature (°C)	pH	Time (min)	Specific energy consumption $kWh\,t^{-1}$	Freeness (CSF)
12.7	48	10.1	5.5	23	260
14.4	60	10.2	4.0	13	270
4.8	47	10.1	10.0	67	210

mill situation. Freeness from high-consistency pulping was higher. At a pH about 10, a small increase in consistency and temperature resulted in a substantial reduction in pulping time and specific electrical energy consumption.

When a wet strengthened liner board (12% wet strength) was pulped to a Sommerville screen flake content of 2%, a similar trend was found; results are given in Table 3.11.

Table 3.11 Effect of increased temperature on pulping of liner board [78]

Consistency (%)	Temperature (°C)	pH	Time (min)	Specific energy consumption $kWh\,t^{-1}$
14.6	73	11.9	13	45
14.9	53	12.0	24	80

In high-consistency pulping trials with news and magazines (80% and 20%, respectively) at a neutral pulping pH, temperature about 50°C, a repulping time of eight minutes resulted in a flake content up to 7%, whereas pilot plant trials with newsprint alone, but with 0.5% NaOH gave a flake content of 1% after two minutes pulping time. Hence pH has a considerable effect on pulping rates [78].

Since current trends are towards both lower temperatures and pH, it is clear that pulpers in existing systems may reach a capacity problem as temperatures and/or pH are reduced. If systems are to be operated under these conditions, pulpers installed will have to be considerably larger than conventional design, unless a soaking tower is used after pulping.

Other pulper additives include deinking chemicals and hydrogen peroxide with associated chemicals to stabilise peroxide. The effect of hydrogen peroxide and stabiliser chemicals is discussed in chapter 5.

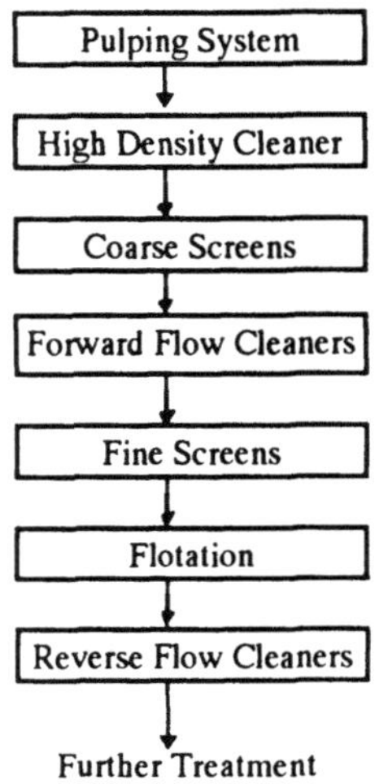

Figure 3.15 Schematic illustration of screening and cleaning systems.

3.4.2 Screening and cleaning

The objective of both screening and cleaning is the removal of non-fibrous contaminants, with minimal losses of useful fibre. Separation depends on physical differences, for example, in screening size and shape are most important whilst in centrifugal cleaning density is the most important difference. There is not a single best way of sequencing cleaning and screening and so many different sequences have been used. Generally, the equipment chosen follows consistency, so that low-consistency equipment is operated after high consistency, to avoid the costs of consistency changes. The sequence used is also in the order of decreasing contaminant size. A typical sequence of screens and cleaning systems is given in Figure 3.15.

3.4.2.1 Screening. Some form of screening has existed for centuries, but the first pressure screens were developed in the late 1930s and had a wooden scraper to keep the face of the screen clear. A mat forms readily on the face of the screen since water passes through the screen plate orifices more easily than fibres. In the 1950s a non-contacting foil was developed, to provide a pressure pulse, which kept the face of the screen clean.

More recent developments in screen design have been mainly in basket or rotor designs, for instance, rotors to provide various intensities and duration of pressure pulses, the development of contoured screens plates and techniques to produce narrow-width slots. Control systems have also developed, to prevent screens blinding. All of these have played a part in maintaining contaminant removal efficiencies as contaminant types changed and concentrations increased.

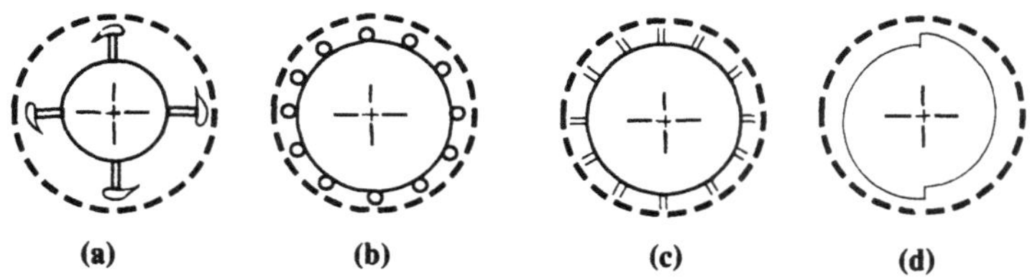

Figure 3.16 Typical pressure screen rotors. (a) Foils; (b) bumps; (c) radial vanes; (d) tapered surfaces.

As well as providing a pressure pulse to clear the mat from the screen, the rotor also fluidises stock, reducing fibre flocculation, which allows improved selectivity, so that fibres are accepted through the screen and debris is rejected. This is especially important at higher consistencies when higher rotor speeds increase the pulse and fluidisation. The clearance between the basket and the rotor has a significant effect on pulse amplitude. Rotor design is also used to try and even out the load on the screen basket - a high proportion of accept stock passes through the feed area of the basket. Rotors used are illustrated in Figure 3.16.

Contoured screens were first introduced during the early 1980s and have largely replaced smooth plates. Stock fluidisation is increased by the contours, increasing the turbulence at the screen surface, as stock flows over the surface. This reduces the blinding tendency, increases screen capacity, allows screens to operate at higher consistencies and reduces fractionation. Although the increased turbulence at the surface reduces screening efficiency, this is more than compensated for by the ability to use narrower slots, due to the increased capacity of the contoured screen. They can wear more quickly - in abrasive stocks the contours can be effectively worn away within a few months, so that capacity falls rapidly. Cross-sections of screen plates are illustrated in Figure 3.17.

Fine perforations and slots cannot be machined with accuracy in screen baskets. Lasers are used, and can produce very fine slots, down to 0.15 mm. Machining even large size slots produces a substantial range in size; for example, tests on two slotted screens with a nominal slot width of 0.45 mm showed the range was 0.40–0.55 mm, and averages for two screens were 0.45 and 0.48 mm [79].

Screen control systems can be very complex - one system described had one supervisory and seven regulatory loops for each screen [80]. Feed control should be implemented - both feed pressure and consistency. Feed pressure has an effect on screen runnability, but within a wide range, little effect on efficiency. In one installation it was found that low feed pressure resulted in operating problems, but an increase in feed pressure gave an increase in capacity without a need for an increased pressure differential between the feed and accepts [81]. Feed pressure can be controlled via

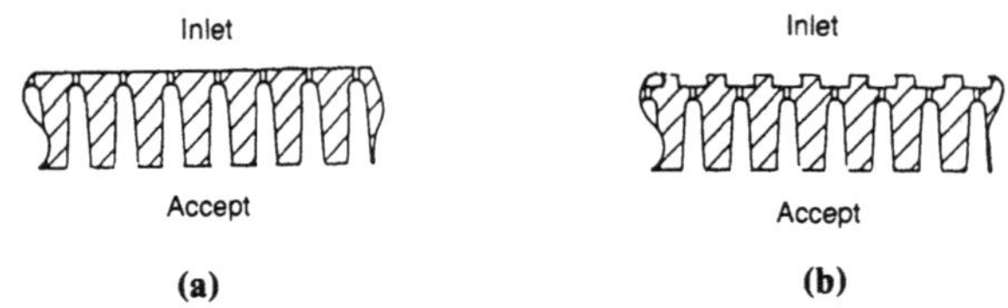

Figure 3.17 Cross-sections of smooth and contoured screen plates. (a) Smooth surface slotted screen plate, (b) contoured surface slotted screen plate.

a simple feedback loop, either to a control valve or variable frequency drive on the feed pump. Feed pressure control also helps to prevent valve cavitation. Consistency control is more important, since screen efficiency and runnability can be affected. Trim dilution can be supplied to pump suctions.

Control systems are commonly based on the pressure differential between feed and accepts; if this is higher than a set point, then the accept valve closes. This allows the screen to be cleared into the rejects stream by the rotor pressure pulses. When the rejects pressure reaches the same pressure as the inlet (zero differential) the accept valve opens and normal operation resumes. Blockage of the rejects valve by debris can be monitored by a flow meter. The rejects valve position changes to provide a constant rejects flow.

The combination of these developments has helped to maintain and improve screening efficiency despite increased contamination loads in wastepaper. They illustrate how complex screening is; efficient screening is dependent on a number of factors, some of which are inter-related. Factors include:

- Screen basket design - opening shape and size, profile, width between openings;
- Velocity through the screen basket opening;
- Stock consistency;
- Size and shape of contaminants;
- Physical properties of contaminants - softening temperatures, conformability, etc.;
- Rotor design - speed, shape, distance from screen plate;
- Power input and pressure differential between feed and accepts;
- Reject to inlet ratios for both flow and mass;
- Stock temperature.

Of the factors given above, only a few can be controlled at mill level, following screen installation, which is why a control system is so important. Maintenance is also important - of instrument sensors, to ensure accuracy and of screen baskets and valves, to clean, repair and replace as necessary. Consistency plays an important role in determining efficiency and as mills put pressure on their recycling systems to produce more tonnes, consistency and temperature are frequently increased. These may reduce screening efficiency, but this is dependent on screen type. In trials using a centrifugal type screen, with a four foil-shaped rotating element on the inlet side of the screen cylinder, slot width, slot velocity and consistency were varied, using a newsprint furnish spiked with pressure sensitive adhesive address labels. Results showed that screening efficiency was more dependent on slot width than any other factor, but that a consistency increase from 0.75–1.25% substantially improved efficiency, especially of the large size slots (0.20 mm and 0.25 mm). Velocity through the slots was of less importance, though at the higher consistency it appeared to be of more significance with larger size slots [82].

These findings were contrary to others, where slot width was reported not to be the most important determinant of screening efficiency. Optimal screening efficiency was reported to be dependent on minimising kinetic energy applied to conformable contaminants, given by low rotor velocity, low-pressure pulse rotor design and screen basket design. These tests were on different screens from those above, with a different rotor design, though in other respects the screens were similar, the centrifugal type, with the rotor on the inlet side of the screen basket. Low-consistency screening (0.8–2.5%) was preferred to thick stock (3–5%) screening, due to the higher forces required to fluidise the stock [83]. This difference illustrates the difficulties involved in choosing and operating screens, since advice is often contradictory.

Although the efficiency of a screening system is generally assumed to be dependent on the primary screen, it is frequently the final stage, or reject disposal, which has the greatest impact on overall efficiency. For many years, the final treatment in a sequence of screens was a vibrating screen, but in this position they are usually very inefficient, and their use contributes significantly to the creation of a heavily contaminated loop. If the final stage rejects only a small proportion of contaminants they will be recycled until they are reduced in size and fed forward with the accepts from the primary screen. Although vibrating screens are now rarely used, recirculation can be caused by the selection of inappropriate final cleaning stages, for example, using centrifugal cleaners as a final cleaning stage. Cleaners may be used, for example, to reduce wear of fine screen baskets, as illustrated in Figure 8.6 but are not suitable for use as a final stage, since the separation principle is different and they will thus not be efficient. If final stage rejects are dewatered but separation of contaminants and water is inefficient, then contaminants will be recycled with water.

Screening is also influenced by treatment prior to the screening stage; for example, if contaminants are reduced in size by the input of a large amount of energy, via pulping, deflaking, dispersion, etc. then the number of contaminant particles will increase, whilst the average size falls. This would reduce screening efficiency, irrespective of the number and types of screen used. Pulping temperatures, pH and chemistry also affect dispersion of contaminants.

3.4.2.2 Coarse screens. Removal of debris whilst still relatively large reduces the load on fine screening systems; for example, a 1 cm^2 piece of polythene could be degraded to 100 pieces, each 1 mm^2. Coarse screening thus helps to prevent fine screens from being overloaded. Perforated pressure screens are usually used in coarse screens, and if used as soon as practical, at a consistency in the range 3–5%, they offer maximum opportunity for large size contaminant removal, as illustrated in Figure 3.12. Rigid particles which are larger than the screen perforation in all three dimensions are removed, as are some long, stiff fibres, such as shives.

Extraction plates in a pulper can be regarded as a first coarse screening stage, but if small diameter holes are used, there is a danger that some debris will be reduced to a size which makes subsequent removal difficult. In the absence of wet strength grades, which are more common in the recycling of board grades (chapter 8) good fibre separation is achieved in high-consistency pulping and so there is no need to retain material in the pulper tub for size reduction. Large diameter extraction plate holes (or extraction from the pulper tub) is preferable, with the screening function provided by a relatively low energy pulper dump screen, which normally has perforations in the range 12–20 mm. Subsequent coarse screening can be by holed pressure screens, with perforations in the range 1.2–2.0 mm, and can be at a consistency in between 1% and 5%. In deinking mills these are usually followed by at least one multistage fine screening system.

3.4.2.3 Fine screens. These are always slotted screens, which can be operated in the consistency range 0.8–5%. Some mills have two multistage fine screens systems, one operating at the upper end of this range, probably with slot widths of 0.25–0.40 mm, with a second low-consistency system (0.8–1.5%) normally having finer slot widths, in the range 0.15–0.25 mm. Even finer slots, down to 0.10 mm, are being introduced. Other mills may have only one stage of fine slots, though some may have a further fine slotted stage in the paper machine approach flow system. In this position screens need to have low pressure pulses. A difficulty with screens in this position is that they increase the machine vulnerability; if screens block, the machine may have to cease production. In general, very fine screens are not used in machine approach flow positions, and slot sizes are in the range 0.25–0.40 mm.

Numerous reports emphasise that slotted screens are the most efficient stage for removal of stickies and other contaminants [20, 28, 84, 85]. In one of these studies, a wide variation in stickie removal efficiency was observed [28] frequently due to poor reject handling systems, following an efficient primary screen.

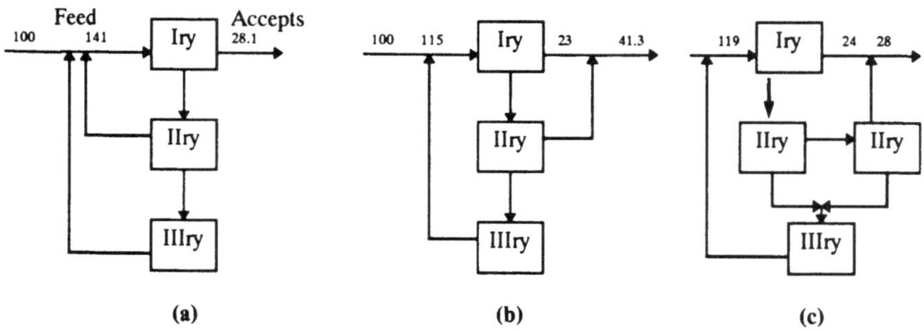

Figure 3.18 Schematic illustration of screen installations. (a) Cascade system, (b) forward feed system, (c) forward feed, series.

There is a wide range of screens available and choice is often made more difficult by conflicting reports, as already noted. In two studies evaluating two alternative screens, different results were found, even though screens evaluated were the same. Screen types were:

- Inward flow, with the foil on the outside of the basket (feed side);
- Outward flow, with the foil on the inside of the basket (feed side);

In one set of trials smooth baskets were used [81] whilst in the other contoured screen baskets were used. This difference may explain the different results obtained. In one case the inward flow screens (with smooth baskets) were found to be less efficient, with lower capacity and more difficult to operate, though fractionation was lower [81]. The other study found no efficiency differences, though the outward flow screen was less prone to blinding [84]. The former evaluation included mathematical modelling of screen efficiencies to assess different options for the complete system. They found that secondary screen efficiency was lower than primary, 68% + 78% stickie removal, respectively. The overall system efficiency was predicted to be 73%, and after installation was complete, was found to be 70% [81]. Although narrower slots were used in the later installation (0.2 mm, compared with 0.3 mm) stickie removal efficiencies were very similar, averaging 69% with 0.2 mm slots [84]. This underscores the lower efficiency of contoured baskets, as indicated above.

There are a variety of different screen layouts possible, some of which are illustrated in Figure 3.18. There are many other possibilities.

A feed forward system, Figure 3.18(b), has the advantage that it avoids recirculation of contaminants which will occur in the cascade system. Another advantage is that since the accepts of the second stage are fed forward, the system capacity is effectively increased. However, there is a cost – the quality of the accepts fed forward from this are lower. If each screen efficiency is 80% and 100 contaminants are in the feed, then at equilibrium the cascade system accepts have 28 contaminants, versus 41 in the feed forward system. This simple model does not include an allowance for size reduction during recirculation, which reduces the quality of the accepts from the cascade system, but neither does it include lower efficiencies for secondary screens, reducing the quality of the feed forward system. A compromise is illustrated in the modified feed forward system, Figure 3.18(c), which feeds second stage accepts forward, but after two screening stages in series. Giving the same feed stock and using the same simple model gives an accepts contaminant count of about 28, similar to the cascade accepts. Both of these systems have higher capital costs than the feed forward, but given the prominent role of fine slotted screening in stickies removal, capital cost savings in these installations are difficult to justify. As discussed above, many factors influence overall efficiency, such as reject rates, the use of other types of cleaners, etc. Some of these are explored in a study describing mathematical modelling of screening systems. This concluded that the reject rate in the primary stage should be as high as possible, whilst that of the tertiary should be as low as possible, to avoid excessive fibre losses [86].

The selection of the final screening stage is important; a poor choice, such as a vibrating screen, will severely reduce the overall system efficiency. As a minimum, it should have the same size slots as the earlier stages. Purging pressure screens, which incorporate a wash cycle and operate on a batch basis (to reject collected debris) are successful in minimising fibre losses, whilst maintaining system efficiency.

One of the most common operating problems is blinding. Studies on different types of basket and slot spacing have examined this in depth. One study found that if fibres or contaminants were caught in a slot, a 'pile' of material grew rapidly, and if other piles were formed, they very rapidly formed a mat. If the height grew to 2–3 mm it was broken up by the main flow stream. It was suggested that this problem was reduced by the use of a basket with a wavy surface [87]. However, slot spacing at 7.5 mm was longer than most fibres and this has been shown to be very significant in the avoidance of blinding. If slots are close together, then screen capacity falls due to stapling and the screen can blind. Blinding and screen capacity were found to be dependent on the percentage of fibres longer than the slot spacing; for short fibre pulps, screens did not blind and capacity was highest. In this study, no blinding occurred when $<7\%$ of the fibres were longer than the slot spacing; limited blinding occurred when 10–20% of the fibres were longer than the slot spacing and severe blinding when 30% or more of the fibres were longer than the slot spacing [79]. Although these percentages will probably vary with different types of screen, it appears that different furnishes should have different slot spacing for maximum screening efficiency. This suggests that mills with blinding problems should try screens with fewer slots in the basket to increase spacing, and not more slots.

Other actions to reduce blinding problems include a good control system to react automatically to partial blinding. Routine testing of debris levels can help to avoid blinding - if contaminant levels are high, reject rates can be increased. The screen should be examined at regular intervals and if necessary cleaned. As water systems are closed up, scale may become a problem, but this can usually be easily removed by an acid wash.

Where air bleeds are installed, these should be always open. Lightweight contaminants can concentrate at the air bleed and so these should be piped to appropriate reject treatment systems. Air in stock can reduce screening capacity significantly. If air bleeds are not open air can accumulate and in extreme cases, can cause damage to the screen basket, since the pressure pulse created by the rotor will be transmitted by air. Rotors may also be damaged, due to partial immersion in stock.

Excessive wear on contoured screen plates is usually due to inefficient removal of grit, sand, etc. Operation under acidic conditions may also increase the rate of erosion of contours. An increase in blinding may be a symptom of wear of the contours on the screen plate.

3.4.2.4 Centrifugal cleaning. There are usually at least two cleaning stages, using centrifugal forces to separate fibre and contaminants, but frequently three or more are used. Each stage is designed to remove different sizes and types of contaminants, for example:

- high density, high-consistency cleaners - removal of medium size contaminants with a high specific gravity, such as paper clips, glass, grit, etc.;
- high-density, low-consistency cleaners (also known as forward flow cleaners) - removal of small contaminants with a high specific gravity (>1.0), such as fine grit, adhesives, sand, coating flakes, some inks, etc. (Figure 3.19);
- low-density, low-consistency cleaners (also known as reverse or through flow cleaners) - removal of small contaminants with a low specific gravity (<1.0), such as waxes, some plastics, some adhesives, etc. (Figure 3.19).

All these operate with the same separation principle - contaminants can be separated from fibre if they have a different density. Although solid cellulose has a density of $1.57\,g\,cm^{-3}$, the specific gravity of various pulp fibres in aqueous suspension is in the range 0.88–$1.08\,g\,cm^{-3}$ [88]. The lower density is due to occluded air, especially that in the lumen or central canal of the fibres. Other tests showed loblolly pine fibres had a specific gravity of $1.53\,g\,cm^{-3}$ [89]. When particles of different specific gravity are subjected to centrifugal forces they separate and form a specific gravity (density) gradient, though size and shape also play an important role in separation. In centrifugal cleaners, stock is injected tangentially into the

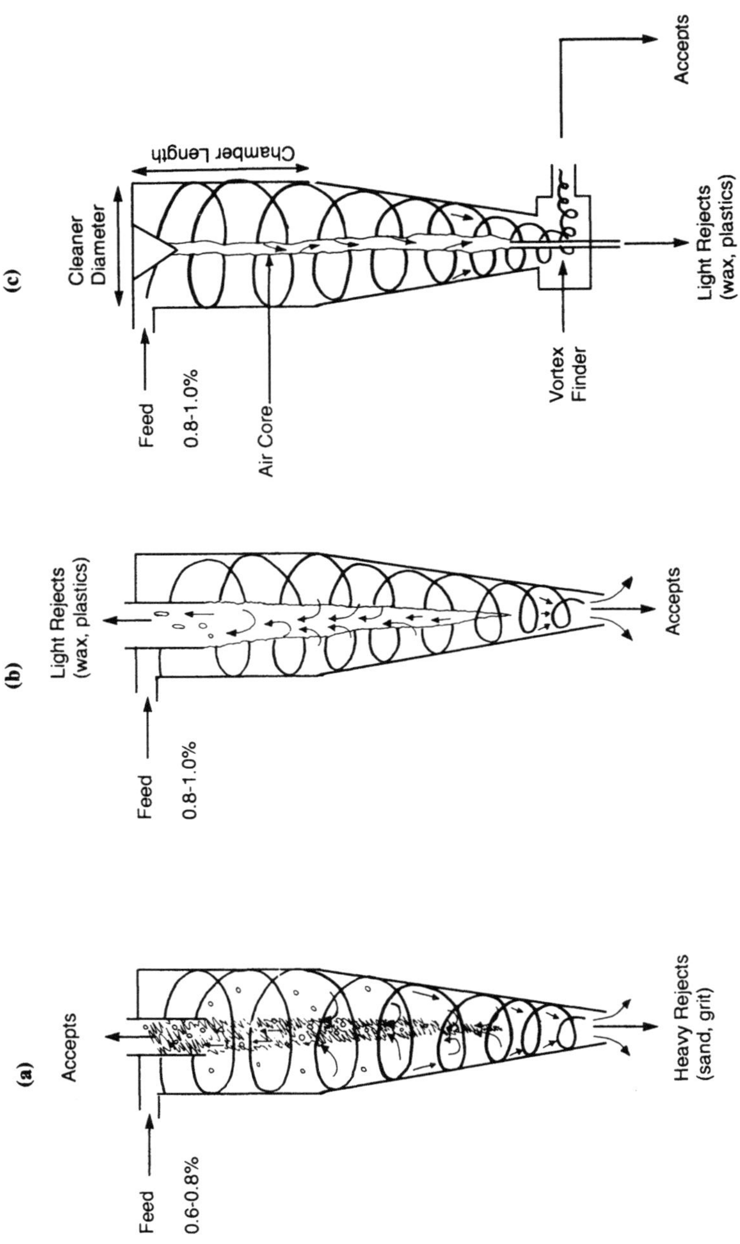

Figure 3.19 Types of cleaner. (a) Forward, (b) reverse, (c) through flow.

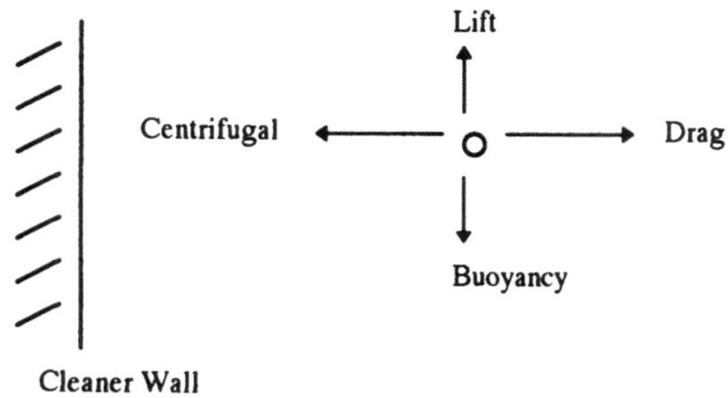

Figure 3.20 Forces acting on a particle in a centrifugal cleaner.

cleaner body, giving a high rotational speed which creates high centrifugal forces on particles, with low-density particles becoming concentrated at the centre of the cleaner, around an air core, and high density particles at the outside, near the cleaner wall.

The top part of the cylinder is usually cylindrical (see Figure 3.19) and the internal diameter is the reference size. When the outlet is located at the axis of the cleaner, stock is forced to spiral down the conical section of the cleaner towards the centre. Complex mathematical models can be developed to describe the flow patterns and forces acting on particles, which are:

- centrifugal force - which results from the rotary motion of the particle, acting outwards;
- drag force - results from the drag force of the inward moving fluid;
- buoyancy force - due to the submergence of the particle in water and unbalanced surface forces;
- lift force - transverse force on a solid spinning in a liquid, generally directed inwards.

These are illustrated in Figure 3.20.

Forces are dependent on the physical properties of the particles, their size, shape and density, cleaner geometry, etc. As a consequence of varying forces cleaners designed to achieve, for example, shive removal may not be the most effective for speck removal. In this case lift forces of these particles would be very different, much higher for shives than for specks.

High-density cleaners are generally the first cleaners after the pulping system, and operate at 'high' consistency - more accurately it is at medium consistency, between 2.5% and 5.0%. They have larger diameters than low-consistency cleaners, due to the presence of relatively large pieces of debris, so that the outlet orifice has a large diameter to prevent blockage. Typical inlet diameters are in the range 150–400 mm. Factors which influence their efficiency are similar to those given below for low-consistency cleaners. If fine screening is to be used before low-consistency forward flow cleaners, then special attention should be paid to high-density cleaner selection and operation; lower diameter, lower consistency and high-pressure drop will give maximum cleaning efficiency. These are sometimes called medium density cleaners.

In some high-density cleaners the centrifugal forces are sometimes reinforced by a rotor. Both types usually reject debris to a reject box, that is, they do not have open continuous rejects. Reject boxes are either emptied automatically, by a controlled time cycle, or manually. Automatic emptying is preferable, since if reject boxes are not emptied, rejects can build up in the body of the cleaner and rapidly wear the cleaner. Pressure drops are usually quite low, in the range 30–50 kPa. Another approach to improving efficiency is to operate with a continuous reject, treated by a second stage, at lower consistency.

Low-consistency, high-density cleaners (forward flow cleaners) were first developed to remove sand and grit in pulp systems, but have been extensively modified for use in paper recycling systems. Selection of cleaners is difficult since many are tagged with arbitrary numerical efficiencies, which frequently bear no relation to their operation in recycling mills. Factors which affect their efficiency include:

- Particle size, shape, surface roughness, density;
- Type of cleaner;
- Cleaner diameter and geometry;
- Reject to inlet ratios - mass and flow;
- Pressure drop across the cleaners;
- Stock freeness, long-fibre content;
- Stock consistency;
- Maintenance.

Many of these are inter-related. One study looked at the efficiency of ten different forward cleaners, with diameters ranging from 24 to 152 mm, in the removal of large flat ink specks, in the approximate size range 0.1–0.5 mm. Actual efficiencies were compared with those predicted by a computer model which introduced the term, cleaning index I

$$\text{where} \quad I = GT/D$$

where G is centripedal acceleration, in the cleaner head zone, produced by the injection velocity, T is residence time, the duration of the centrifugal separation - the calculation is based on the real and a standard design - and D is the cleaner head diameter.

The cleaning index was roughly proportional to the pressure drop across the cleaner, that is, the difference in pressure between the inlet and accepts outlet. Trials showed that small cleaners had a higher efficiency; highest was 24 mm, which is not used in the paper industry. There were some exceptions to this general rule: one 150 mm cleaner was more efficient than a 75 mm, due to the low injection velocity and residence time in the 75 mm cleaner.

The effect of consistency was marked, but could be compensated for by other changes. At high consistencies (1.4–1.8%) small diameter cleaners (60–80 mm) were more efficient than large diameter cleaners (150 mm) at low consistencies (0.7–0.9%), provided higher pressure drops were used, 2 bar (206 kPa) vs. 1.2 bar (120 kPa), respectively. The 60 mm cleaner gave a significantly better performance than the 80 mm diameter cleaner. Highest efficiencies were at a reject flow ratio of 15–30%. These results were in line with predicted efficiencies from the model used. However, the model predicted an increase in efficiency with increased temperature, which was not observed [90].

Other reports on speck removal appear to be contradictory, since they suggest good speck removal is achieved only at low consistency. One report was that a dramatic reduction in efficiency was observed when consistency was above 0.7%; specks were from electrostatic inks. In this case cleaners were operated at 0.5% to achieve good speck removal. However, no details of cleaner diameter, pressure differentials, etc. were given [91].

Another model was developed, based on laboratory studies of forward cleaners with different diameters, 75, 125 and 150 mm. The type of dirt evaluated was not specified. Pressure drops were 35–245 kPa and consistency was 0.5%. Hydraulic relationships were also examined, the feed flow rate being shown to be proportional to the square root of the pressure drop and the reject rate (volume) proportional to the pressure ratio, defined as:

$$\text{pressure ratio} = P_a/(P_f + P_r)$$

where subscripts a, f and r refer to accepts, feed and rejects, respectively.

In this study, dirt removal was not found to be related to cleaner diameter. Slightly cleaner accepts were given by the cleaners with a very high reject rate [92].

Although these data appear contradictory, it is not possible to make comparisons, since dirt was defined and measured differently. This was shown to have a major bearing on data evaluation [90]. The study which examined the widest range of cleaners and variables probably provides the most useful information, that is, low-diameter cleaners and high-pressure drops can afford good cleaning efficiencies, even at higher consistencies. However, there is a limit to cleaner diameter, since cleaners with a narrow rejects orifice are very prone to blockage. Cleaners with a diameter of 75 mm have the smallest diameter practical in most wastepaper recycling applications.

Normally in deinking mills, three or four stages of forward flow cleaners are required, with final stage rejects being discharged to some type of water treatment system. The arrangement of these can be similar to those illustrated in Figure 3.18 for fine screens - feed forward or cascade. The cascade system is normally used, even though it has higher capital and operating costs. Some installations feature two cleaning stages in series. Since specks are best removed by small diameter cleaners and shives by larger diameter cleaners, different diameters for each stage ensure maximum contaminant removal. In this case, to ensure cleaning efficiency is maintained, the reject systems should also be separate, so that there would be two primary, two secondary, etc. stages of forward cleaners.

The influence of air on centrifugal cleaners has been assessed. Air entrained in stock migrates to the centre of the cleaner, giving a stable column of air along the axial centre line. Residence time of air was very low, < 1 s. When viewed in a transparent cleaner, the air cone was twisted and rough in appearance, with a diameter from 2 to 4 mm up to 15 mm, dependent on the pressure differential, absolute pressures and the amount of air in stock [93]. Since the diameter of the air core can restrict the reject flow, it must be compensated for by increasing inlet pressure. Whilst this is possible in long-term situations, it is much more difficult with short-run variations in air content, such as are found when the proportion of coated paper grades vary, especially in woodfree deinking. Rejects discharging to an open trough can entrain air which is then recycled with rejects. This can be avoided by discharging rejects via submerged tips or into a reject header, for pressurised rejects. Vacuum systems can also be used to strip air from rejects, which prevents it exiting with accepts. This would also help to stabilise the air core.

Air was deliberately injected into some cleaners during the 1960s and 1970s, sometimes with chemicals, in order to remove lightweight contaminants. This has been developed with gas sparged cyclones designed to remove ink as well as other hydrophobic materials.

Mill surveys examining removal of stickies found rather low efficiencies, with system efficiencies in the range 10–40%. Although individual stages had higher efficiencies, up to 60% removal, the presence of blocked cleaners or low pressure drops in one or more stages reduced overall efficiencies [28]. This highlights the need for continuous maintenance of cleaning systems, both to provide routine cleaning of blocked cleaners, but also to check for wear. When cleaners are blocked, if a high concentration of grit is present, cleaner bodies can be damaged very quickly. As cleaners wear, irregularities develop at the surface and adversely affect efficiency. High-wear areas such as the lower cone and the rejects tip should be inspected regularly and worn parts replaced as required. Other parts should also be inspected, but this can be at less regular intervals. One of the difficulties inherent in canister systems is the inability to check for blocked cleaners during normal operation.

Several studies have shown there is a specific gravity below which forward flow cleaners are not efficient though this is dependent on particle size and shape. The limit found was $1.05\,\mathrm{g\,cm^{-3}}$ [28, 94]. To overcome this, cleaners which removed lightweight particles were developed in the late

1960s – reverse flow cleaners. As illustrated in Figure 3.19(b) rejects were from the overflow and accepts from the underflow. Reverse flow cleaners are geometrically similar to forward flow cleaners, though there are differences in the outlet diameters of the underflow and overflow. In reverse flow cleaners the underflow diameter is larger and the overflow smaller. The consistency of the accepts is higher than the feed, the reverse of the situation with forward flow cleaners. However, reverse flow cleaners required high-pressure drops and high reject rates for maximum efficiency. One study which examined efficiency found that it increased with increased pressure drops, up to a maximum of 620 kPa, when efficiency started to fall. As the pressure drop increased, the capacity per cleaner increased, but so did losses [95].

Through flow cleaners were introduced in the early 1980s. In these, accepts and rejects are at the bottom of the cleaner, with the rejects picking up the air core at the axial centre line, as illustrated in Figure 3.19(c).

Reject rates are low, typically 1–10%, which means that usually only two stages of these cleaners are required. Operating consistencies recommended are $<1.0\%$, with relatively low pressure drops, typically 70–110 kPa. Cleaning efficiencies are slightly lower than for reverse flow cleaners and due to low operating pressures have a greater tendency to plug. Generally, lightweight cleaners are inefficient in removing particles with a density $>0.95\ g\ cm^{-3}$ [28, 94].

Control of cleaning systems is simple and is usually manual, by throttling valves in feed, accepts and rejects, to set pressures. Sometimes a closed loop system is used automatically to control either the feed or accepts pressure.

Final stage rejects sometimes include fibre recovery systems to minimise fibre losses, but these may have a significant effect on system efficiency. If final stage rejects are discharged to a sewer, there should be no possibility of recirculation without water treatment. Ideally, final stage rejects would be pumped directly to a sludge blending tank, provided the volume, relative to other sludge sources, is low.

Another lightweight cleaning device is the Gyroclean, in which the body of the cleaner rotates; it is a rotating centrifuge. This allows very high centrifugal forces to be developed, which with longer retention times (5–10 s) than in centrifugal cleaners allow effective separation of lightweights. One mill reported very effective removal of lightweight plastic, with a very low final concentration of plastic [96].

3.4.3 Dispersion

Asphalt dispersion units were first researched in the USA in the mid 1940s and a number of patents were granted in the period 1946 to 1962. Most of these expired in the period 1970–1973 [97]. These early systems were used to deal with asphalt treated, waxed and bitumenised boards. Temperatures up to 150°C were used to melt contaminants, which were then dispersed; there was no removal of contaminants. This gave recycled board an acceptable appearance. Even though consistencies were high, up to 40%, their energy needs resulted in high operating costs.

In 1973, two papers were published describing 'new' dispersion units, one a high-speed, single shaft, refiner disc unit and the other a double shaft, relatively slow unit [98, 99]. These and similar units were first used in board grades, especially waste-based liner, then in deinking lines for tissue and then in newsprint. The emergence of polymeric inks, used in laser printers and copiers, gave further impetus to the use of dispersion and kneading units, since these break down large ink flakes to a size which allows more efficient removal of ink. Dispersion has also been used to replace deinking - ink is dispersed, so that there is no specky appearance, but ink is not removed. There is a bewildering choice in units, often with contradictory information regarding the effects of kneading and dispersion units on fibre properties and contaminants. Part of this arises from the complexity of the process; for example, if a unit is also being used as a mixer for the addition of chemicals for bleaching, then final properties are the result of a number of different effects, including:

- ink release by bleaching chemicals;
- ink release from fibres as well as ink particle size reduction by forces from kneading or dispersion;
- bleaching of fibres from chemicals present;
- effect of mechanical forces on fibre properties;
- effect of curl induced by the kneading or dispersion unit on fibre properties;
- effect of temperature (if at elevated temperatures) on fibre properties.

Although mechanically it is a simple process these effects can be difficult to interpret. Reasons for using dispersion have changed with time and with applications. Originally, they were to ensure that visual contaminants were reduced in size, to improve the visual appearance of the sheet, as well as to disperse large pieces of contaminant which could cause operating problems. As problems with inks grew, emphasis switched to the reduction in size of ink speck and so its location in deinking plants changed. Its use has advantages and disadvantages, and the decision on use will depend on the balance between these, as well as the visual property requirements of the finished product.

3.4.3.1 Types of dispersion and kneading units

Low-speed units. Low-speed units, usually known as kneaders, can have single, double or triple shafts. They have been used extensively in Asia. The action is a gentle shearing, with a twisting and compressive action on fibres, with no cutting, since forces are generated by inter-fibre interactions, similar to a high-consistency pulper. Operating temperatures are usually ambient, though steam can be added, if required. Due to the friction forces generated, exit temperatures can be quite high, 70–80°C. Units are almost never pressurised. One evaluation of a twin shaft kneader found that speck reduction was dependent on its nature. Bitumen was difficult to disperse and 2% bitumenised paper content required an energy input of 120–130 $kWh\,t^{-1}$, with steam addition - input temperature minimum of 85°C - for effective bitumen dispersal. If a lot was present, then it was rolled into balls. Wax dispersal was easier. When the furnish contained 10% waxed paper, it was dispersed by 75 $kWh\,t^{-1}$, if the input temperature was 45°C [100]. These results suggest that dispersal of some contaminants such as hot melts may require quite high energy inputs and that under normal conditions, they are neither totally dispersed nor rolled into balls, but some dispersion probably occurs. The same study found that when a blend of 50% mechanical and 50% chemical pulp was treated by a kneader followed by low-consistency refining and properties were compared with low-consistency refining alone, strength properties were similar, but the kneading system required less total energy to reach a specific strength value; for example, to reach a specific tensile of 40 $Nm\,kg^{-1}$, low-consistency refining alone required about 35% more energy. Total energy use was high; in kneading plus refining it was about 280 $kWh\,t^{-1}$. In this case fewer fines were created, which gave better drainage than the refining treatment on its own [100].

Part of the reason for the higher tensile value was probably to do with the introduction of curl into chemical fibres. This was examined using two low-speed kneaders, one a single shaft, the other a double shaft, using sulphite pulp of varying yields, 50–88% yield. Results showed that curl was dependent on pulp yield; low-yield pulp (50%) was highly curled, whereas high-yield (>80%) pulp did not retain curl.

The effects included an increase in the wet-web stretch at low solids content, a small reduction in wet-web tensile but an overall increase in work to rupture, which would give better machine runnability when using fibres which retained curl. Little or no change was seen in wet-web properties of high-yield pulps. Effects were related to energy input; for example, with low-yield pulp treated by a single shaft kneader, wet-web stretch increased from about 9% to 15%, with an energy input of about 50 $kWh\,t^{-1}$. The response given by twin shaft treatment was slightly different; at 50 $kWh\,t^{-1}$, wet-web stretch was about 13% and increasing this to 15% required considerably more energy, 140 $kWh\,t^{-1}$. The dewatering unit (a screw press) used prior to kneading introduced considerable curl; wet-web stretch increased from about 5% to 9%.

Dry sheet properties followed a similar pattern: little or no change with high-yield pulps, but low-yield pulps showing appreciable changes. The 50% yield pulp showed a significant reduction in breaking length and an increase

Table 3.12 Properties before and after kneading [101]

	Input (after screw press)	Kneader type	
		Single shaft	Double shaft
CSF (ml)	646	630	650
Bulk ($cm^3 g^{-1}$)	1.62	1.5	1.6
Stretch (%)	3.0	3.0	3.6
Burst index ($kPa\, m^2 g^{-1}$)	4.3	3.4	3.4

in tear for both units. In the case of the twin shaft unit, with an energy input of about $50\,kWh\,t^{-1}$, breaking length fell by about 25%, whilst tear increased by about 25%, whereas with the single shaft unit, these were less pronounced, about 15% reduction in breaking length and a 10% increase in tear. Again, the dewatering screw press had a significant effect on these properties, which illustrates that they are due to curl not changes in fibre during the kneading process. Some results for the 50% yield pulp are reproduced in Table 3.12, at about $55\,kWh\,t^{-1}$ energy input [101].

The reduction in bulk is unexpected, since curled fibres would be expected to have higher bulk. However, the effect may be due to the effects of kneading on fibres; the high-yield pulp (88%) which did not retain curl had a substantially greater loss in bulk. This illustrates difficulties in interpreting results; the effects of kneading on fibres can be evaluated only after compensating for the effects of curl. If curl is desired in high-yield pulp it can be set by heat treatment; for instance, one study on TMP showed curl would be set by storage at 65°C, at high consistency. Fibres developed residual latency which could not be removed [102].

The action of kneading on ink speck size reduction has been studied extensively and results show it is effective in reducing ink speck size [98, 103–106]. Most of these papers have concentrated on ink speck size reduction during one or two kneading stages and subsequent removal by washing and flotation, and they show this is an effective approach for deinking when large specks are created, for example, laser printed, xerographic, UV coated, offset, etc. However, few details have been given about the effects on fibre properties. One report which did give details used a mixture of news and magazines. Although this furnish does not normally cause large ink specks, when offset printing of newsprint was first introduced into Europe in the mid 1970s the inks initially used caused severe deinking problems, especially due to the production of large specks from newspapers which were more than a few weeks old. Thus there was interest in using dispersion (kneading) to reduce these in size.

A slow-speed single shaft kneader was used and this reduced ink speck size successfully. Stock consistency was 22%, inlet temperature 33°C, exit

Table 3.13 Fibre properties before and after single shaft kneading [103]

	Before kneading	After kneading
Freeness (CSF, ml)	177	172
Bulk ($cm^3\,g^{-1}$)	1.97	2.02
Stretch (%)	1.83	2.03
Burst ($Kp\,cm^{-2}$)	1.38	1.18
Breaking length (km)	3.18	2.87
Bendsten ($ml\,min^{-1}$)	252	325
Brightness (% ISO)	58.5	58.0

Table 3.14 Effect of kneading on fibre length [103]

	Before kneading	After kneading
Long fibres (%)	34.9	31.8
Short fibres (%)	21.8	23.3
Fines (%)	43.3	44.9

temperature 43°C at an energy input of $52\,kWh\,t^{-1}$. Some results are reproduced in Table 3.13.

A comparison with results in Table 3.12 shows a similar pattern, very little change in freeness, an increase in stretch and a reduction in burst. However, bulk has increased. Tear increased by almost 13% and it was suggested this was due to curl [103]. Tensile breaking length fell by about 10%. Brightness fell very slightly, which suggests ink had been dispersed. Smaller ink specks have a greater surface area and so absorb more light, which results in a lower brightness. If residual ink is high and ink dispersion is effective, brightness can fall substantially, by up to 8 units. Although there was only a small change in freeness, it was found that kneading reduced the long-fibre content and created some fines. Results are given in Table 3.14; fractionation was by the Brecht–Holl method.

Other reports [105, 106] have suggested that there is no fibre degradation during low-speed kneading, which is the reverse of these results.

High-speed units. High-speed units are known as dispersers or dispergers. They have single shafts, similar to refiners, with two plates, one fixed, the other rotating at high speed. A number of different types are available and these vary in plate patterns, operating temperatures and control systems. Plate patterns include multistage concentric rings, with intermeshing teeth, a bar pattern, from very coarse to fine and pyramidal tooth configurations. Operating temperatures vary from 50°C to 125°C, with the latter achieved in a pressurised system. Control systems range from simple to complex – in these, disc gaps are set to very fine tolerances. Different plate designs and operating disc gaps are used to achieve varying degrees of dispersion, but normal operation is due to internal friction created in a fibre mat

Table 3.15 Fibre properties before and after high-speed dispersion [108]

	Before dispersion	After dispersion
Freeness (CSF, ml)	145	97
Bulk ($cm^3\,g^{-1}$)	1.9	1.8
Tensile (km)	3.8	3.93
Tear factor ($mNm^2\,g^{-1}$)	7.8	7.6
Bendsten ($ml\,min^{-1}$)	223	240
Brightness (% ISO)	57.5	54.8

between two plates; the action is primarily brushing and flexing, whilst trying to avoid cutting, crushing and bruising of fibre. One study looked at the influence on ink speck dispersion of different plate patterns, ranging from very coarse to a combination of coarse (centre of disc) and fine (periphery of disc). Inlet temperature was about 50°C and applied energy was between 53 and 330 $kWh\,t^{-1}$. Two consistencies were used, 15% and 30%. Ink dispersion was better with line plates at 30% and little benefit was gained by power inputs >70 $kWh\,t^{-1}$ [107].

It is probable that high-speed dispersion also introduces curl into some types of fibres, so interpreting data is difficult. One study looked at the effects of dispersion on a furnish of 60% newsprint and 40% magazines, sampling after deinking by flotation. Inlet temperature was 77°C, outlet 89°C, consistency 25–27% and energy input 52 $kWh\,t^{-1}$. Disc clearance was set at 2.4 mm. Effects on fibre properties are given in Table 3.15.

Freeness was low initially, partially due to a high ash content, 15%. When compared to values in Table 3.13, the reduction in freeness and a reduction in tear are the two substantial differences. In the case of high-speed dispersion the reduction in tensile is smaller than with low-speed kneading. As the furnishes used were different, it would be wrong to draw detailed conclusions from this comparison. The reduction in brightness shows effective ink dispersion and laboratory work showed that ink removal following dispersion was efficient. A Brecht–Holl fractionation showed no change in fibre length, that is, no fines were created by dispersion [108], unlike slow-speed kneading. Since the freeness fell significantly this is surprising.

Other studies have shown that fines can be created with high-speed dispersion. One of these used fibre which was derived from domestic trash, using a wet processing system [109]. An extensive series of tests was carried out using a pressurised disperser. The fibre blend was a mixture of mechanical pulp (35–45%), unbleached chemical (<10%) and the remainder bleached chemical pulp. Disperser variables examined included:

- temperature, ranging from 90°C to 115°C;
- disc gap, using narrow gaps, ranging from closed (0.01 mm) to 0.4 mm.

Table 3.16 Effects of high-temperature dispersion on fibre properties [109]

	Before dispersion	After dispersion, at conditions given					
Temperature (°C)	-	90	105	115	115	115	106
Disc gap (mm)	-	0.4	0.4	0.4	0.07	0.01	0.01
Freeness (CSF, ml)	390	168	292	310	283	257	191
Breaking length (km)	3.18	4.40	3.43	3.31	3.38	3.47	4.04
Tear index	8.02	7.86	8.24	8.28	8.05	8.15	8.12
Burst index	1.8	2.56	2.07	1.81	1.81	1.97	2.45
Brightness (% ISO)	37.1	35.5	37.5	37.4	37.4	37.1	35.5

Table 3.17 Effect of high-temperature dispersion on contaminants [109]

	Before dispersion	After dispersion, at conditions given					
Temperature (°C)	-	90	105	115	115	115	106
Disc gap (mm)	-	0.4	0.4	0.4	0.07	0.01	0.01
Dirt count (mm m^{-2})	1008	270	277	306	-	209	169
Stickies (#/100gbdf)	180	93	61	121	12	19	43

Results from these tests are reproduced in Tables 3.16, 3.17 and 3.18. Energy input varied, according to temperature. At high temperatures it was in the range 40–50 kWh t^{-1}.

These results show clearly that dispersion is a complex process. If results in Table 3.16 are compared with those in Table 3.15 (low-temperature, wide gap) it is clear that under some conditions, for example high temperature (115°C) and a disc gap of 0.4 mm, effects are similar, i.e. relatively minor effects on strength and a drop in freeness. Under other conditions - at low temperature (90°C) and wide gap (0.4 mm) - effects are very different: the loss in freeness is very high and strength properties (with the exception of tear) increase substantially, tensile breaking length by almost 40% and burst by >40%. Under these conditions the disperser is acting as a high-consistency refiner. The reduction in brightness is a consequence of the breakdown of contaminants, which is detailed in Table 3.17.

Contaminants are not removed by dispersion; in the case of dirt specks, they were reduced in size so that they were no longer visible to the naked eye. Stickies were reduced in size so that they passed through the slots of a 0.15 mm screen plate used to separate fibre from stickies. The reduction of both follows the same general trend - higher temperatures and narrow disc gaps are most effective. Fibre length analyses given in Table 3.18 reveal the intensity of dispersion [109].

The series of results reveals some of the contradictions in dispersion - average fibre length was reduced significantly, dependent on dispersion conditions, but tear strength did not fall. High temperatures reduce the

Table 3.18 Effects of high-temperature dispersion on fibre length [109]

	Before dispersion	After dispersion, at conditions given					
Temperature (°C)	-	90	105	115	115	115	106
Disc gap (mm)	-	0.4	0.4	0.4	0.07	0.01	0.01
Average fibre length (mm)	1.34	1.18	1.31	1.25	1.21	1.11	1.05
Bauer McNett							
+28	18.8	3.5	0.3	0.1	5.6	12.9	7.4
−200	10.8	22.5	11.8	18.8	24.0	10.2	15.4

effects on fibres, probably due to the reduction in viscosity and possibly softening of some of the fibre components; modified lignins soften at about 90°C and native lignins at about 130°C, dependent on lignin species. At higher temperatures, the reduced viscosity reduces frictional forces on fibre and increased fibre flexibility may also limit fibre damage. High temperatures, but with low disc gaps appear to be best for contaminant reduction. Results demonstrate clearly that dispersion conditions can be varied, dependent on the effects desired.

Modelling can be used to predict dispersion effects. Using the results, for example, for freeness, a model can be developed to predict freeness at any combination of temperature and disc gap. The correlation coefficient (r^2) for the plot is 0.89, which is an acceptable fit. It is illustrated in Figure 3.21.

Another study examined the dispersion of toner inks from laser printed and xerographic papers. Similar physical effects to those above were found. Dispersion of ink was better at higher temperatures and appeared to be independent of disc gap. Subsequent laboratory deinking by flotation, or washing, or both, showed that ink removal was better following high-temperature dispersion. Best results, in terms of both residual speckiness and brightness were achieved after both washing and flotation; some results are summarised in Table 3.19 [110].

These results suggest considerable benefits from dispersion followed by deinking: a higher brightness (and better cleanliness) is obtained, especially when high dispersion temperatures are used. The brightness of the unprinted edge of the papers used in the furnish was 81.5% ISO, and high-temperature dispersion followed by flotation and washing gave a brightness well in excess of this value.

A study on the effect of low-speed kneading concluded that high-temperature kneading was ineffective in the dispersion of laser toner inks due to the softening of inks or reduced friction. Temperatures used were 38°C and 82°C [111]. The effectiveness of deinking after high-temperature, high-speed dispersion suggests that it is not due to softening of toner, but is due to reduced viscosity, which would reduce frictional forces. The glass transition temperature for toner particles varies. Although components of thermoplastic resins, for example, copolymers of styrene and butylacrylate,

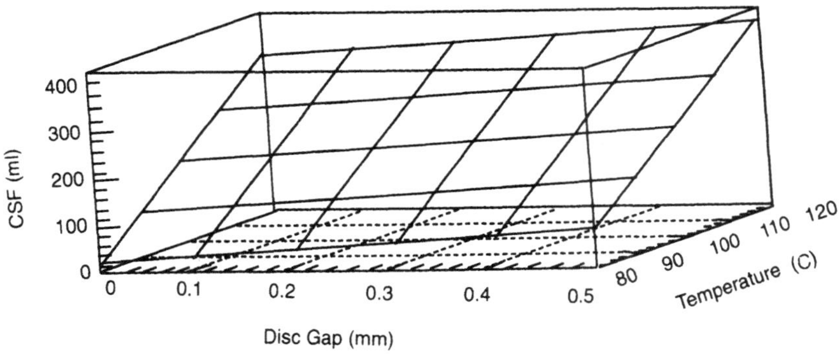

Figure 3.21 Three-dimensional plot relating dispersion temperature (°C), disc gap (mm) and freeness (CSF, ml).

Table 3.19 Deinking laser printed and xerographic grades after high-temperature dispersion [110]

Dispersion conditions		Brightness (% ISO)			
Temperature (°C)	Disc gap (mm)	Initial	Flotation	Washing	Flotation and washing
90	0.5	60.2	68.0	77.6	79.7
105	0.5	61.9	70.0	82.1	82.8
125	0.5	61.9	72.7	83.2	84.2
No dispersion	-	62.8	70.8	74.6	77.7

soften at 60–100°C, fusing temperatures used are higher; in radiant fusing typically it is 150°C and in roll fusing 180–200°C. However, when toner is melted onto paper, cross-linking occurs between copolymers, which would increase softening temperatures. In the high-speed dispersion study, high temperatures were clearly better in terms of subsequent ease of deinking. Irrespective of the type of unit, the variables which control dispersion intensity are:

- temperature;
- stock consistency;
- disc gap (high-speed units), or clearance in slow-speed units;
- plate design (high-speed) or type of low-speed unit;
- time within the unit;
- pH.

The dewatering system can have a significant effect on paper properties, due to the introduction of curl [101], as can temperature and residence time at high temperature [77].

Advantages of dispersion and kneading include:

- dispersion of visual contaminants;
- improved deinking, for high-quality products, from laser printed papers;
- in the case of high-speed units, they can eliminate the need for refining, since strength properties are enhanced;
- very efficient use of bleaching chemicals can be achieved (cf. section 5.6);
- in high-temperature units there is a reduction in microbiological activity;
- in high-speed units, contaminants are reduced in size, so that fewer problems from large contaminants are likely;
- curl can be introduced, which increases wet-web strength and so can improve wet end runnability.

Disadvantages include:

- high operating cost, due to energy requirement;
- dependence on unit and conditions - there can be a loss of freeness and the production of fines. Both low-speed and high-speed units can produce fines;
- there is no contaminant removal, just dispersal.

3.4.3.2 System design and applications. In Europe, up to the end of the 1980s, high-speed dispersion units were installed at the end of a wastepaper processing line. In this location they were very successful in contaminant dispersion, which substantially improved paper machine operation, due to less deposition problems with stickies. Visual quality was improved due to speck size reduction. A reduction or elimination of refining helped to limit increased power costs, but brightness could fall, due to ink speck dispersion. As paper machine white water circuits were closed up, the need to remove dispersed ink and contaminants led to the installation of a second stage of flotation deinking following dispersion, initially in newsprint deinking mills, but followed in woodfree mills, for both tissue and printing and writing papers. The use of dispersion as a means to protect the paper machine has thus diminished and it is now viewed as an integral component of a deinking system, especially in woodfree deinking mills which may be exposed to laser printed and xerographic toners. Typical systems are illustrated in Figure 3.22.

In systems shown in Figure 3.22(b) and (c) removal of ink and other contaminants after dispersion is possible by both washing and flotation. The system shown in Figure 3.22(c) includes two dispersing units; one could be a low-speed unit for ink dispersion, allowing removal by flotation and washing, followed by a high-speed unit to give a final contaminant dispersal stage, for final protection.

Some installations had high-speed dispersion units following pulping, in high-quality printing and writing as well as OCC recycling. Dispersion

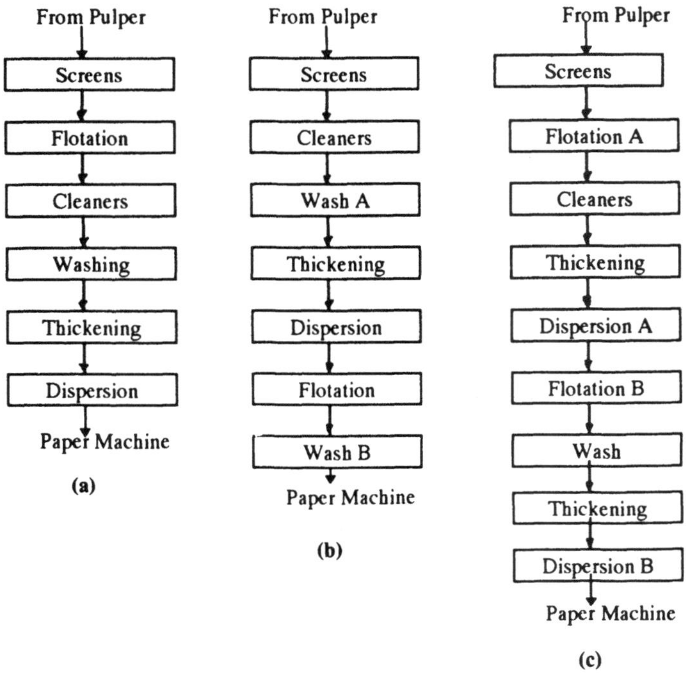

Figure 3.22 Possible location of dispersion units. (a) Dispersion as protection, (b) tissue, (c) high-grade printing.

was used to deflake, but since screening and cleaning followed dispersion, contaminant removal efficiencies were reduced.

In Japan and other Asian countries, low-speed dispersion was introduced early in the system, illustrated in Figure 3.23. Although kneading is said to roll up contaminants such as foil and polyethylene, its effect on stickies is less clear. Since dispersion of asphalt by low-speed kneaders is not efficient other than at high temperatures and high-energy inputs, it is probable that dispersion of most hot melts would be minimal. However, pressure sensitive stickies are easy to disperse; this occurs in high-consistency pulpers and so is likely in low-speed kneaders. Although low-speed kneading may increase the efficiency of removal of some contaminants by cleaning and screening, there may be a reduction with others, such as pressure sensitive adhesives. Nevertheless, this type of installation has been successful in Asia, which has led to proposals for two kneading stages to allow the recycling of laser and xerographic papers into high-quality printing and writing papers [111].

There has been considerable debate over which approach is best in assisting the deinking of laser printed papers [112], but it appears probable that either is satisfactory.

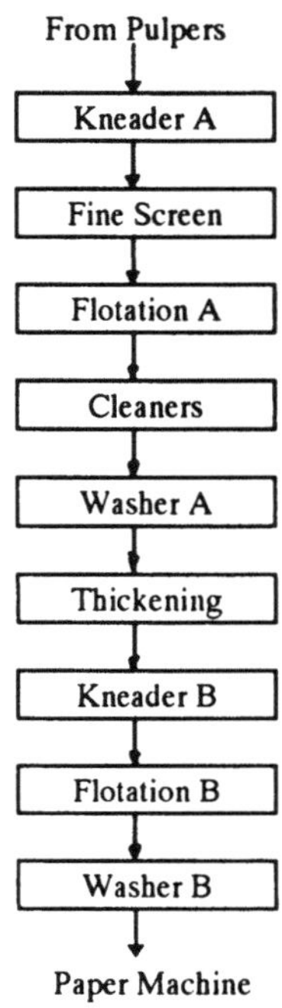

Figure 3.23 Low-speed kneading system for high-quality printing and writing paper [111].

3.4.4 Deinking

Deinking is a very complex subject that could be the subject of a separate text. This section is a brief review. Deinking chemistry is discussed separately, in sections 4.2 and 4.3.

Deinking is by two methods, washing and flotation. Up to the late 1970s they were regarded as competitive systems, rather than complementary. There is now a widespread recognition that better results are achieved by a combination system, since wash deinking is less effective in the removal of large ink specks ($>20\,\mu m$), whilst flotation deinking is less effective in the removal of small ink particles ($<20\,\mu m$). However, small particles can be removed by flotation, if they are agglomerated to form large particles.

When printed papers are repulped, ink is broken down into particles with a wide size range which are suspended in the pulping liquid. Deinking is the separation of these particles from fibres, ideally with no loss of useful fibres. Since recycled fines contribute little to strength properties, their loss and that of fillers is acceptable in grades such as tissue. There is a link between brightness and ink content, provided the size distribution of ink particles is weighted towards small and medium size particles. Large specks ($>40\,\mu m$) contribute to visual unacceptability but have only a small effect on brightness. Very small ink particles would be required to have no effect on brightness, about 1/10th of the wavelength of light.

As wastepaper is pulped, fibres become wetted whilst simultaneously, ink particles are produced. At the fibre surface a layer of water is strongly bound and it is inevitable that this contains some ink. In irregular areas, for instance, near fibrils, the bound water layer may extend enough to 'trap' small ink particles. Tests with cotton fibres have shown that very small particles (< 50 nm) of carbon black have a strong affinity with fibres [113] and it has been suggested that ink is lodged in fibrillated surfaces [114]. Since complete ink removal does not occur during either flotation or wash deinking, either or both of these mechanisms help to explain why, but there are many other phenomena in deinking which are difficult to explain. Much about deinking is unknown; for example, some individual fibres become very heavily coated with ink particles. This is not because they are of a specific fibre type but their affinity for some ink particles is very strong. It is not known why. Nor is it known exactly what role surface chemistry plays in ink removal. Although there is considerable emphasis in the literature on the dependency of removal on particle size, there are many exceptions to this general rule - for example: washing and flotation does not remove all ink specks irrespective of the number of stages; some very large flakes of a specific toner can be removed very efficiently by flotation; it is difficult to remove all small ink particles from water-based inks by washing; the mechanism by which coated papers improve flotation is not clear, but is not due to size considerations; the role of base papers in improving deinking is not clear, etc.

Although some ink is removed by screens as well as centrifugal cleaners, most ink is removed by either or both flotation and washing. Other methods of ink removal have been researched, for example, removal by screening and cleaning following chemical agglomeration but these methods have not been widely adopted.

3.4.4.1 Wash deinking. At its most simple, ink is broken up on pulping and some of these ink particles are removed by subsequent wash stages with the washer filtrate. This can be quite effective even without the use of chemicals. Even if no pulping aids such as caustic soda are used, a high proportion of ink can be removed by wash deinking following mechanical dispersion of ink, which in many applications, for example tissue, will give an acceptable quality.

Methods for wash deinking are thickening processes, using a mechanical separation process: ink particles are separated from fibres by a separating element, frequently a synthetic (plastic) wire. Ink specks need to be of a size which can pass through the mesh of the separating element. Many models for wash deinking are based on ink removal being proportional to the amount of water removed during the thickening stage [115] but these do not accurately predict efficiency, even if the model is based on colloidal particles. As fibres are separated from water, by flow through a mesh, ink particles are removed with water. At first large particles may be removed, dependent on mesh size, but as dewatering continues, fibres are laid down on the mesh and a mat forms which is a very effective filter. It can trap even colloidal particles; for example, the retention of soap was shown to be high when pulp was dewatered through a filter - this retained colloidal calcium soap particles or fines with calcium soap adhering to them [116]. Thus the efficiency of wash deinking is determined by the speed of mat formation and the thickness or basis weight of the mat. Even in washing devices where a mat is not formed, for example a screw press, water is pressed through an effective mat. Efficient washing is only realised in that portion which is next to the dewatering element. Hence washing is inefficient in the removal of large ink particles, since they are the first to be trapped in the mat as it forms, and very small particles are the last. Separation is purely based on size, not on surface properties, so efficient washing also has a high solids loss, of mineral fillers, fibre fines and crill. Hence washing is usually accompanied by an increase in freeness and strength properties such as tear. If all the solids removed are inert then there is a general improvement in strength properties, but if fines removed are 'active', that is, they contribute to inter-fibre bonding, properties such as burst will fall.

Washing efficiency is also affected by variables such as pH, temperature and treatment given before the washing stage, usually during pulping.

3.4.4.2 Pulping conditions. Chemical concentrations during pulping have a considerable impact on washing efficiency, since these influence the size of ink particles. A study which used newsprint showed that high pH (>11) increased washing efficiency due to the smaller size of ink particles. The high pH is probably effective in breaking down ink and conferring a high degree of stability to the particles so that there is no agglomeration. Very high levels of caustic alkalinity reduced brightness. The addition of other chemicals such as surfactants and dispersants also improved subsequent deinking (cf. sections 4.2 and 5.2). Optimal pulper temperature was between 50°C and 60°C [117]. Although these results were specific for the type of ink present, they provide a general rule, but with exceptions, such as flexographic inks. Pulping consistency is also important; higher consistencies produce smaller specks which give a lower pulper brightness, but ink removal by washing is more effective, so the final brightness is usually higher.

3.4.4.3 Wash conditions. Counter-current washing (in which dilution water for wash stage x is the backwater from wash stage $x + 1$) is common in wash deinking, since this enables water consumption to be reduced.

However, this reduces washing efficiency, when measured by brightness improvement. Washing efficiency is sensitive to water hardness or the presence of aluminium ions - it is significantly higher when soft water is used [117]. This may be due to the agglomeration of ink by calcium or magnesium. When aluminium ions are present, ink is precipitated onto fibre, so that washing under acid conditions appears to reduce efficiency. In the absence of aluminium, the best results when washing newsprint were in the range pH 4 to 6, with an optimum at pH 5.0 and least effective washing at pH 9.0 [118]. If washing is carried out under acid conditions, the efficiency will vary, according to the amount of aluminium present; its source is predominantly from acid sized papers.

In counter-current wash systems, especially using deckers, the nature of the solids in the filtrates changes from stage to stage. When relatively clean water is used as dilution water for the final wash stage (x), solids present in the wash backwater include mineral fillers and fibre fines. When this is used as dilution water for the preceding wash stage ($x - 1$), the ash and fines concentration in the thin stock feeding the washer increases significantly beyond that of the thick stock which was diluted, due to the filler and fines removed during the final wash stage (x). As the mat on the wash unit ($x - 1$) forms, an efficient filter is formed more rapidly, due to the increased fines concentration, but due to their small size mineral filler particles can pass through this until a very efficient filter is formed. Thus the backwater from this stage contains more mineral filler and less fines than the dilution water which was used. If the backwater is reused as dilution water (wash stage $x - 2$), a similar mechanism occurs. Although the solids concentration may not change substantially between wash stages, the nature of solids in the backwater does. This explains why counter-current washing is more efficient in limiting fibre losses, but less efficient in actual ink removal; some ink removed is trapped by the more efficient filter mat, increasing overall ink retention.

Although models for washing predict that washing efficiency is increased by increasing the outlet consistency, at best this has a marginal effect. The inlet consistency is much more important. At low consistencies water removal can be considerable before an effective filter mat forms. One study showed that when washing through a mesh the optimal feed consistency was 1.0–1.5%. If it was in the range 0.5–1.0%, fines loss increased by up to 33%, but with little additional ink removal, measured indirectly by brightness [118].

3.4.4.4 Equipment used in wash deinking. Almost any type of thickening device functions as a washing unit, though there are exceptions. A two stage thickener, which passes first stage filtrate back through the filter mat will not function as a wash unit. If backwater from a washing stage is reused without ink and solids removal, effective washing will not take place. A wide variety of devices have been used as washers, including screw presses, gravity deckers, sidehill screens and belt washers. Some details of these are given below.

Screw presses have been used for many years in deinking, both as wash units and thickeners. They have a pitched screw rotating in a fixed perforated screen barrel. The pitch of the screw, its length, speed and perforation size can be varied. Extracted water drains through the perforations. Typical inlet consistencies are in the range 2–10%, whilst outlet consistencies are up to 35%. They are very efficient in ash removal, but usually with a high associated loss of fibre and fines. Fibre losses can be reduced by the use of a fibre recovery stage prior to backwater clarification or discharge. Their effect on fibre properties by the introduction of curl is not widely understood, nor is the effect of curl on subsequent deinking.

One report on screw press washing suggested that water is removed without being drained through a filter mat [119]. However, water extracted from inside a pulp mass is effectively filtered through the pulp mass, which functions as a filter. This effect is shown in the same report by results obtained by varying the screw speed. As the rotating speed increased from 15 rpm to 22.5 rpm, with a 4.2% consistency feed, the volume of water removed fell, but ash removal increased, which suggests more dewatering occurred without a filtering effect. This study also examined the effects of varying the inlet consistency. At a feed consistency of 3.4%, ash removal was about 70%, at 6%, ash removal was more than 50% and at 8.3% feed consistency, about 40% ash was removed. Fines and fibre losses show a similar pattern: at a feed consistency of 3.4%, fines and fibre losses were >10%; at 6% inlet consistency, losses were about 5%; and at 8.3% < 5% of fines and fibres were lost. In most cases this fibre loss would be unacceptable, so fibre recovery is essential, for example, using an inclined screen such as a hydrasive, which functions as a fractionating device, dependent on the mesh or slot size. This study, using newsprint (70%) and magazines (30%) also found some evidence of ink speck breakdown in the screw press; the number of ink particles with a diameter of about 100 μm was reduced by about 30%. The effects of varying hole size in the screen barrel were also examined, and it was shown that reducing the diameter from 1.4 mm to 1.0 mm altered the performance at low consistency. When the basket with the smaller perforations was used the filtrate solids concentration was quite constant, between 5 and 6 mg l^{-1}, over a range of inlet consistencies, from 2% to 8.3%. However, with the larger diameter perforations, at 2% inlet consistency, the filtrate solids were almost 10 mg l^{-1} falling to just over 6 mg l^{-1} at a feed consistency of 4% [119]. This suggests that if low consistencies are to be dewatered by a screw press, small diameter holes would help to reduce fibre losses.

Ash reduction in a screw press is also influenced by stock freeness, pH, temperature and condition of the screw press. High freeness, pH and temperature increase the washing effect. As the edge of the flight of a screw press wears, efficiency falls. In some, the perforations are kept clean by the action of a brush, which also wears and efficiency falls off as this wears, since the perforations are not kept clean. Scale can build up and begin to close perforations. Usually the scale is calcium carbonate, which can be removed by an acid wash.

Drum washers, or gravity deckers have become less popular with the introduction of belt washers. They are usually plastic or metal mesh covered cylinders, rotating in a vat, and water passes through the mesh, forming a mat of washed stock on the mesh and the filtrate or backwater is collected after passing through the mesh. The mat is picked up by a couch roll and doctored off. Mesh cleanliness is maintained by showers.

Variables which affect washing efficiency on a decker include:

- Stock freeness;
- Head difference between the stock in the vat and the filtrate;
- Mesh size and cleanliness;
- Peripheral speed of the drum;
- Water system and quality;
- Temperature, pH, etc;
- Inlet consistency.

Some of these have been outlined above. Higher peripheral speed usually results in better washing, since the basis weight of the filtering mat formed on the mesh falls. However, this also results in greater fibre and fines losses. Increased temperature increases capacity, and entrained air reduces capacity. Typical performance is 40–60% ash removal with an inlet consistency of about 1% and a mat consistency of 4–8%. Operating problems include the blinding of mesh, with fine mesh (40 mesh) blinding faster than a more open mesh (80 mesh). With a fine mesh solids losses are lower, but so is washing efficiency. Maintaining a good seal between the vat and the filtrate is also difficult, a poor seal resulting in higher fibre losses. High-speed operation to improve washing efficiency may lead to stress cracking in the cylinder body, especially if combined with high couch roll pressures.

Disc and vacuum filters are used as thickeners, rather than washers, but the concept is similar. Most thickeners are operated in closed loops, with no associated water clarification. In many cases a two stage cycle is used; a cloudy filtrate is produced as the mat builds up on the disc segments, which is collected separately from the clear filtrate - collected after the mat is formed, when it is an effective filter. Cloudy filtrate can be recycled, or used as consistency dilution water. Filter mesh is usually a finer mesh than used on a decker. Vacuum assistance to dewatering may be provided, either by a vacuum pump or by the use of barometric legs. These units are very effective thickeners, with discharge consistencies of up to 30%, so they are sometimes used to thicken stock before dispersion.

Sidehill screens are curved inclined frames, covered with a mesh. Stock overflows a head box at the top of the mesh and baffles in the head box ensure there is an even distribution over the screen. As stocks flows down, the mesh water drains through. This forms a mat at the surface of the mesh, but this is moving and tumbling down the screen. Some water removal is from the inside of a pulp mass, but ash and ink removal can be efficient, though solids losses are also high. The low capacity of sidehill screens has restricted their use in large deinking systems, but their low capital costs makes them attractive in small mills. Capacity is in the range 13.5–16.7 $t\,day^{-1}\,m^{-1}$ width [115]. With inlet consistencies in the range 0.6 to 1.0% ash removal should be in the range 50–70%, with discharge consistencies of 3–6%. Screens can be severely affected by blinding, which affects performance and capacity.

Belt washers are relatively new wash units, with the first introduced in the early 1980s, and are very efficient. Since tissue requires low ash concentrations in recycled pulp they have been widely applied in deinking for tissue. There are two in common use and both are high speed, with maximum speeds of about $1000\,m\,min^{-1}$. One of these is similar to a paper machine wet end; stock is dewatered in the nip formed by a large diameter roll and a wire. Dewatering is very rapid and is assisted by the high centrifugal forces developed, as well as shear forces of the wire on the mat which forms on the roll. It is doctored off the roll by a blade. Washing efficiency is influenced by similar factors to those already discussed, including:

- inlet consistency;
- stock freeness, pH, temperature;
- speed;
- type of wire (porosity);
- jet speed to wire ratio.

These influence the basis weight of the mat formed and the speed of its formation; the lower the basis weight, the more efficient the washing, but the higher the total solids loss. Ash removal is at its highest with a low basis weight mat, given by a low inlet consistency, high speed and an open wire. Under these conditions high ash removal efficiencies can be achieved, up to 80%, but with high associated losses of up to 35% total solids including ash and fines. Losses are dependent on the type of stock and the level of mineral filler. They can be reduced by increasing the basis weight, by reducing the speed, increasing the inlet consistency, etc. At 40% ash removal, losses can be $<15\%$ of total solids. Typical inlet consistencies are in the range 0.7–3%, with outlet consistencies of 6–12%.

The other type of belt washer operates with similar principles, but with some variations. Dewatering is between a large diameter roll and a wire, but in this case the roll is grooved and dewatered stock sticks to the wire, which travels around another roll which can assist in additional water

Table 3.20 Yield reduction with washing and flotation [120]

	Inlet	1st	2nd	Flotation	
		Wash		6 min	12 min
Brightness (% MgO)	46.1	54.0	59.2	57.7	59.5
Ash	16.5	5.1	3.1	14.0	12.4
Long fibres (%)	27.2	46.6	50.4	-	31.2
Short fibres (%)	21.3	27.4	31.1	-	21.4
Fines	51.5	26.0	18.5	-	47.4
Yield (%)	100	67.6	60.2	-	89
CSF (ml)	115	375	505	115	135

removal. The mat is picked up by the second smooth roll and is doctored off. High ash removal efficiencies are achieved, with typical feed consistencies in the range 0.7–3% and discharge consistencies 8–15%.

In some deinking systems there is a desire to produce a recycled fibre with minimal ash content, <2%. This is difficult to achieve and requires multiple washing stages and is inevitably associated with high total solids losses, since these are much greater in moving from 5% to 1% ash than, for example, 25% to 21%. In the latter case, total losses may be 8% solids including ash and dissolved solids, whilst in the former total losses would be 16% or more including ash and dissolved solids. Achieving very low ash levels is thus very costly. This is illustrated by the yield figures given in Table 3.20, which show that in the first stage wash, ash reduction was from 16.5 to 5.1, which resulted in total losses of 32.4%, but a further 2% ash reduction, to 3.1%, increased total solids losses by 7.4%. It also shows the significant effect that effective washing has on the fibre length distribution and hence on many properties of those fibres. Two stages of belt washing were used with a furnish of 70% news and 30% magazines. These were also compared with flotation, given in Table 3.20 [120].

3.4.4.5 Flotation deinking. In 1903 it was found that air bubbles would pick up chemically coated mineral particles in a weak acid solution. This discovery led to the development of flotation in mineral ore dressing. Patents were filed for deinking applications in the early 1930s and the first reports on pilot plant trials, in a mill in the USA, appeared in 1952. Flotation was reported as achieving better brightness results than washing and it was suggested that the best approach was a combination of washing and flotation [121]. The first cells used in deinking were similar to those used in mineral ore flotation. Although these are still in use in both mineral ore flotation and deinking, there has been considerable development in both industries; for example, in the separation of molybdenum and copper ores, deep column (15 m) double aerated flotation tanks, using nitrogen gas as a carrier, have been introduced. A similar design has been used successfully in flotation deinking on a pilot plant scale. A high efficiency of ink removal (93%) was claimed [122]. Comparisons with conventional deinking have yet to be published.

There has been considerable development in flotation units for deinking, from the Denver cell initially used, with many modifications and new designs. In the period 1972–1993 there were many changes with at least ten equipment suppliers offering flotation systems. Developments have been extensively reviewed [123].

Although there have been many changes in flotation cell design, these have not dramatically increased performance. In the UK, flotation cells which were originally used in the mid 1970s produce a brightness increase of 10–12 units, whereas redesigned cells (from the same supplier) which were installed in the late 1980s give an increase of 12–14 units using similar wastepaper grades (news and magazines). Although performance has not radically altered, other aspects have, for example, the area required and energy consumption. In 1966, flotation alone was reported to require 84 $kWh\,t^{-1}$ [124] whereas in newer designs power consumption is in the range 40–65 $kWh\,t^{-1}$.

With the exception of water-based (including flexographic) inks, ink particles are more hydrophobic than paper fibres, which is the basis of separation by flotation. However, some fibres can be hydrophobic and some inks can be hydrophilic [125]. Chemicals are added to increase the difference, but even in the absence of added ink collector chemicals, some ink is still removed by flotation. Chemical aspects of flotation deinking are reviewed in section 4.3. A great deal is not known about flotation deinking, for example, a chemical explanation of why magazines improve the deinking of newsprint. Flotation deinking is complex with chemicals introduced by wastepaper considerably increasing its complexity [126]. Ink removal efficiencies are usually in the range 50–70%. The efficiency of ink removal is influenced by a wide range of factors, and for ink removal to occur a number of interactions must occur which include:

- ink particles must be 'free', that is, not attached or bound to fibre surfaces, or trapped in fibrillar areas or in bound water;
- ink particles must collide with an air bubble, that is, ink particles must not follow streamlines around the bubble;
- following collision, ink particles and air bubbles must form a complex, that is, the collision or an interaction must provide sufficient energy to overcome repulsive forces;
- the air bubble/ink particle complex must rise to the surface;
- the air bubble/ink particle complex must be removed and not recycled back into the system.

Each of these steps is dependent on a variety of mechanisms and chemical conditions, some of which are inter-dependent, so that flotation deinking efficiency is affected by many factors, which include:

- stock consistency. As consistency increases the drag on the air/ink complex increases, since the fibres form a network. This reduces the rise velocity and in addition collisions with fibre may dislodge ink particles, especially if these are large. The optimal consistency varies with the design of individual units, but is usually in the range 0.8–1.2%. However, lower consistencies may give better flotation efficiencies. One study which compared laboratory and pilot scale deinking used different consistencies, 0.6 and 1.0%. Ink removal efficiency was better at the lower consistency; for example, using the pilot scale unit at 0.6% gave ink removal efficiency of about 80% with particles in the size range 50–150 μm, whereas at 1.0% efficiency was about 50%. With ink particles of about 300 μm, at 0.6% consistency, removal efficiency was about 50%, but only 20% at 1.0% consistency. Increased ink removal efficiency at the lower consistency in the laboratory scale unit was also shown [127].
- stock temperature and pH. One laboratory study showed that as temperature increased, flotation efficiency fell, though this may be relevant only to the chemical system studied (soap) [128]. The same study showed that as pH increased above 9, efficiency fell, and in the range 6–9, it was constant. However this is very dependent on specific chemicals used and on the presence of other chemicals, for example, some soaps precipitate out at a neutral pH and under acid conditions aluminium can 'fix' ink to fibres. Chemical type determines pH dependency, etc. which is illustrated by the conclusions drawn from another study. This used paper printed with UV cured ink and a commercial surfactant. Laboratory flotation runs led to the conclusion that the flotation cell temperature was optimal between 40°C and 45°C and a pH of 6.0 gave the best brightness [129].

When using laboratory trials to determine optimal conditions and chemical types for flotation the difficulties experienced in transferring results to a full scale operation must be continually considered. The chemistry of mill backwater is very complex and inevitably many types of chemical are present. Reproducing the complexity of a full scale mill in a laboratory or pilot plant is extremely difficult, since many interactions between wastepaper-derived chemicals may take place. Some of the chemicals which could contribute to interactions, even when present at very low concentrations have been identified as lignin derivatives. Two groups are believed to be present, those which contribute to foam generation and those which give hydrophilic characteristics to some ink particles [126].

- water hardness. This has been studied extensively [116, 125]. With fatty acid soaps hardness must be not less than about 10°DH (equivalent to 179 mg l^{-1} of $CaCO_3$) or efficiency falls. However, increased calcium concentrations increase fibre losses [125] (cf. section 4.3.3).

- collector chemical concentration. The optimal level depends on the chemical type - with fatty acids it is usually in the range 0.4–0.8% on fibre. High concentrations can have negative effects due to chemical carry-over and in the case of fatty acids, may affect the coefficient of friction of the paper being produced.
- ink particle size. Many studies have shown that ink particle size is important in determining ink removal efficiency [130, 131]. Usually the size range reported as being optimal for removal by flotation is 10–100 μm. However, smaller particles can be removed if they agglomerate. Very large toner ink specks were observed by the author to be removed very efficiently; in this case the surface chemistry of the ink overcame problems due to large size including reduced buoyancy of the bubble/particle complex and separation of the complex components by collisions with fibres in the network. Prior chemical or mechanical treatment has a significant effect on ink particle size and so affects flotation efficiency.
- ink particle chemistry. Different inks are removed with varying efficiencies, not only due to differences in sizes, but also to their surface properties. Treatment which increases the hydrophobic nature of the ink particles will tend to improve removal efficiencies. One study of flotation of papers from six different copiers found that the dirt count after deinking ranged from 10 to 90 per gram of pulp. The low count was visually acceptable. There were no clear reasons for the differences observed, though in some cases toner was found to be still attached or even encasing fibres [132]. Some of this ink is probably released by treatments like dispersion or kneading.
- air bubble diameter. Some ink particles are drawn past bubbles by fluid stream lines. The probability of collision is higher with increasing ink particle size and reduced bubble size. However, the production of very small air bubbles, as in dissolved air flotation, may tend to increase fibre losses. Minimal air bubble size has been estimated at 0.3–0.5 mm diameter [133].
- hydraulic flow patterns. These are a function of the design of the unit and production rates. Forces acting on ink particles are complex, but they determine if collisions result in the formation of an ink particle/air bubble complex. Surface chemistry is also important. Good mixing is essential, but air bubbles must also be allowed to float to the surface, for removal with attached ink. Large particles reduce the rise velocity and increase drag. Microturbulence increases collision frequency, without redistributing ink.
- air-to-stock ratio. Increased air injection can lead to increased turbulence which can prevent ink removal by not allowing ink and bubble complexes to rise to the surface, or by separation of ink particles from the air bubble, but these problems can be eliminated by the design. Japanese flotation cells have been designed to allow much higher air-to-stock

ratios than were common in Europe, for example, with a ratio of 10:1 compared to about 4:1 used in Europe. This increases the number of air bubbles and so increases the probability of a collision of an ink particle with an air bubble. High retention times also achieve this - typical retention times in European cells are 5–10 min, whereas in Japan up to 20 min is used.

- foam removal system. Several systems are in use, overflow, vacuum assisted, etc. If foam is not formed or is very unstable, ink removal is inevitably poor, unless a hydraulic overflow is used. Foam generation and stability can be unpredictable, especially with woodfree papers. Where hydraulic overflows are used, treatment of these primary rejects by another stage of flotation is essential to limit fibre losses and restrict the generation of sludge. The presence of coated wastepapers increases foam generation, which is also influenced by added chemicals, base paper characteristics, ink type, etc.

3.4.4.6 Other deinking techniques. Dry deinking has been extensively researched, with several USA patents granted in the mid 1980s [134], but the system was not successful commercially. More recently, properties of toner-based inks have been utilised to develop a simple system for the removal of these inks, without relying on traditional washing or flotation. In these, chemicals are added which agglomerate the ink and which can also result in a density change. Large, relatively stiff globules of ink are removed by centrifugal cleaning and fine screening [63, 64, 135]. A system which was recommended for deinking by this method is illustrated in Figure 3.12.

Solvent deinking has also been evaluated and solvent recycling was used for a period [136] but due to difficulties in solvent recovery and containment, is unlikely to be adopted.

Using ultrasonics to break down inks for subsequent conventional removal has also been evaluated. Using gloss printed and overvarnished paper, good ink dispersion was achieved in 2–3 min, at low consistency and temperature (26°C). Higher temperatures gave better results. No chemicals were used. Treatment after ink dispersion employed washing through three stages of sidehills. Power consumption (to achieve fibre and ink dispersion) was high, at 830 $kWh\,t^{-1}$, but on the same laboratory scale, low-consistency pulping required 570 $kWh\,t^{-1}$ [137].

3.4.4.7 The role of the base paper in deinking. There have been several reports on the importance of base paper in deinking, which must be partly due to the presence of chemicals which have an impact on the deinking process. One report was of a significant difference in the ease of deinking due only to the type of base paper [138]. Another study looked at some physical paper properties, to see if these could be related to ease of deinking, but no relationship was found. However, if papers were coated with cationic starch, in general they were easier to deink when printed using toners, though this varied according to the type of toner, which had a much greater effect. It was not the presence of a coating which improved deinking, since in some cases a coated paper gave the worst results [139]. Alkyl ketone dimer (AKD) sized papers also generally gave a better deinking result [140].

3.4.4.8 Development of deinking. It is clear that ink manufacturers can have a significant effect on the economics of recycling. Systems being built to use 'problem ink' grades such as laser or flexographic inks are more complex and hence more costly to operate. Yet there is enough evidence available to show that these inks can be made to be easier to remove, thus improving the economics of recycling. This importance was demonstrated very clearly during the mid 1970s in Europe, as offset printing of news was introduced. Initial inks were very difficult to remove, but when paper makers highlighted problems, new ink formulations were developed which were easier to deink. Since then, unfortunately, there have been few examples of cooperation, but it is still of great importance. One other example of improving recycling by ink producers was the development of easy-to-deink inks to print telephone directories in Germany. One study in the UK developed inks with good printing properties which were easy to deink, but these were not developed by printing ink companies as they were slightly more costly [141]. Inks can be better engineered for recycling. So can base papers, though little is known on how to achieve this, but it is in the paper industry's best interest to understand how base paper properties affect subsequent deinking, since this would ultimately lead to better recycling economics.

If wastepaper disposal imposes a burden on society, then all parts of the paper chain could ease that burden by making recycling easier. This can be done - the final consumer by sorting before disposal; the paper industry by collecting and using wastepaper, as well as enhancing its ease of recycling when new paper is made; printing companies, including original equipment suppliers to offices, by developing inks and toners which are easier to remove; and the adhesive industry by producing adhesives which cause fewer problems during recycling. If all parts of the chain played their part, the high levels of recycling which are now demanded would become much easier to achieve.

References

1. Rose, A. (1981) The development of wastepaper as a raw material for the paper and board industry, *CEPAC Conference*, Brussels.
2. Moreland, R.D. (1986) Stickie control by detackification, *TAPPI Pulping Conference*, TAPPI Press.
3. Scott, W.E. (1989) A survey of the various contaminants present in recycled mill white

water systems, *TAPPI Contaminant Problems Recycling Seminar*, TAPPI Press, pp. 69–80.
4. McKinney, R.W.J. (1987) Test methods for assessing stickie contamination - a review, *TAPPI Pulping Conference*, TAPPI Press, pp. 725–729.
5. Anon, (1993) Methods to quantify stickies - a mill survey, *Prog. Paper Recycling*, **2**(3), 80.
6. Pira Report 32/TENQ/811/349 (1981) Pira, Leatherhead.
7. Dunlop-Jones, N. and Allen, L.H. (1989) The influences of washing and defoamers and dispersants on pitch deposition from unbleached kraft pulps, *J. Pulp and Paper Sci.*, **15**(6), 225.
8. McKinney, R.W.J. and Currie, P.G.C. (1986) Stickie pacification - new additive shows promise, *Paper Technol. Ind.*, **27**(4), 182.
9. Sjöström, J., Holmbom, B. and Wiklund, L. (1987) Chemical characteristics of paper machine deposits from impurities in deinked pulp, *Nordic Pulp & Paper Res. J.*, **4**(2), 123.
10. Dunlop-Jones, N. and Allen, L.H. (1988) A rapid method for the qualitative analysis of plastic and sticky contaminants by pyrolysis-gas chromatography, *Tappi J.*, **71**(2), 109.
11. Berben, S. (1992) Sticky detective work, *Prog. in Paper Recycling*, **3**(1), 66.
12. Cathie, K., Haydock, R. and Dias, I. (1991) Understanding the fundamental factors influencing stickies formation and deposition, *First Research Forum on Recycling*, CPPA Technical Section.
13. Gunderson, D. (1993) Friction properties of paper and paperboard, *TAPPI Recycling Symposium*, TAPPI Press.
14. Ling, T.F. (1991) Modifying surface properties of stickie materials through polymer/ surfactant adsorption, *1st Research Forum on Recycling*, CPPA Technical Section.
15. Allen, L.H., Cavanagh, W.A., Holton, J.E. and Williams, G.R. (1992) The use of talc for pitch and deposit control in the modern kraft pulp mill, *TAPPI Pulping Conference*, Book 2, TAPPI Press, p. 509.
16. de Jong, R.L. (1984) The influence of zeta-potential on the agglomeration of stickies on a tissue machine using waste, Pira Symposium *Stickies - an overall view*.
17. de Jong, R.L. (1989) Deinking problems with woodfree wastepaper, *TAPPI Contaminant Problems Recycling Seminar*, TAPPI Press.
18. Alince, B. and Robertson, A.A. (1978) Colloidal aspects of the retention of positively charged additives, *Tappi J.*, **61**(11), 111.
19. Sigman, M.A. and Rohlf, E.V. (1993) The effect of system charge on stickies deposition, *TAPPI Pulping Conference*, TAPPI Press, p. 507.
20. Burrows, P.R.M. and Small, J.D. (1982) Removal of sticky contaminants from waste based papermaking furnishes, *European Seminar on Technological Progress in Recycling of Paper & Board*, Brussels.
21. Hassler, T. (1988) Pitch deposition in papermaking and the function of pitch control agents, *Tappi J.*, **71**(6), 195.
22. McKinney, R.W.J. (1987) Stickies removal, *PITA Paper Week Conference*.
23. McKinney, R.W.J. (1989) A review of stickies control methods, including the role of surface phenomena in control, *TAPPI Contaminant Problems Recycling Seminar*, TAPPI Press.
24. Doshi, M.R. (1991) Properties and control of stickies, *Prog. Paper Recycling*, **1**(1), 54.
25. Doshi, M.R. (1992) Quantification, control and retention of depositable stickies, *Prog. Paper Recycling*, **2**(1), 45.
26. Lacy, M. and Cawein, D. (1993) Contaminant Control at Arkansas kraft, *Prog. Paper Recycling*, **4**(2), 81.
27. Hisey, B.A. (1986) Quality control of recycled newspaper, *TAPPI Pulping Conference Proceedings*, TAPPI Press, p. 197.
28. McKinney, R.W.J., Cathie, K. and Voss, G.P. (1986 and 1987) Efficiency of stickie removal systems, *Pira Project Report*, PB/MC/85/3.
29. Prein, M. (1992) Contaminant challenges at Stone Container - Snowflake mill, *Prog. Paper Recycling*, **1**(4), 59.
30. Andrews, W.A. (1992) Stickie situation at Garden State, *Prog. Paper Recycling*, **1**(2), 67.

31. Wigglesworth, A. (1983) *Pira Project* PB/3F/81/2 Final Report.
32. McBride, D. (1993) High density kneading: an alternative to dispersion, *TAPPI Recycling Symposium*, TAPPI Press, p. 173.
33. McKinney, R.W.J., Cathie, K. and Biddlecombe, J. (1988) *Disperger Trials on Waste Derived Pulp*, Pira report 32/TENQ/803/120.
34. Fetterly, N. (1992) The role of dispersion within a deinking system, *Prog. Paper Recycling*, **1**(3), 11.
35. Wade, D.E. (1987) Sticky pacification with synthetic pulps, *TAPPI 1987 Pulping Conference*, TAPPI Press.
36. Goldburg, J.O. (1987) Use of zirconium chemical in stickie contaminant control, *Tappi Papermakers Conference*, Vol **3**, TAPPI Press, p. 585.
37. Hoekstra, P.M. and May, O.W. (1987) Developments in the control of stickies, *TAPPI Pulping Conference*, TAPPI Press, p. 573.
38. Williams, G.R. (1987) Physical chemistry of the absorption of talc, clay and other additives on the surface of stickie contaminants, *TAPPI Pulping Conference*, TAPPI Press, pp. 563–572.
39. Ormerod, D.L. and Hipolit, K.J. (1987) Aluminum control prevents stickie deposition, *TAPPI Pulping Conference*, TAPPI Press, pp. 597–604.
40. Duffy, R.J. and Aston, D.A. (1988) *Annual Meeting*, Canadian Pulp and Paper Association, pp. B81–B84.
41. Miller, P.C. (1988) Chemical treatment programs for stickies control, *TAPPI Pulping Conference*, TAPPI Press, pp. 345–348.
42. Dykstra, G.M. and May, D.W. (1989) Controlling stickies with water soluble polymers, *TAPPI Contaminant Problems Recycling Seminar*, TAPPI Press, pp. 97–100.
43. Inyarn, F.H. (1991) Cleaning of secondary fibre stickies may necessitate monitoring for VOCs, *Paper Recycling*, (ed. Patrick, K.L.) Miller Freeman, San Francisco, p. 105.
44. Fogarty, T.J. (1992) Cost effective, common sense approach to stickies control, *Tappi 1992 Pulping Conference*, TAPPI Press, Vol. 3, p. 573.
45. Doshi, M.R. (1989) Additives to combat sticky contaminants in secondary fibres, *TAPPI Contaminant Problems Recycling Seminar*, TAPPI Press, pp. 81–90.
46. Coenen, T.J. (1993) Problems with stickies/contaminants and some simple ways to improve on existing systems, *Prog. Paper Recycling*, **2**(3), 91.
47. McAlpine, I. (1984) Zirconium chemicals for stickies control, *Pira Symposium*, SPB/8.
48. Schyrench, L. (1984) Talc for stickies control, *Pira Symposium*, SPB/8.
49. Withiam, M.C. (1990) The effects of fillers on paper friction properties, *Tappi J.*, **73**(4), 249.
50. Paraskeva, S. (1988) Theoretical considerations on reject disposal, *Tappi Pulping Conference Proceedings*, TAPPI Press.
51. Bliss, T. (1988) Reject handling in secondary fiber systems, *Tappi J.*, **71**(6), 87.
52. McBride, D. (1993) Reject handling and sludge pressing in recycling and deinking systems, in *Secondary Fiber Recycling*, TAPPI Press (ed. Spangenburg, R.J.).
53. Smith, W.E. and Orr, T.W. Maintaining forming fabrics and wet felts – a surface chemistry approach, *Tissue Issues*, Niagra Lockport.
54. Hawes, J.M. (1992) Design considerations for paper machine clothing in secondary fiber applications, *Prog. in Paper Recycling*, **1**(2), 38.
55. Anon, Repulpability of splices/splicing tape, *TAPPI Useful Methods*, UM 213.
56. Anon, Dispersability test for adhesives, *TAPPI Useful Methods*, UM 666.
57. Stegink, D.W. (1992) The recyclability of Post-it^{TM} brand notes in office wastepapers, *Prog. Paper Recycling*, **1**(2), 30.
58. Clewley, J.A. (1983) Bridgewater – A new mill for the eighties and beyond, *Paper Technol. Ind.*, **24**(1), 5.
59. Patterson, J.H. (1987) Deinking at Bridgewater, *Resource Utilisation* – for Profit or Loss, PITA Annual Conference.
60. Galland, G. and Verna, Y. (1991) Deinking of wastepaper containing waterbased flexoprinted newsprint, *1st Research Forum on Recycling*, CPPA.
61. Clement, J.M. (1993) Application and Development of the JMC double-loop recycling process, *TAPPI Pulping Conference Proceedings*, TAPPI Press, p. 1025.

62. Fetterly, N.W. (1991) Deinking ONP for directory, the low lost approach, *TAPPI Pulping Conference*, TAPPI Press.
63. Darlington, W.B. (1988) A new process for deinking electrostatic printed secondary fibre, *TAPPI Pulping Conference*, TAPPI Press.
64. Olson, C.R., Hall, J.D. and Phillipe, I.J. (1993) Laser print deinking using chemically enhanced densification and forward cleaning, *Prog. Paper Recycling*, **2**(2), 25.
65. Dexter, D.J., Orth, L. and Peil, J. (1992) The chemistry of deinking 'laser' office waste at Prime Fiber corporation, *Tappi Pulping Conference*, TAPPI Press.
66. Hamilton, F.R. (1987) Pulping systems, in *Secondary Fibres and Non-wood Pulping*, (eds Hamilton, F. and Leopold, B.), The Joint Textbook Committee of the Paper Industry.
67. Siewert, W.H. and Kebs, J. (1986) Trends in pulping - A new pulper concept, *Tappi Pulping Conference Proceedings*, TAPPI Press.
68. Heinbockel, W. (1979) *The twin-Pulp System*, Paper presented at Escher Wyss Seminar, Constance.
69. Anon (1979) In *Fibreflow Wastepaper Processing System*, Paper presented at Escher Wyss Seminar, Constance.
70. Siewert, W.H. (1983) High consistency pulping - new technology development and future aspects, *Tappi Pulping Conference Proceedings*, TAPPI Press.
71. Bahr, T. (1981) Recent developments in helical-rotor pulpers, *Tappi J.*, **64**(7), 43–46.
72. Anon, (1989) Pulping of secondary fibre, *TAPPI Contaminant Problems Recycling Seminar*, TAPPI Press.
73. Stradel, M. and Roberge, N. (1979) Upgrading of mixed wastepapers by dry sorting, *TAPPI Pulping Conference Proceedings*, TAPPI Press, pp. 281-287.
74. Mason, W.H. (Jan 1928) US Patent Number 1,655,618.
75. Maners, H. (1978) The Siropulper - a new concept in wastepaper recovery, *Appita*, **32**(2), 124.
76. Mentz, J.R. (1991) Steam explosion technology, *TAPPI Pulping Conference Proceedings*, TAPPI Press, pp. 83-84.
77. Reynolds, H.H. and Reuben, M.D. (1966) Effect of cooking variables on properties of deinked fibres, *Tappi J.*, **49**(7), 64.
78. Merrett, J.K. (1987) Repulping at high consistencies, *Appita*, **40**(3), 185.
79. Gooding, R.W. and Craig, D.F. (1992) The effect of slot spacing on pulp screen capacity, *Tappi J.*, **75**(2), 71.
80. Dundie, D.P. (1991) A systems approach to pressure screen control in the pulp mill, *Tappi J.*, **74**(11), 97.
81. Eck, T.H., Rawlings, M.J. and Heller, P.A. (1985) Slotted pressure screening at Southeast Paper manufacturing company, *TAPPI Pulping Conference*, TAPPI Press.
82. Vitori, C.M. (1991) Stock velocity and stickies removal efficiency in slotted pressure screens, *1st Research Forum on Recycling*, Technical Section, CPPA, pp. 133-142.
83. Heise, O. (1992) Screening foreign material and stickies, *Tappi J.*, **75**(2), 78.
84. Lerch, J.C. and Audibert, S.W. (1989) Slotted screen system evaluation at Garden State Paper, *TAPPI Contaminant Problems Recycling Seminar*, TAPPI Press.
85. Cruea, R.D. (1983) Ultrafine cleaning of deinked stock, *Tappi J.*, **66**(8).
86. Doshi, M.R. and Prein, M.G. (1986) Effective screening and cleaning of secondary fibres, *TAPPI Pulping Conference*, TAPPI Press.
87. Chi, J.Y. and Defoe, R.J. (1992) Fundamental hydraulic study of screening - Part I, *TAPPI Engineering Conference*, TAPPI Press, pp. 439–461.
88. Yiannos, P. (1964) *Tappi J.*, **47**(8), 468.
89. Clark, Jd'A. (1978) Pulp technology and treatment for paper, 1st Edition, Miller Freeman Publications, San Francisco.
90. St Amand, F.J. and Perrin, B. (1991) The effect of particle size on ink and speck removal efficiencies of deinking steps, *1st Research Forum on Recycling*, Technical Section, CPPA, pp. 39–47.
91. Dexter, R.J., Orth, R. and Peil, J. (1992) The chemistry of deinking laser office waste at Prime Fiber corporation, *TAPPI Pulping Conference*, TAPPI Press, pp. 653-657.
92. Wright, M.D. (1992) Performance of forward cleaners, *Prog. Paper Recycling*, **1**(4), 38.

93. Bliss, T. (1993) What is the effect of entrained air on the performance of hydrocyclones, *Prog. Paper Recycling*, **2**(2), 89.
94. Wise, E.M. and Arnold, J.M. (1992) The role of specific gravity for removal of hot melt adhesives in recyclable grades, *Tappi J.*, **75**(9), 181.
95. Diehl, D. (1982) Inlet pressure key to efficient operation of reverse flow cleaners, *Pulp and Paper*, **58**(3), 134.
96. Marson, M. (1989) New lightweight removal cleaner solves mill plastic problem, *TAPPI Pulping Conference*, TAPPI Press.
97. Schall, H.E. (1986) Asphalt dispersion systems, past and present, *TAPPI Pulping Conference Proceedings*, TAPPI Press.
98. Siewert, W.H. (1973) The preparation of printing, coated, wet strength, waxed, laminated and bituminous wastepapers, *Paper*, 18th July, pp. 98–100.
99. Brauns, D. (1973) A new method of reclaiming wastepaper, *Svensk Papperstichnung*, **76**(5), 192.
100. Lundberg, per R. (1989) Jan/Feb Le défibrage a forte concentration, *Papier, Carton et Cellulose*, pp. 66–73.
101. Pelletier, L.J., Seth, R.S., Barbe, M.C. and Page, D.H. (1987) The effect of high consistency mechanical treatment on the properties of sulphite pulps, *J. Pulp and Paper Sci.*, **13**(4), 121.
102. Luman, W.E. (1986) Curl setting during storage of TMP at high consistency, *J. Pulp and Paper Sci.*, **12**(4), 108.
103. Ortner, H. (1977) Dispersion: a means of improving the quality of deinked pulp, especially from offset printing newspapers, *Wochenbl Papirfabr*, **105**(23/24), 981.
104. Gilkey, M., Shinohara, H. and Yoshida, H. (1988) Cold dispersion unit boosts deinking efficiency at Japanese tissue mills, *Pulp and Paper*, **62**(11), 100.
105. Ferguson, L. and McBride, D. (1992) Deinking sorted office waste, *CPA Annual Meeting*.
106. Kotani, Y. (1993) Modern deinking method for recovered papers. The efficiency of high density kneading and flotation, *Prog. Paper Recycling*, **2**(4), 110.
107. Rangamannar, G. and Silveri, L. (1989) Dispersion - an effective secondary fiber treatment process for high quality deinked pulp, *TAPPI Pulping Conference*, TAPPI Press, pp. 381–390.
108. Ortner, H. and Fischer, H.S. (1990) The application of dispersion for quality improvement of deinked pulps, *TAPPI Pulping Conference*, TAPPI Press, pp. 749–756.
109. Johnson, A. (1989) Quality of recovered fibres from municipal waste, *Pira Secondary Fibre Conference*.
110. Cathie, K., Biddlecombe, J., Staves, J. and McKinney, R.W.J. (August 1989) *Pira Research Project* 32/PB/88/8.
111. Kotani, Y. (1992) Deinking office wastepaper in Japan, *OPTEC Seminar*.
112. McBride, D. (1992) High density kneading offers alternative to conventional deinking, *Pulp and Paper*, **66**(5), 149.
113. Compton and Hart (1951) quoted by K. Durham in *Surface Activity and Detergency* McMillan (1961).
114. Turvey, R.W. (1984) Chemical aspects of deinking, *PITA Paper Week Conference*, pp. 293–303.
115. Horacek, R.G. and Forester, W. (1993) Washing, in *Secondary Fibre Recycling* (ed. Spangenberg, R.J.), TAPPI Press, pp. 163–181.
116. Larsson, A., Stenius, P. and Odburgh, L. (1985) Surface chemistry in flotation deinking - part 3, *Svensk Papperstidning*, **88**(2), 2.
117. De Ceuster, J. and Papageorges, G. (1981) Physiochemical aspects of wastepaper deinking by washing, *Appita*, **35**(2), 145.
118. Cruea, R.P. (1978) Deinking laboratory evaluations and total system concepts, *Tappi J.*, **61**(6), 27.
119. Egenes, T.H., Helle, T., Bendiksen, P.B. and Hegstad, G. (1993) Removal of water and contaminants from ONP Stocks in a screw press, *2nd Research Forum on Recycling*, CPPA Technical Section, pp. 71–78.
120. Linck, E., Mayr, H. and Siewert, W.H. (1983) The Variosplit, a new machine for improving wastepaper stocks, *TAPPI Papermakers Conference*, TAPPI Press.

121. Jelks, J.W. (12th July 1952) Deinking paper pulp by flotation, *Paper Mill News*.
122. Petri, B., Dobby, G. and Whiting, P. (1993) *Deinking Wastepaper in a Flotation Column*, unpublished paper.
123. McCool, M.A. (1993) Flotation deinking, in *Secondary Fiber Recycling* (ed. Spagenburg, R.J.), TAPPI Press, pp. 141-162.
124. Ortner, H. (1966) Present state of technical knowledge of deinking paper by flotation, *Paper Technol. Ind.*, **7**(5), 431.
125. Turvey, R.W. (1991) Why do fibres float, *1st Research Forum on Recycling*, Technical Section CPPA, pp. 123-131.
126. Turvey, R.W. (1990) Chemical aspects of flotation deinking, *4th PTS Deinking Symposium*, Munich.
127. Vidotti, R.M., Johnstone, D.S. and Thompson, E.V. (1992) Comparison of bench scale and pilot plant flotation of photocopied office wastepaper, *TAPPI Pulping Conference*, TAPPI Press, pp. 643-652.
128. Larsson, A., Stenius, P. and Odberg, L. (1984) Surface chemistry in flotation deinking - part 1, *Svensk Papperstidning*, **87**(18), 158.
129. Forester, W.F. (1986) Deinking of UV cured inks, *TAPPI Pulping Conference*, TAPPI Press, pp. 419-423.
130. Larsson, A., Stenius, P. and Odburgh, L. (1984) Surface chemistry in flotation deinking - Part 2, *Svensk Papperstidning*, **87**(18), 165.
131. Peel, J.D. (1985) The importance of ink particle size in deinking processes, *PTS Deinking Symposium*, Munich (1985).
132. Quick, T.H. and Hodgson, K.T. (1986) Xerography deinking - a fundamental approach, *Tappi J.*, **69**(3), 102.
133. Schulze, H. (1991) The fundamentals of flotation deinking in comparison to mineral flotation, *1st Recycling Forum*, Technical Section CPPA, pp. 161-167.
134. United States Patent Numbers, 4,615,767 (1986) and 4,668,339 (1987).
135. Olson, C.R., Richmann, S.K., Sutman, F.J. and Letscher, M.B. (1993) Deinking of laser printed stock using chemical densification and forward cleaning, *Tappi J.*, **76**(1), 136.
136. Aldrich, L. (1977) A new look at deinking with solvents, *Tappi J.*, **60**(8), 114.
137. Turai, L.L. and Chung-Haw, T. (1978) Ultrasonic deinking of wastepaper, *Tappi J.*, **61**(2), 31.
138. Kübler, R. (1987) Inks and Deinking, *3rd PTS Deinking Symposium*, Munich.
139. Cathie, K. and Mallouris, M. (1993) Use of surface energy measurements and other parameters to predict the deinkabilily of laser printed papers, *2nd Research Forum on Recycling* Technical Section, CPPA, pp. 179-199.
140. Cathie, K. and Burdett, S. (1993) Understanding why laser printed papers deink differently, *TAPPI Pulping Conference*, TAPPI Press, pp. 619-642.
141. Burstall, M. (1985) Inks and deinking - problems and prospects, *Paper Technol. Ind.*, **26**(7), 327.

4 Chemical use in recycling

R.W. TURVEY

4.1 Introduction

Currently, about one third of the world's paper and board making fibre is recycled [1]. The major portion of this is simply pulped, mechanically cleaned, and the resulting stock is then used directly on the paper, board or tissue machine. Waste containing wet-strength chemicals is often pulped with appropriate chemicals to break down the wet-strength resins, for example alkaline hypochlorite for Kymene types. Dyed waste is sometimes bleached, but most recycled fibre is not treated with chemicals.

When a printed paper is pulped it usually gives a dark stock, or a stock which is specky. It is clear that the print on the paper is responsible for this behaviour, because unprinted paper gives a bright, speck-free stock. Dark stock is produced by print which breaks up into small particles, with typical examples being newspapers, especially old ones. Specks are larger print particles which are visible to the naked eye, produced from, e.g. photocopy papers and papers printed with UV-cured inks. Print contains solid pigment particles which cannot be bleached, dissolved or destroyed, and in order to make such stock bright and speck-free it is necessary to remove the print particles which are generated during pulping.

About 15% of all recycled wastepaper is treated to remove print particles, so the final stock is bright, speck-free, and can replace virgin fibre in newsprint , tissue and printing grades [1]. This treatment is called deinking, and chemicals are usually used to improve the mechanical removal of the print particles. Deinking probably accounts for most of the chemical use in paper recycling, and this chapter will therefore concentrate on the use of chemicals in deinking. The first part gives an overview of the main chemicals used in deinking, the second part describes their use in greater detail, and the third part examines particular problems in deinking and how chemicals can solve them. The final part considers the future of chemical use in recycling.

Table 4.1 Estimated amount of wash deinking carried out worldwide*

Year	1978	1980	1988	1989	1990
Amount ($\times 10^6$ tonnes)	~1.2	1.68	1.3	<3.3	1.5

*Source: literature values and personal communications

Table 4.2 Estimated amounts[a] of wash deinking chemicals used worldwide

Chemical	Assumed addition level (% of waste weight)	Amount used (tonnes/day)
NaOH	0.3	12
Surfactant/solvent	0.1	4

[a]Calculated assuming 1.5 million tonnes/year worldwide are wash deinked

4.2 Chemical use in deinking - an overview

4.2.1 *Wash deinking*

Wash deinking was the earliest method used to improve the quality of recycled paper [2]. The process consists simply in pulping, followed by one or more dewatering steps, possibly using countercurrent water flow. Deckers, disc filters, belt presses, screw thickeners and similar devices are used to dewater the stock. Washing will brighten stock if the print particles present are between ~1 and 10 μm in diameter, as these particles will then wash out [3]. Washing can also significantly reduce filler levels, although visible print specks are not removed efficiently. They have a diameter greater than ~45 μm and the dewatered stock pad acts like a filter and retains most of them.

Wash deinking can be practised in any mill which has a pulper and a dewatering device. As these are standard mill items it is very difficult to know which mills wash deink, and how much waste is wash deinked. One estimate suggests that about 1.5 million tonnes of paper is wash deinked per year worldwide, and that this amount has remained approximately constant for the last decade (see Table 4.1). It is probably true to say that the majority of wash deinking is carried out in the USA, and on mainly woodfree wastepapers.

4.2.2 *Types and amounts of chemicals used in wash deinking*

Most chemicals used in wash deinking are proprietary mixtures of surfactants and solvents. Chlorinated solvents were common in early formulations but are not currently used. Sometimes phosphoric acid is included in the formulation. Sodium hydroxide is almost universally used during pulping. Hydrogen peroxide and sodium silicate are sometimes used, as is sodium hypochlorite. Table 4.2 gives estimates of certain chemicals used in wash deinking.

4.2.3 What wash deinking chemicals do

In wash deinking the aim is to detach the print from the fibre surface and break it up into particles with diameters of ~1 to 10 μm. These particles should have a surface which suspends them in the water phase, and which has no affinity for the fibres. Deinking chemicals can help detach print, weaken and break the print vehicle network holding the pigment particles, and suspend the resulting small print particles in water.

4.2.4 How wash deinking chemicals work

The print on woodfree wastepapers typically consists of pigment particles held in a strongly cross-linked polymer. Surfactants probably wet the print, allowing the solvents/alkali/oxidising agents to penetrate micro-cracks. Their swelling action and chemical attack weakens the vehicle network in the print, and the mechanical action of the pulper and subsequent equipment break it down more easily into small particles. The surfactants suspend these smaller particles in the water phase and they are then washed out in the dewatering step.

Newspaper is successfully wash deinked by several companies, but often the print particles generated on pulping are below 2 μm in size, with or without wash deinking chemicals being present. Such print particles are difficult to wash out and only low-brightness paper can be made from this stock. Newspaper printed by the water-based flexographic process is a good example of this behaviour. Very small print particles are formed on pulping because the network enclosing the pigment particles completely disintegrates. Many of these particles lodge in the surface of fibres and fines and cannot be removed.

4.2.5 Problems caused by wash deinking chemicals

The surfactants used in wash deinking can cause problems in the system. Large volumes of dirty water are generated in the deink plant and typically this water is cleaned by using clarifiers and flocculating chemicals. The dispersive nature of the surfactants in the water can upset the flocculating chemicals, while conversely, any flocculating chemical going into the deink plant can upset washing. Surfactants can upset papermaking if they are not washed from the fibre, and sizing chemicals and retention aids can also be affected [4].

Table 4.3 Estimated amount of flotation/combination deinking carried out worldwide[a]

Year	1978	1980	1982	1986	1988	1989	1990	1991
Amount ($\times 10^6$ tonnes)	2.58	3.13	4.7	7.5	9.7	11.5	13.5	15.4

[a]Source: literature values and personal knowledge

Table 4.4 Estimated amounts of wastepaper flotation deinked per day in 1991[a]

	Wastepaper (tonnes)	
Region	Newsprint	Woodfree
Japan	6 044	1 064
North America	7 615	5 430
Europe	11 005	4 717
Other	4 045	2 406
Total	28 709	13 617
	(42 326)	

[a]Source: literature values and personal knowledge

4.2.6 Flotation and combination deinking

In flotation deinking, pulped wastepaper is fed into cells. Stock consistency is typically 1%. Air is injected with the stock; the print particles adhere to the air bubbles, rise to the surface and are removed. In order to stick to the air bubbles, the print particles need to have a mainly hydrophobic surface and to be ~10–150 μm in diameter [3]. Flotation deinking will increase the brightness of pulp by removing the small print particles which cause darkening; it will also partially remove specks.

Flotation can be followed by washing, to increase brightness and to remove filler. Flotation itself is a poor remover of filler. Most flotation is carried out in alkaline conditions, and initial post-flotation washing was also alkaline [5]. In 1984, Bridgewater Mill pioneered the use of an acid wash [6,7]. Some hydrophobic organic materials can dissolve or become suspended during the alkaline pulping and flotation stages. Acid conditions destabilise these materials, causing them to precipitate as sticky particles. An acid wash after flotation can remove them, and reduce any stickies problems on the paper machine. Following Bridgewater, the expression *combination deinking* nowadays usually refers to alkaline flotation/acid washing.

Table 4.3 gives estimated figures for the amounts of wastepaper deinked by flotation and combination methods. Similar estimates have appeared in the literature [8, 9]. All estimates show that flotation deinking is the largest volume deinking method, and that its use is growing rapidly. Table 4.4 gives

Table 4.5 Major chemicals used in flotation deinking

Chemical	Typical addition levels (%)	Usual addition points	Estimated worldwide addition amount (tonnes/day)
NaOH	0.25–1.0	Pulper/bleach tower/disperger	252
H_2O_2	0.5–1.0	Pulper/bleach tower/disperger	198
Sodium silicate	0.25–1.5	Pulper/bleach tower/disperger	141
Chelating agents	0–0.5	Pulper/bleach tower	-
Soap	0.25–1.0	Pulper/cells	167
Surfactants	0.01–0.2	Pulper	22
Talc	0–1.0	Pulper/cells/final stock	-
Calcium salts	-	Cells	-
H_2SO_4	-	Final stock	-
Sodium hydrosulphite	-	Final stock	-
Sodium hypochlorite	-	Final stock	-
Flocculating chemicals	-	Backwater	-

The following assumptions are made:
1 wastepaper consumption figures taken from Table 4.4;
2 amount of NaOH is 0.75% on news, 0.3% on woodfree;
3 amount of H_2O_2 is 0.75% on news, 0% on woodfree;
4 amount of silicate is 0.5% on news 0% on woodfree;
5 amount of soap is 0.7% on news, 0.4% on woodfree;
6 amount of surfactant is 0.2% on news, 0.1% on woodfree;
7 in Japan 100% waste deinked with surfactants, none with soap;
8 in North America 15% waste deinked with surfactants, 85% with soap;
9 in Europe 10% waste deinked with surfactants, 90% with soap;
10 in rest of world, 50% waste deinked with surfactants, 50% with soap.

a picture of the application of deinking by region. Flotation deinking generates a small amount of waste in the form of a concentrated scum. This can be thickened by centrifugation or pressing, and used as landfill or burnt in mill furnaces. Less useful paper-making fibre/fine/filler is lost in this scum than in a comparable wash deinking system.

4.2.7 *Types and amounts of chemicals used in flotation deinking*

Table 4.5 lists the major chemicals used in flotation deinking, and also gives an order-of-magnitude estimate for the amounts used. The difficulty of estimating such figures is illustrated in Table 4.6 which compares some of the calculated values with those reported in the literature.

4.2.8 *What flotation deinking chemicals do*

In order to be removed from the fibre/water mixture in the flotation cell, a print particle must adhere to an air bubble and float. It will do this best when it is ~10–150 μm in diameter, has a mainly hydrophobic surface, and is not stuck to a fibre. Deinking chemicals added to the pulper help the removal of the print from the fibre surface. Chemicals added to the pulper

Table 4.6 Deinking chemical usage in North America

Chemical	Estimated amount used (tonnes/day)		
	a	b, c	b, d
NaOH	57	36	-
H_2O_2	53	15	13–27
Silicate	38	32	-

a, Calculation for North America in 1991; assumptions as in Table 4.5
b, Presumably include wash deinking figures
c, *Pulp and Paper*, March 1987, p. 41, quoting *Chemical Marketing Reporter*
d, *Pulp and Paper*, July 1990, p. 45. Information from peroxide manufacturers

and to the flotation cell increase the hydrophobic nature of the surface of the detached print particles.

Newsprint is extensively deinked by flotation deinking (Table 4.4). Some chemicals increase the size of print particles in a pulped newsprint stock [10–14].

4.2.9 *How flotation deinking chemicals work*

Since the first commercial flotation deinking plant was installed in 1950 [15], considerable thought has been devoted to understanding how certain chemicals improve the performance of such plants. A picture has emerged which has gained wide currency in the technical literature. According to this picture, fibres are swollen by NaOH, and this action breaks the bond between fibre and print, because the print does not swell. The mechanical action of the pulper then knocks this loosened print from the fibre.

The print on magazines and similar waste papers is composed of pigment particles held together by cross-linked networks. This network is typically an alkyd resin which polymerised when the ink, after being applied to the paper, set to a print. Newspapers printed by the web offset process have a similar print structure. Sodium hydroxide is believed to hydrolyse ester groups within this network, an action which causes the print to break into smaller particles. The result is a pulp in which there are few visible specks, and in which the print particles are in the size range most suited for flotation.

Soap is commonly added to the pulper or the flotation cell. The actual materials used are the sodium salts of fatty acids such as palmitic, stearic and oleic, i.e. materials similar to normal household soap. To improve the effectiveness of flotation deinking, enough calcium ions must be present in the water to react with the soap and produce the insoluble calcium salt. This wax-like material is believed to coat the print particle surface, making it hydrophobic. The print particle, with its newly acquired surface, then adheres to an air bubble and floats out of the stock. Because soap increases the tendency of print particles to collect at air bubble surfaces, soap is called a 'collector', a term which comes from mineral flotation.

Hydrogen peroxide is believed to bleach fibres (especially mechanical), while silicates and chelating agents are also present to inactivate metal ions in the waste and in the backwater, which might otherwise catalytically decompose the peroxide.

Some comments concerning this mechanism will be made later, in the sections dealing with individual chemicals. The actual role of surfactants in flotation deinking will also be described later.

4.2.10 Problems caused by flotation deinking chemicals

Chemicals added in the deink plant can adhere to fibres or remain in the water associated with the fibres, and then be carried to the paper machine. Calcium ions and calcium soap salts can cause deposit problems on the machine. Sodium silicate is sometimes said to slow drainage on dewatering devices, but many mills never encounter such problems. Both soap and surfactants are known to disturb sizing if they are carried over.

4.2.11 Disperger chemical addition

Pulp from woodfree waste often contains visible print specks. One way to improve the quality of the stock is to run it through a mechanical attrition device to try to break the print specks into particles below the size where they can be seen as specks. Such mechanical devices usually run at high temperatures and high consistencies, and sometimes chemicals are added to improve their performance. Typically these are NaOH, surfactant, hydrogen peroxide and silicate. Where they are found to be effective, they presumably weaken the print by chemical attack. Colour stripping and bleaching can also be achieved by addition of an appropriate chemical to dispergers or kneaders.

4.3 Specific chemicals used in deinking

4.3.1 Sodium hydroxide

Alkali is probably the oldest chemical used in deinking. Koops used it around 1800 to wash deink, while Strachan [16] suggests that Koops's work was based on German deinking techniques of perhaps some 50 years earlier. The original reasons are unrecorded. Possibly it was used because it was known to improve the washing of cloth, where it is suggested that it forms soap by saponifying the fats which are naturally associated with dirt and cloth [17]. This soap suspends the dirt particles in the water phase and allows them to be rinsed away [18].

In turn-of-the-century patents describing deinking formulations, several authors attempted to explain what was happening in the hot, strongly alkaline pulping liquors [16]. The general idea was that the caustic alkali saponified and dissolved the oily vehicle of the print, thus liberating small pigment particles. These became suspended in the water and thus could be washed out. Later workers also strongly suggested that alkali saponified the vehicle in the print. The papers being wash deinked at that time contained inks which were typically based on modified linseed oil. The belief that soap was formed from print by sodium hydroxide, and that this soap helped suspend and remove print particles during wash deinking, is echoed in a 1956 comment [19]: "one mill reported recently, however, that a few pounds of soap had been included as a permanent part of its deinking chemical furnish, in order to promote better ink removal. It was believed that the decrease in proportion of linseed oil and other saponifiable ink constituents was a cause of poor ink removal, and the added soap was found to be helpful." Other comments in the same reference point to high levels of sodium hydroxide addition in previous times. Levels of 1.5–8% on wastepaper weight are recorded.

The belief that sodium hydroxide hydrolyses ester groups in print particle networks, causing the print to break up into small particles, is accepted by the majority of writers on deinking. Most quote it as fact and give no supporting evidence or references. Occasionally, some evidence for saponification is given, usually as an inference, rather than a proof [20]. A small number of writers have suggested that saponification does not occur. Strachan [16] remarks that oxidised vegetable oil varnish is only saponifiable to a very limited extent. Mattingley [21] writes that it appears highly unlikely that any significant degree of ink vehicle saponification occurs, even at pH values up to 11.5. Bassemir [22] comments that print which has dried by oxidation can only be partially saponified by strong alkali at moderate to high temperatures. Kübler [23] could detect no soap generated by the action of alkali on print, and concludes that saponifiable ink constituents are not actually saponified during deinking.

When used in a flotation deinking system on a news/magazine waste, sodium hydroxide increases final brightness. Exactly why these alkaline conditions give better flotation is not really evident from the technical literature. The general feeling seems to be that the presence of alkali aids the removal of print from the fibres and breaks it down into particles suitable for flotation by the processes described above, and that these actions allow more print to be removed from the system.

Currently, some flotation deink plants are experimenting with neutral pulping and deinking, for a variety of reasons. With some woodfree wastes, speck removal improves at neutral pH values. Often no chemicals at all are used; soap is difficult to work with at neutral pH as it will form the corresponding fatty acid and this has little deinking effect. Much of the waste being treated in this way is printed with laser and similar toners, which do not react chemically with NaOH.

Another reason to operate with less or no NaOH is the COD level in the backwater. High addition levels of NaOH will alkali-extract lignins and other materials from fibres, especially from mechanical fibres [24]. With increasing demands to reduce COD levels in effluent waters, some deink plants are trying to work close to neutrality. Both wash deinking and flotation deinking plants are investigating neutral deinking.

A well-known effect of sodium hydroxide is that it increases the strength of papers made from dried fibres. This can be particularly useful when deinking mechanical grades [25]. This effect may be caused by the opening up of the inside of the dried fibre, or by hydrolysis of ester bonds within the fibre wall. The amount of water in the fibre wall increases, making it more flexible. Increased inter-fibre bonding follows, and this increases paper strength. Ester hydrolysis in mechanical fibres is known to occur [26].

4.3.2 Soap

Soap was used in wash deinking in the early part of the century, but this use has disappeared because synthetic surfactants give better results.

Soap is currently one of the major deinking chemicals used in flotation deinking. It was one of several chemicals tried as collecting agents in early flotation deinking experiments. One feature of the use of soap is that its calcium salt form is well known as a defoamer. Waste containing magazines can create considerable volumes of foam which can overflow from the flotation cell and disrupt the deink plant. Soap addition to a flotation cell gives a foam which is deep enough to be separated from the stock by controlled overflow or by vacuum suction, but which does not grow uncontrollably. In addition to this foam-controlling effect, it was found that soap addition also gave brighter stock.

Originally, soap was created *in situ* by running liquid fatty acid into the pulper, where it was saponified by the sodium hydroxide present. The addition level of this soap was $\sim 1.2\%$ on a news/magazine furnish. Liquid fatty acid is mainly oleic acid, but soaps made from most fatty acids have been evaluated for deinking performance [27]. Ricinoleic acid (from castor oil) possibly gives the highest brightness [28], while stock deinked using stearic acid is said to cause least problems on the paper machine [29]. Mixtures of C(16)–C(18) fatty acids are said to show synergy, and to give the best deinking [30]. Cost and availability, however, mean that all deinking soaps are based on commercial blends of fatty acids. Tallow-based soaps are widely used. Such soaps contain stearic/palmitic/oleic acids in an approximately 1/1/1 mixture and will adequately deink news/magazine waste at addition levels of ~0.7% on waste weight. A modern trend is to reduce the level of oleic acid still further; e.g. some current soaps contain ~14% of this component. Exceptions to this trend are deink plants situated in integrated mills which produce tall oil. The latter contains a lot of oleic acid, and is used because it is available at low cost.

Deinking soaps are usually supplied to deink plants as soaps, and not as fatty acids to be saponified on site. Saturated acids are slow to saponify, and adding them to a pulper can result in waxy spots in the final paper, due to unreacted fatty acid being carried over to the machine. Even adding solid pre-made soaps to the pulper can result in waxy spots in the final paper, because the soaps are slow to dissolve. Many mills purchase such soaps as 20% solutions in water, and store them above 45°C. Others dissolve the soap in hot water before use. Cooled solutions of soap turn into gels, and care must be taken not to allow spills to run undiluted into drains, where they cool and can cause blockages. Spilled soap solution is also very slippery.

In the deink plant, soap solution is added to the pulper or to the flotation cell. To function as deinking chemicals there must be calcium ions present in the water [27, 31–33]. A level of at least 10 German degrees hardness (180 ppm as $CaCO_3$) is often recommended. For mills with soft water, calcium salts must be added (see below).

The technical literature contains several suggested mechanisms to explain how the calcium soap salt improves deinking (Figure 4.1). Most of these suggest that the soap salt sticks to the print particle surface in some way. This action can give the print particle a waxy coating, which will increase its adherence to an air bubble. The addition of calcium soap salts to a deinking system has also been reported to cause agglomeration of print particles, typically giving a doubling of size [10, 11]. The other reported mechanism suggests that the precipitating calcium soap salt removes material extracted from the fibres which would otherwise go onto the print particles and give them a hydrophilic surface, thereby stopping them from adhering to air bubbles and floating.

Studies have shown that calcium soap salts can be carried by the deinked stock to the paper machine. The amount involved has been variously stated as 10%, 33% [38], and up to 50% [39] of the added soap. One suggested cause of this carry-over is that the pulp pad which is formed on dewatering devices filters out the calcium soap salts which have not floated [40]. Because they are not removed, they go to the paper machine. Problems can be caused by this material, for example deposition, cylinder adhesion [29] and interference with sizing [4].

4.3.3 Calcium salts

Calcium salts are added to flotation deinking systems which have low calcium ion levels in order to make fatty acid soaps function as deinkers

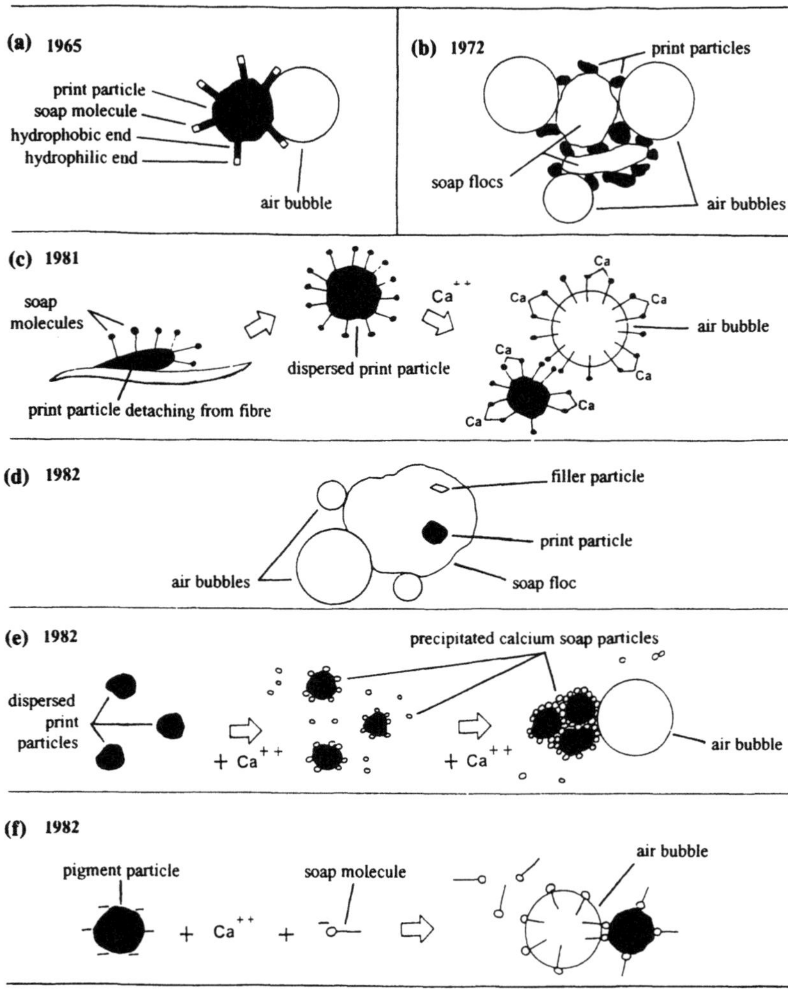

(g) 1987 *Natural surface active materials can be extracted from paper. In a flotation deinking plant these can adsorb onto print particles, give them a hydrophilic surface and stop them floating. Recirculation of water in a deink plant will increase their concentration, unless they are removed. As soap precipitates by reacting with calcium ions, it offers a fresh surface onto which these disruptive materials can attach. They are then precipitated out of the system as the calcium soap floc grows. (Silicate may act in the same manner. Peroxide may destroy these disturbing materials. Conventional deinking chemicals may be simply cleaning deink plant backwater.)*

Figure 4.1 Suggested mechanisms by which calcium soap salt formation can improve flotation deinking. (a) Redrawn after Schweizer [34] and Ortner *et al.* [109]. Note that calcium ions are *not* involved in this mechanism. (b) Redrawn after Bechstein and Unger [35]. (c) Redrawn after Ortner [36]. (d) Redrawn after Fischer [37]. (e) Redrawn after Larsson *et al.* [10, 39]. (f) Redrawn after Hornfeck [27]. (g) Hypothesis by Turvey [12, 44].

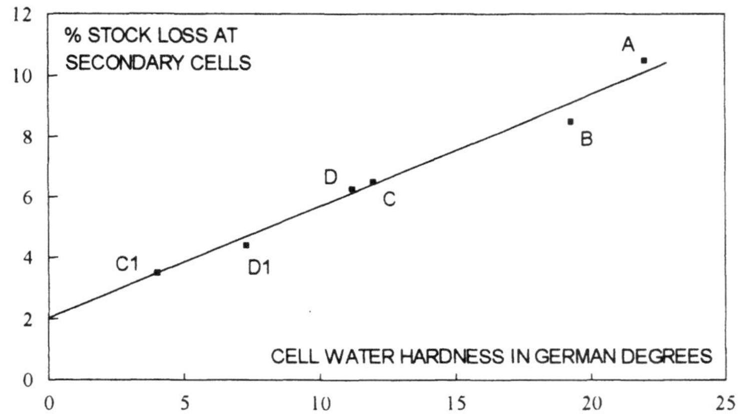

Figure 4.2 Mill stock loss values across flotation deinking cells as a function of cell water hardness for a number of industrial deink plants: (A) waste German news/mags (60/40), $NaOH/H_2O_2$/silicate plus soap, Escher-Wyss cells; (B) waste UK news/mags (60/40), $NaOH/H_2O_2$/silicate plus soap, Voith cells; (C) waste Swedish news/mags (70/30), $NaOH/H_2O_2$/silicate plus soap, Voith cells; (D) waste North American news/coated paper (80/20), $NaOH/H_2O_2$/silicate/DTPA plus soap, Voith cells; (C1) as C, but soap replaced by proprietary deinking emulsion; (D1) as D, but soap replaced by proprietary deinking emulsion. Source: personal knowledge and technical literature.

(see preceding section). The most commonly added salts are calcium chloride and calcium hydroxide, which are added to give hardness levels of 10 German degrees and above, and are typically added to the first flotation cell. Chloride ion can give corrosion problems in deink plants, and the addition of calcium hydroxide means that higher levels of sulphuric acid are needed to reduce the pH of the final stock.

During any flotation deinking, a certain amount of stock floats to the surface and is removed with the scum. The amount of stock lost has been shown to depend on several factors, a prime one being water hardness. Figure 4.2 plots stock loss values against flotation cell hardness for several industrial deink plants.

The way in which calcium ions increase stock loss is the subject of current research. One suggested mechanism is that calcium ions form a complex with the print particles, which then sticks to the fibres [41]. This makes them more hydrophobic, and therefore they float.

The increasing use of calcium carbonate as a filler means that magazine waste and similar papers carry this material into the deink plant. Very little is removed in the flotation cells, and so most of it goes to the paper machine. Machines running under acid conditions can dissolve it. Deink plants often run on paper machine backwater, and the calcium ion content can build up to very high levels in such deink plants. Levels as high as 100 German degrees have been observed, and such levels can give high stock loss, and lead to deposit formation. Some European mills which use deinked stock to make newsprint have changed from acid to neutral papermaking to stop calcium ion buildup in their deink plants.

4.3.4 Surfactants

A surfactant is a molecule with two distinct parts; one part will dissolve in water, while the other part will not. This latter part usually has an affinity for oils and similar materials, and also prefers air to water. Put into water, surfactants go to surfaces; hence, with a slight change of spelling, their name, **surf**ace **act**ive **a**ge**nts**.

In conventional cloth washing, surfactants are believed to work by reducing the surface tension of water, thus allowing it to penetrate between the dirt particle and the cloth [17, 18]. Combined with mechanical agitation, this action lifts the dirt particle into the water. Dirt particles are usually covered with an oil film, and the oil-loving part of the surfactant adheres to this surface. The other end of the surfactant molecule remains in the water. The resulting complex is a dirt particle with a hydrophilic surface, which remains suspended in the water phase and can be removed by dewatering and rinsing.

It is quite likely that this is the mechanism by which surfactants remove print particles in wash deinking. The original surfactant used was soap. One problem with soap as a washing surfactant is that it reacts with calcium ions to give an insoluble salt which cannot suspend particles in a water phase. Soap has now been replaced by synthetic surfactants, which give much better results. There are thousands of synthetic surfactants to choose from, and trial and error has identified those which give the best results.

Surfactants have also been used in flotation deinking from the earliest days. Most surfactants give poor results in a flotation cell, probably because they both disperse the print and also give the resulting small particles a hydrophilic surface. Such particles will not float. Surfactants used in flotation deinking cannot work in the way that surfactants probably work in wash deinking; larger print particles with hydrophobic surfaces are the particles which float.

A number of surfactants have been identified which improve flotation deinking. These usually contain ethoxylated chains as the hydrophilic end, with a variety of different structures as the hydrophobic end. Typical molecules which have been used in deinking are shown in Figure 4.3.

The ratio of the hydrophobic to the hydrophilic part of a surfactant molecule can be given a numerical value which is called the hydrophile–lipophile balance (HLB) value. This number can be evaluated by experiment on individual surfactants. The higher the HLB, then the more soluble the surfactant. By increasing or decreasing the length of either part of a surfactant molecule, families of surfactants can be made whose members have

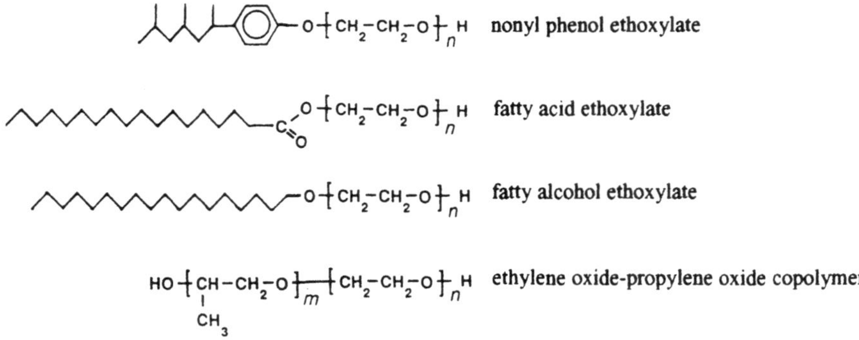

Figure 4.3 Typical surfactants used in deinking.

HLB values which cover a wide range. Studies on such molecules show optimum flotation deinking effectiveness around HLB values of 15 [42, 43].

In laboratory pulpings with certain waste papers these surfactants can cause considerable agglomeration of print particles [12]. Presumably these molecules are going onto print particle surfaces and then causing some agglomeration. They can also make the surface more hydrophobic, but how this is done is not known. Attempts have been made to explain surfactant behaviour in flotation deinking, but these studies do not explain why only certain surfactant structures and certain HLB values improve flotation deinking. Ethoxylated surfactants are known to come out of solution as the temperature is raised [42], and possibly this behaviour is involved in their flotation deinking abilities.

To successfully use a surfactant in flotation deinking requires very careful selection. Early use of surfactants led to foam buildup and a drop in stock brightness. This behaviour was ascribed to these surfactants generating foam and causing dispersion [33]. Alternatively, it may have been because soap was replaced by surfactant, and the system thereby lost the defoaming ability of the calcium soap salt, plus the ability of this salt to remove the dispersive materials which were being alkali-extracted from the wastepaper [12, 44].

Surfactants can be made to work well in a flotation deink plant at addition levels typically between 0.01 and 0.2%. They can give higher brightness values and also lower speck counts than soap. Since they do not need calcium ions in order to deink, stock loss can be reduced by deinking in soft water. To use surfactants in flotation deinking requires careful choice, which is based on experience. Formulated products often work best, and these sometimes include fatty acids. Incorporating surfactants in emulsions is another good way in which to use them.

Surfactants can leave mill deink plants in backwater, and injudicious choice can lead to foam problems in downstream treatment plants. Early surfactants, such as nonyl phenol ethoxylates, are now banned in some countries because they can generate phenols as they are biologically broken down. The environmental impact of surfactants should therefore be considered carefully before they are used.

4.3.5 Displector™ deinking chemicals

The trade mark Displector was coined by Leif Elsby of Berol Kemi AB in 1982 [45], and is now used by Eka Nobel and its subsidiaries. It was used initially in Europe, and then became well known in North America [46].

Displector is derived from the words **disp**ersing co**llector**. The first Displector was a chemical which was designed to disperse print (i.e. detach it from the fibre), and then collect together the detached print particles to give an agglomerate with a hydrophobic surface which could then be removed by flotation. This chemical was a fatty acid alkoxylate. The conjecture was that this molecule would act initially like a washing surfactant, and disperse/detach print particles. It was thought that the molecule would then hydrolyse in the alkaline pulping medium, and break into two parts; one part, the fatty acid, would react with calcium ions and behave as a classical soap collector, causing the print particles to agglomerate and have a hydrophobic surface, while the other part of the molecule, the alkoxylate chain, would be completely soluble in the water phase and play no further role. Subsequent Displector deinking chemicals are formulated products based on this original idea.

It is not known if Displector deinking chemicals hydrolyse during pulping. Some fatty alcohol alkoxylates have structures similar to the original Displector chemical, and give similar results in flotation deinking. Since these cannot hydrolyse, this suggests that Displector chemicals do not hydrolyse either.

Displector chemicals probably give their best performance in the flotation deinking of woodfree wastepapers.

4.3.6 Emulsions

Emulsions are useful ways to deliver chemicals which are otherwise difficult to handle. Mixtures of fatty acids and surfactants can give better flotation deinking than either material on its own. Liquid mixtures can sometimes be made, but when using saturated fatty acids, such as stearic acid, intractable solids often result. Some surfactants which are useful in flotation deinking are also solid, and emulsions can be made in which these surfactants suspend fatty acid particles in water. The result (technically a dispersion, but invariably called an emulsion) is a liquid which is easier to handle than its active components. Such materials can be pumped at any rate required into a flotation deink plant. Well-formulated emulsions can combine the best of both soap and surfactant deinking, giving higher brightness and a lower speck count at lower addition levels than soap. Surprisingly, although they contain fatty acid, they usually show little need for calcium ions to be present. Lower stock losses can therefore be achieved by deinking in soft water, and in addition, they can also be used successfully in neutral deinking conditions.

It has been suggested that oleic acid, while not a sticky itself, helps carry over stickies to the paper machine [47]. Emulsion deinking products, however, can be formulated to contain no oleic acid.

Ammonium stearate is a self-emulsifying chemical which has been used to a small extent in commercial deinking, despite the obvious drawback of its odour. Emulsified calcium stearate has also had some use as a deinking chemical, but experience suggests it is not quite as effective as soap.

4.3.7 Hydrogen peroxide

Hydrogen peroxide is widely used in deinking, especially in flotation deinking, where it is commonly added to the pulper. It is also sometimes added to the bleach towers between the pulper and the cells. Addition to bleach towers after deinking is practised, and in some mills addition to disperging equipment is also carried out.

The initial reason for addition of hydrogen peroxide to deink plants was almost certainly to act as a bleach and to increase the brightness of stock. It is known to be effective at stopping the darkening of mechanical fibres caused by sodium hydroxide [48]. Much of the technical literature describing the use of peroxide in deink plants assumes that its only role is that of a bleach [42] and much of the technical work performed has been to find the optimum conditions to bleach deinked stock. Analogy is often made to mechanical fibre bleaching, and to the importance of the correct total alkali hydrogen peroxide ratio [49]. Even for pulper addition of hydrogen peroxide, it is usually assumed that its main role is in bleaching.

Some workers have suggested that peroxide plays a role in deinking, in addition to the bleaching which undoubtedly occurs [50]. Peroxide addition to the pulper can give higher final stock brightness than peroxide addition to a post-flotation bleach tower [51]. One suggestion to explain this is that peroxide destroys material, possibly alkali-extracted from fibres, which may give print particles hydrophilic surfaces and stop them sticking to air bubbles during flotation [12].

It is also suggested that peroxide breaks bonds in print networks [30, 52–54], which can help detach print from fibres, and also create smaller print particles. These smaller particles may float better, which will decrease speckiness. The action of dispergers can sometimes be improved by peroxide addition.

One problem which occurs with the use of peroxide is catalase decomposition. Catalase is an enzyme generated by bacteria to protect themselves against peroxide attack [55]. It catalytically destroys peroxide very rapidly. The effect in a deink plant is obviously to reduce brightness sharply. It can be easily identified by measuring how fast a backwater sample destroys peroxide; if it is rapid, then either catalase or metal ions are present. To check which species is present, a sample of backwater is boiled and then cooled. If this treated water does not destroy peroxide, then catalase is present. (Boiling denatures and destroys the enzyme, while metal ions are unaffected by such treatment.) It is difficult to free a system from catalase. Stressing the bacteria by either raising or lowering the temperature sometimes works, and adding large amounts of peroxide can also resolve the problem. The use of peracetic acid has been suggested as a way to clean the system, as has hypochlorite addition. Draining the system and cleaning mechanically also helps, while the use of biocides offers little help, as most of these work best in acid or neutral environments.

4.3.8 Sodium silicate

Sodium silicate was probably first used in deinking because it is the time-honoured additive to use with hydrogen peroxide. In this application it inactivates metal ions, especially transition metal ions, which can catalytically decompose peroxide. It also buffers the system at around the pH where peroxide works best. Sodium silicate comes in a variety of SiO_2/Na_2O ratios; in deinking a common ratio which is used is 3.3/1.

Sodium silicate is known to have a powerful effect on deinking performance, particularly in newsprint deinking [56]. If brightness falls in a deink plant, one of the first things to be checked is the silicate pump, as the nature of the silicate solution means it is prone to blocking.

The literature contains suggestions that silicate does more in deinking than just create a stable environment in which the peroxide can work [13]. One reason for these suggestions is that chelating agents can also provide a stable environment for peroxide, but if they replace silicate totally in deink plants the deinked stock final brightness is usually not as good [49]. The addition of silicate when peroxide is not present can also give better deinking.

It has been suggested that silicate acts as a dispersing agent, and helps remove print from fibres [42]. In wash deinking, it is suggested that silicate suspends the print particles in water and this detergent action increases brightness.

One aspect of silicate behaviour seems to have been little commented on, namely it will react with calcium ions almost instantaneously to give a precipitate [29, 57]. This behaviour is similar to that of soap. In flotation deinking, perhaps this gives an agglomeration of print particles in the same way that soap is believed to work [13]. It may also give a more hydrophobic surface. It has been suggested that this precipitate can adsorb the natural surface active chemicals that are alkali-extracted from fibres, which would result in better flotation of print particles [12].

The addition of silicate can also reduce stock loss in flotation deinking. A reasonable explanation for this behaviour is that it reduces the calcium ion level in the system [57].

4.3.9 Chelating agents

Chelating agents have been used in deinking for many years. Their main use is to sequester metal ions, and so give hydrogen peroxide a stable environment in which to work [42]. DTPA is probably the main chelant used, because it is not oxidised by alkaline peroxide [58–60]. The use of other chelants, such as EDTA, glucoheptonate, phosphonates, etc. has been reported. Chelants have not been able to replace sodium silicate completely in deinking, and where used they are usually used with silicate [49]. Possibly the large amounts of calcium and magnesium ions often found in deink plant waters are chelated in preference to other metal ions. The use of chelants in deinking may be reducing.

4.3.10 Phosphate, carbonate, borate, etc.

These chemicals have been recorded in early deinking literature [16, 61] but are probably not widely used. They can reduce water hardness by precipitating calcium salts. Sodium carbonate has been used to partially replace sodium hydroxide and sodium silicate. Sodium phosphate can exist in such various forms as metaphosphate, orthophosphate and polyphosphate. It sometimes appears in formulated deinking products, and even in patented mixtures [62, 63]. Phosphoric acid is a component of wash deinking formulations, usually combined with surfactants and solvents. It may improve wash deinking in the same way that phosphates are believed to improve cloth washing, by removing calcium ions. In cloth washing, it is suggested that calcium ions hold dirt particles to the cloth by calcium half-salt bridges which link carboxylic acid groups on both the cloth and the dirt surface. By sequestering/removing calcium ions this method of retention is stopped [17].

4.3.11 Clay

The addition of clay to flotation deinking is probably seen more in the literature than in the deink plant itself. The literature typically tells us that a point or two higher brightness can be obtained if clay is added, but the addition amounts are usually high, and the brightness increase can sometimes be due to the brighter clay [64, 65]. Descriptions of addition in deink plants do exist [38, 66], but clay addition is generally uncommon in these plants.

The reason for the interest in clay addition may be connected with early observations that the addition of magazines will significantly improve the flotation deinking behaviour of newsprint waste [67]. Brighter stock is obtained, and brightness fluctuations are smoothed out. Early thoughts were that clay in the magazines was the agent responsible for these improvements. This does not seem to be the case, and the reason why the presence of magazines usually improves newspaper deinking is still a mystery [30].

4.3.12 Talc

Talc is added in deink plants, usually to pacify stickies [68]. It is often added to the pulper in a flotation deink plant, and sometimes to flotation cells, where it can help to reduce foam levels. Most commonly, it is added to the deinked stock.

4.3.13 Sulphuric acid

Sulphuric acid is usually used to reduce the pH of deinked stock, especially flotation deinked stock, before it goes to the paper machine. This action reduces the 'pH shock' on the paper machine. Combination mills wash the deinked stock at this point in order to remove stickies which have dissolved in the alkaline deink plant, and which precipitate in acid conditions. Some mills are now only reducing the pH to a value of 7, instead of ~5, because magazines and similar waste contain increasing amounts of calcium carbonate. This dissolves in acid conditions, and can subsequently cause deposits in the system. Calcium sulphate is a common deposit in these situations.

4.3.14 Bleaching chemicals

A variety of bleaching chemicals are added at various stages to stock, either during or after deinking. The most commonly used agent is probably hydrogen peroxide, which has been discussed earlier, but other chemicals, such as oxygen and ozone, may become important bleaches for deinked stock in the future.

Sodium hydrosulphite (also known as dithionite) is a reductive bleach, which is commonly used as a bleach tower agent to bleach deinked stock after the pH has been adjusted to acid or neutral. The literature reports that it bleaches wash deinked stock best at a pH of ~6.2 and flotation deinked stock at ~7.2 [69]. A typical use of hydrosulphite is in a deink plant using a newsprint/magazine furnish. The brightness of deinked stock is monitored on a continuous basis, and hydrosulphite is dosed in as necessary to maintain a target brightness.

Formamidine sulphinic acid (FAS) is a relatively new bleaching agent which is used to some extent in deink plants. It is a reductive bleach working in a similar manner to hydrosulphite [70], but advantages over hydrosulphite have been claimed [71].

Sodium hypochlorite is still employed in deink plants, where its main role is not to bleach fibres but to discharge the colours of dyes. Red and yellow dyed papers will typically not be bleached by peroxide or hydrosulphite. Waste containing chemical fibres is usually treated, as mechanical fibres can yellow in the presence of hypochlorite.

4.3.15 Flocculating chemicals

Water recirculating in a deink plant is often passed through a clarifier system and treated with flocculating chemicals. Standard flocculating chemicals are usually used, and in some flotation deink plants polyacrylamide and bentonite work well. Cleaning backwater helps a flotation deink plant run more evenly, possibly by removing the disruptive chemicals which can reduce flotation print or cause excessive foam. However, flocculating chemicals can be affected by some of the surfactants used in wash deinking.

4.3.16 Solvents

The main current use of solvents is probably in wash deinking, and in the deinking of laser print and photocopying paper. In wash deinking, the solvent is often paraffin. The literature is vague about how solvents work, but typically they are said to 'emulsify' the print. Probably the solvent swells the vehicle network holding the pigment particles, and weakens or dissolves it so that surfactants and alkali can break the print into small particles and suspend them in the water. In laser print deinking, the print is held together by a plastic structure which can take in solvents such as turpentine, and on heating these become soft enough to flow into spheres, which harden on cooling and can then be screened out.

Solvents will dissolve waxes, stickies and other contaminants. Some waste paper, if washed in solvent with little or no water present, will give extremely clean fibres. The solvent will remove all print and all contaminants. However, it is difficult to economically run such processes – the main problem is recovering all of the solvent from the fibre. Any solvent retained means a financial loss, and is also a possible source of pollution. A commercial deink plant using a water-free solvent wash did exist at one time [72], and the literature reveals an ongoing interest in such processes [73]. The literature also describes solvents such as cyclohexyl pyrrolidone

[74], which can be used in two-phase mixtures with water, and which can also remove print completely. Practically, however, solvents are used at low addition rates in substantially 100% water systems, and their use is probably quite low when compared with other deinking chemicals.

4.4 Some specific recycling problems

4.4.1 Stickies

Stickies are the small glue-like particles which come from self-adhesive labels, hot melt adhesives, etc. [75, 76]. They cause problems by sticking to the hot drying cylinders and tearing holes in the sheet and can also cause printing presses to blind. They have been causing problems for many years. A 1952 report stated that two pounds of rubber-like thermoplastic material can ruin a hundred tons of fibre [61].

The combination process is one way to reduce stickies problems [6, 7], and talc is the time-honoured additive [68]. Other chemicals have been suggested to pacify the surface of stickies, to disperse them, or to otherwise stop them causing problems. These have been as diverse as surfactants [77–80], zirconium salts [81–83] and fibrillated polypropylene [84]. Few of these, however, seem to be in use in deink plants. Stringent wastepaper quality control, increased use of mechanical removal devices (especially fine slotted screens of typically 0.2 mm slot width [85, 86]), plus talc where necessary, can significantly reduce stickies problems.

To stop stickie deposits in deink plants (and also on paper machines), some mills add cationic chemicals to the water system [87]. These coat surfaces, and repel stickie particles.

It has been suggested that in flotation deinking systems oleic acid can increase the carry-over of stickies to the paper machine [47]. However, it is difficult to use fatty acids or soaps containing no oleic acid, although one way is to use fatty acid emulsions.

4.4.2 Laser print

Laser and photocopy print usually give specks when pulped [88]. These thin flat particles are difficult to remove as they are too large to float out, have the wrong shape to be removed in cyclones, and will turn edge on and pass through fine slotted screens. They can be reduced in size by running stock through disperging units. Chemically they are unlikely to be affected by sodium hydroxide or by peroxide.

Laser and xerographic prints soften above ~70°C, so one treatment is to pulp at high temperature. The aim is to soften the print particles so that they fuse together to give large print balls, which can be more easily removed. Proprietary chemical formulations are added to the pulper to enhance the action of heat, and these consist typically of solvent plus surfactant [89], solvent plus solid thermoplastic particles [90–93], or all three materials [92, 93]. Solvent mixes with the heated print to make it softer. Surfactant possibly carries the solvent to the print by emulsifying it, or reduces the surface tension between the print particles. The solid thermoplastic particles become soft in the pulper, mix with the solvent and become tacky towards the laser print particles. The result is a semi-liquid print particle which flows into a spherical shape. It can also join with other print particles or the added thermoplastic particles to form large spherical agglomerates which harden as the temperature drops. Their shape then allows them to be easily removed by screens or cyclones.

There are many chemical formulations used to make laser and photocopy inks, and a solvent/surfactant mixture which will improve the deinking of one particular print type will often fail to affect another.

4.4.3 Carbonless copy paper

Carbonless copy paper, also known as NCR paper [94], typically comprises two or more sheets of paper; one sheet has a coating of acidic clay, while the other has a coating of capsules which contain a dye. The dye is colourless until made acid; in use, pressure ruptures the dye capsule, and the acid clay turns it into a coloured line. Early carbonless copy papers contained polychlorinated biphenyls (PCBs) as part of the oil component used to dissolve the dye, but current ones do not contain such materials.

When such paper is recycled, the pulp looks wonderfully clean and bright in the alkaline pulping, washing and flotation stages, but then turns highly coloured as it is made into paper at neutral or acid pH. The problem is the dye in the capsules. The capsules rupture during pulping, and the dye appears to fix onto the fibres, and cannot be washed or floated out. Peroxide seems to have little effect on the usual dyes that are used.

Currently, carbonless copy paper is a difficult waste to recover in a clean condition. Some is deinked, but much is burnt or used in dark products, which is a loss because the paper usually contains high-quality chemical fibre. However, reports in the literature do suggest possible progress in the reuse of such papers [95].

4.4.4 Water-based flexographic print

Water-based flexographic inks were developed in order to reduce oil and solvent vapours in print rooms and thus be more environmentally friendly. Typically, they are used to print newspaper. Such inks are made by grinding carbon particles in water with surfactants present. This gives a suspension of extremely small carbon particles coated with acrylate or styrene acrylate polymers. These polymers are then treated with a volatile amine, such as ammonia. When applied to paper the amine evaporates, and the carbon particles bind together through the polymers on their surfaces. This gives a print which will not smear under pressure.

When pulped in alkaline solution, the original polymeric surfactants are reformed, and suspend the extremely small carbon particles in the water. These particles are usually below the limit at which they will wash out, and so they probably lodge in the fibrillar surface of the fibre. Neither, of course, will they float, as they are too small and have a hydrophilic surface. The result is a black stock which is very difficult to brighten [96].

Several ways have been suggested to raise the brightness of stock consisting of pulped water-based flexo news. Pulping at pH < 8.5 does not regenerate the water-soluble surfactant on the carbon particle surface, and the print particles tend to stay together [96, 97]. Several specific chemicals have been proposed to improve the deinking of such wastes [98–101]. Their current use in deink plants seems limited.

Practically speaking, deink plants appear able to tolerate ~5–10% of water-based flexo printed news without seeing a significant brightness drop [102].

4.4.5 Filler removal

To use recycled fibre in tissue and some other types of paper, fillers need to be removed, and the usual way to do this is to wash them out. The addition of surfactants and phosphates will sometimes improve the removal procedure.

The technical literature reports that various chemicals will increase the removal of fillers in laboratory flotation deinking [100], although there seems to be little or no actual use of these chemicals in deink plants.

4.5 Future deinking chemicals

4.5.1 General

As legislation and commercial pressure causes more wastepaper to be deinked, the quality available will decrease, mainly because the 'easy-to-deink' waste is already being collected and deinked. This extra wastepaper will be more mixed, and older/dirtier. It will also include print which is difficult to deal with, such as carbonless copy paper, laser, thermal, water-based flexo, etc. This will mean that more mechanical and chemical treatment will be needed to allow this waste to be used, with improved deinking chemicals being needed to deal with some of these wastes.

The effluent from deink plants is increasingly becoming an object of scrutiny by those local authorities charged with monitoring air and water purity and regulating landfill activity. The role that deinking chemicals play in the effluent will therefore become increasingly important. The effect of sodium hydroxide on COD has been commented on previously, as has the fact that some countries ban the use of nonyl phenol ethoxylates. The next generation of deinking chemicals will not only have to work better than the current ones, but will have to comply with a myriad of anti-pollution regulations. Where deinking chemicals actually end up, in what amounts and in what forms, will one day be required information from suppliers.

4.5.2 Enzymes

Enzymes are chemicals, albeit complex ones. The addition of enzymes to deinking systems has recently been the subject of several technical papers, and is attracting the interest of some large deinking companies. The reported work describes the addition of enzymes to Japanese or Korean newspapers [103–105], or to waste with coloured print [106, 107]. After flotation, the stock is said to be brighter and less specky. The authors claim that the enzymes improve the detachment of print particles from the surface, and also cause the print to break up into smaller particles, which then can be floated more easily. The mechanisms by which the enzymes do this are not specifically described. Enzymes have also been reported to increase the strength of paper made from recycled fibre [25], and to improve drainage [108].

Cellulase enzymes will degrade cellulose, so perhaps this is the mechanism that detaches print from fibres [104, 105]. Lipases will hydrolyse ester groups, so this is possibly the mechanism that causes print to break into smaller particles [105, 107]. The print on Japanese newspapers is known to be different to the print on European and American newspapers, consisting of acrylate and other polymeric binding materials, which can make it adhere to fibres and strongly resist being broken during pulping [60, 109]. Because of the differences in print between Western newspapers, enzyme deinking may possibly have little effect on such papers, although it may work well on coloured prints.

Enzymes are reported to be already in commercial use in deinking [110], and if enzyme deinking becomes a widespread technique it could lead to significant changes in deink plant operation; sodium hydroxide is not needed, and this in turn may reduce peroxide and silicate use.

References

1. Veverka, A.C. (1990) *Pulp Pap.* Sept., pp. 97–103.
2. Wells, S.D. and Spearin, W.E. (1956) *TAPPI Monogr. Ser. No. 16*, pp. 1–10.

3. Harrison, A. (1989) *Pulp Pap.* Mar., 60-65.
4. Moyers, B.M. (1992) *Tappi J.* **75**(1), 111-115.
5. Kalish, J. (1978) *Pulp Pap. Int.*, **20**(6), 56-63.
6. Scott, W.B. (1989) *Pap. Technol.*, May, V/12-V/15.
7. Claydon, P. (1991) *Pap. Technol.*, May. pp. 34-37.
8. Carroll, W.P. and McCool, M.A. *TAPPI Pulping Conference*, 1990, Book 1, 145-152.
9. Estes, T.K. *TAPPI Pulping Conference*, Book 1, 345-350.
10. Larsson, A., Stenius, P. and Strom, G. (1982) *Wochenblatt für Papierfabrikation*, **110**(14), 502-506.
11. Marchildon, L., Lapointe, M. and Chabot, B. (1989) *Pulp & Paper Canada*, **90**(4), T153-158.
12. Turvey, R.W. Paper read at 4th PTS Deinking Symposium, Munich. Condensed version 128-129, Symposium Proceedings, 1990; and in *Wochenblatt für Papierfabrikation*, **120**(8), 308-309, 1992.
13. Ali, T., McLellan, F., Adiwanta, J., May, M. and Evans, T. *1st Research Forum on Recycling*, Toronto 1991. Conference Proceedings, 21-29.
14. Nguyen, N.G., O'Neill, M.A., Jordan, B.D. and Doriss, G.M. (1992) *J. Pulp Pap. Sci.*, **18**(5), J193-J196.
15. Jelks, J.W. (1954) *Tappi J.*, **37**(10), 176A-180A.
16. Strachan, J. (1918) *The Recovery and Re-Manufacture of Waste-Paper*, The Albany Press, Aberdeen, Scotland.
17. Ullmann's *Encyclopedia of Industrial Chemistry*, 1987, 5th edn, Vol. A8, pp. 315-448.
18. Shaw, D.J. (1970) *Introduction to Colloid and Surface Chemistry*, 2nd edn, Butterworths, London, pp. 117-132.
19. Altieri, A.M. and Wendell J.W., Jr. (1956) *TAPPI Monogr. Ser. No. 16*, 68.
20. Okada, E. and Urushibata, H. *TAPPI Pulping Conference*, 1991, Book 2, 857-864.
21. Mattingley, J.T. (1978) *Pulp Pap. Int.* **20**, 64-65.
22. Bassemir, R.W. (1979) *Tappi J.*, **62**(7), 25-26.
23. Kubler, R. Paper read at 3rd PTS Deinking Symposium, 1987, Munich. Symposium Proceedings, 14-18.
24. Berndt, W. (1982) *Wochenblatt für Papierfabrikation*, **110**(15), 539-541.
25. Bhat, G.R., Heitmann, J.A. and Joyce, T.W. (1991) *Tappi J.*, **74**(9), 151-157.
26. Katz, S., Liebergott, N. and Scallan, A.M. (1981) *Tappi J.*, **64**(7), 97-100.
27. Hornfeck, K. (1982) *Wochenblatt für Papierfabrikation*, **110**(15), 542-544.
28. Bechstein, G. (1975) *Das Österreichische Papier*, **4**, 16-22.
29. Mattingley, J.T. (1984) *Tappi J.*, **67**(6), 74-77.
30. Read, B.R. *TAPPI Pulping Conference*, 1991, Book 2, 851-856.
31. Raimondo, F. and Ortner, H. (1966) *Revue ATIP*, **20**(5), 218-227.
32. Weigl, J., Scheidt, W., Phan-Tri, D. and Grossman, H. (1989) *Wochenblatt Papierfabrikation*, **117**(16), 733-738.
33. Ortner, H. (1965) *Tappi J.*, **48**(2), 37A-41A.
34. Schweizer, G. (1965) *Wochenblatt für Papierfabrikation*, **93**(19), 823-830.
35. Bechstein, G. and Unger, E. (1972) *Zellst. Pap. (Berlin)*, **10**, 297-306.
36. Ortner, H. *TAPPI Monograph on Recycling*, 3rd edn, chapter on Flotation Deinking published separately in 1981, TAPPI Press, Atlanta.
37. Fischer, S. (1982) *Wochenblatt für Papierfabrikation*, **110**(14), 511-514.
38. Eriksen, E. (1979) *Pap. Technol. Ind.*, **20**(8), 279-281. See also *Paper*, 4 Aug. 1980, 23-24.
39. Larsson, A., Stenius, P. and Odberg, L. (1984) *Svensk Papperstidning*, No. 18, R158-R164.
40. Larsson, A., Stenius, P. and Odberg, L. (1985) *Svensk Papperstidning*, No. 3, R2-R7.
41. Turvey, R.W. (1993) *J. Pulp Pap. Sci.*, **19**(2), Mar., J52-J57.
42. Ferguson, L.D. (1992) *Tappi J.*, **75**(7), 75-83.
43. Schuster, A. (1990) Flotations-Deinking im schwach alkalischen pH-Bereich mit Zusatz von nichtionischen Tensiden, *Diplomarbeit*, Jul., PTS Munich.
44. Turvey, R.W. (1987) *Pap. Technol. Ind.*, **28**(1), 366-368.
45. Elsby, L. (1992) *Paper Age*, **108**(1), 24-25.
46. Horacek, R.G. and Jarrehult, B. (1989) *Pulp Pap.*, March, 97-99.

47. Lippert, G.V. Paper read at 1st PTS Deinking Symposium, 1981, Munich. Symposium Proceedings, 128–130.
48. Kutney, G.W. and Evans, T.D. (1985) *Svensk Papperstidning*, No. 9, R84–R89.
49. Helmling, O., Suss, H.U. and Eul, W. *TAPPI Pulping Conference*, 1986, Book 2, 407–417.
50. Blechschmidt, J. and Ackermann, Chr. (1991) *Wochenblatt für Papierfabrikation*, **119**(17), 659–661.
51. Opherden, A., Ramatschi, Chr. and Bruning, F. (1985) *Zellst. Pap. (Berlin)*, **6**, 219–222.
52. Mack, H. (1963) *Tappi J.*, **46**(3), 141A–146A.
53. Ducey, M. (1987) *Pulp Pap.* March, 41.
54. Carmichael, D.L. (1990) *Pulp Pap. Can.*, **91**(10), T365–T367.
55. Prasad, D.Y. (1989) *Tappi J.*, **72**(1), 135–137.
56. Helmling, O., Suss, H.U. and Berndt, W. Paper read at 2nd PTS Deinking Symposium, 1985, Munich. Symposium Proceedings, 57–61.
57. Turvey, R.W. Paper read at PITA Conference, Manchester, March 1990. Conference Proceedings, (unpaginated).
58. Mathur, I. *TAPPI Pulping Conference*, 1991, Book 2, 1015–1021.
59. Bambrick, D.R. (1985) *Tappi J.*, **68**(6), 96–100.
60. McCormick, D. *TAPPI Pulping Conference*, 1990, Book 1, 357–364.
61. Forsythe, J.J. and Bragdon, C.R. (1956) *TAPPI Monogr. Ser. No. 16*, 46–47.
62. Calmanti, G., Gafa, S., Dadea, G.M., Gatti, A. and Burzio, F. GB Patent 2 007 252, May 1979, to Montedison S.p.A. Also US Patent 4 231 841, 4 Nov., 1980.
63. Blum, S. International Patent WO 90/10 749, publ. 20 Sept. 1990, to Blumco Detergents Ltd.
64. Raimondo, F.E. (1967) *Tappi J.*, **50**(9), 69A–74A.
65. Letscher, M.K. and Sutman, F.J. (1992) *J. Pulp Pap. Sci.*, **18**(6), J225–J230.
66. Zabala, J.M. and McCool, M.A. (1988) *Tappi J.*, **71**(8), 62–68.
67. Wultsch, F. (1963) *Tappi J.*, **46**(3), 147A–150A.
68. Williams, G.R. *TAPPI Pulping Conference*, 1987, Book 3, 563–571.
69. Hache, M.J.A. and Joachimides, T. (1992) *Tappi J.*, **75**(7), 187–191.
70. Eul, W. and Suss, H.U. Paper read at PIRA Conference, Leatherhead, 28 Feb.–2 Mar., 1989, Conference Proceedings, (unpaginated in Vol. 2).
71. Matzke, W.H. and Selder, H.H. *TAPPI Pulping Conference*, 1988, Book 1, 203–211.
72. Aldrich, L.C. (1977) *Tappi J.*, **60**(8), 114–116.
73. Anon (1991) *Tappi J.*, **74**(9), 42.
74. Mestetsky, T. (1974) *Pap. Trade J.*, 1 July, 20–23.
75. Sjostrom, J., Holmbom, B. and Wiklund, L. (1987) *Nord. Pulp Pap. Res. J.*, **4**, 123–126, 131.
76. Miller, P.C. *TAPPI Pulping Conference*, 1988, Book 2, 345–348.
77. Hamilton, F.R. (1984) *Tappi J.*, **67**(4), 56–58.
78. Elsby, L.E. *TAPPI Pulping Conference*, 1986, Book 1, 187–191.
79. Moreland, R.D. *TAPPI Pulping Conference*, 1986, Book 1, 193–198.
80. Ling, T.F. *TAPPI Pulping Conference*, 1991, Book 2, 1039–1044.
81. Nakamura, S., Nomura, M. and Sakai, K. Japanese Patent, Special Application 1979-25 565, 19 Sept. 1980, to Honshu Paper Manf. Co. (English translation from Magnesium Elektron Ltd.)
82. McAlpine, I. (1984) *Paper*, **202**(3), 24.
83. Goldberg, J.Q. *TAPPI Pulping Conference*, 1987, Book 3, 585–596.
84. McKinney, R.W.J. and Currie, P.G.C. (1986) *Pap. Technol. Ind.*, **27**(4), 182–188.
85. Cruea, R.D. (1983) *Tappi J.*, **66**(8), 67–68.
86. Vitori, C.M. *1st Research Forum on Recycling*, Toronto, 1991; Conference Proceedings, 133–142.
87. Duffy, R.J. and Aston, D.A. *CPPA Annual Meeting*, 1988, B81–B84.
88. Carr, W.F. (1991) *Tappi J.*, **74**(2), 127–132.
89. Richman, S.K. and Letscher, M.B., US Patent 5 141 598, granted 25 Aug. 1992, to Betz PaperChem. Inc.
90. Quick, T.H. US Patent 4 276 118, granted 30 June 1981, to Weyerhauser Company.
91. Quick, T.H. and Hodgson, K.T. (1986) *Tappi J.*, **69**(3), 102–106.

92. Darlington, W.B. (1989) *Tappi J.*, **72**(1), 35–38.
93. Darlington, W.B. US Patent 4 820 379, 1989, granted 11 Apr. 1989, to PPG Industries Inc.
94. Stadelhofer, J.W. and Zellerhoff, R.B. (1989) *Chem. Ind.*, 3 Apr., 7, 208–211.
95. Anon. (1991) *Pulp Pap. Int.*, **33**(4), 21, See also *Paper*, **215**(7), 16, 1991.
96. Clewley, J. (1992) *Paper*, **217**(7), 32–33.
97. Galland, G. and Vernac, Y. *1st Research Forum on Recycling*, Toronto, 1991. Conference Proceedings, 31–36.
98. Jarrehult, B., Horacek, R.G. and Lindqvist, M. *TAPPI Pulping Conference*, 1989, Book 1, 391–405.
99. Bast, J. Paper read at *5th PTS Deinking Symposium*, 1990, Munich. Symposium Proceedings, 98–102.
100. Liphard, M., Schreck, B. and Hornfeck, K. *TAPPI Pulping Conference*, 1990, Book 2, 965–973.
101. Cowman, J.S. European Patent 0 478 505 A2, granted 23 Sept. 1991, to Sandoz Ltd.
102. Anon. (1990) *Paper*, 4 Sept, 14.
103. *Tomorrow's World*, BBC TV, 16 May 1991: BBC 1 film showing enzyme-enhanced flotation deinking of newspaper at Daeduk Science Park, Korea.
104. Kim, T-J., Ow, S. S-K. and Eom, T-J. *TAPPI Pulping Conference*, 1991, Book 2, 1023–1030.
105. Davis, J.L. (1991) *Tappi J.*, **74**(2), 33.
106. Gibson, K. *TAPPI Pulping Conference*, 1991, Book 2, 775–780.
107. Daniels, M.J. (1992) *Paper Technol.*, **33**(6), 14–17.
108. Anon (1991) *Chem. Market. Rep.*, 9 Sept., p. 15. Also 24 Aug. 1992, pp. 7, 15.
109. Ortner, H., Wood, R.F. and Gartemann, H. (1975) *Wochenblatt für Papierfabrikation*, **103**(16), 597–601. (English translation available from Voith, based on paper given at *TAPPI Secondary Fibre Conference* 1974.)
110. Pommier, J-C. (1990) *Paper* 2 Oct., 34–36.

5 Recycled fibre bleaching

A. RENDERS

5.1 Introduction

Given the increasing use of recycled fibres in a wide variety of paper applications, deinked pulp must meet higher quality market requirements. To achieve this, mechanical processes such as deinking, cleaning and screening must be carefully optimised and controlled, and the use of various chemical treatments to modify the molecular structures responsible for light absorption become necessary.

Bleaching or brightening of recycled fibres can be achieved with oxidative or reducing agents. Of the oxidative bleaching chemicals that are available, hydrogen peroxide is most commonly used with recycled fibres. The greater part of this chapter describes the use of hydrogen peroxide at several stages in the deinking line and the parameters affecting its efficiency. Considerable attention is given to optimisation of the chemicals that are used with hydrogen peroxide.

Some particular applications for oxygen and ozone in wastepaper are also described, and a brief overview of recycled fibre bleaching with reducing agents is given.

5.2 Bleaching with hydrogen peroxide

Hydrogen peroxide has the advantage of being compatible with the environment, as it decomposes to water and oxygen. Therefore, strong environmental concerns make the product even more attractive as a bleaching agent for recycled fibres.

Hydrogen peroxide is the most frequently used chemical for high-yield pulp bleaching when high brightnesses are required. For a mixed wastepaper furnish of old newsprint (ONP) and old magazines (OMG), bleaching has some similarity to mechanical pulp bleaching. However, most of the recycled fibres have been bleached before and so they will not necessarily respond in the same way to the bleaching stage as virgin pulps.

The bleaching reactivity of hydrogen peroxide can be attributed to the perhydroxyl anion, OOH^-, which is formed in alkaline conditions, according to the following:

$$H_2O_2 + OH^- \rightarrow OOH^- + H_2$$

The bleaching of recycled fibre with hydrogen peroxide needs the optimisation of both chemical and physical parameters to initiate and support the perhydroxyl anion formation. On the other hand, careful control of contaminants is required to avoid hydrogen peroxide decomposition and this is described further in this chapter.

For the utilisation of hydrogen peroxide in a deinking line, several addition points can be considered, as follows:

in the pulper;
in a bleaching tower;
in the dispersion unit.

These applications are described below.

5.3 Hydrogen peroxide addition in the pulper

Pulping or disintegration of the dry wastepaper is the first operation in a deinking line, and the most commonly used is high consistency pulping (14–18% stock consistency (s/c)). The pulping stage determines the efficiency of subsequent decontamination stages since the size of the contaminants and their bonding with the cellulose fibres is determined in this stage.

Besides the separation of useful material (such as fibres and, to some extent, also filler material) from impurities (such as ink particles, solid material, adhesives, etc), the fibres can be bleached in the pulper. The role of different chemicals, hydrogen peroxide, sodium silicate and caustic soda for both deinking and for bleaching is described elsewhere, as well as the optimum addition levels [1].

The results described hereafter were obtained with an equivalent mixture of newsprint and magazines. Pulping time, temperature and consistency were held fixed at 30 min, 50°C, and 11%, respectively. A 1% fatty acid soap solution was added to all pulpers. Water hardness was corrected to 10 degrees German water hardness (10°dH) for water used in the pulper and for dilution to 0.8% consistency before flotation. Flotation was carried out at 40°C and for 20 min; the flotation pH was corrected to pH 9 before flotation.

Brightnesses were measured on both sides of the brightness pads and reported as mean values. To eliminate the influence of the pH on the brightness measurements, the pads were acidified to pH 5.5. In general, pulping chemistry has to be optimised to obtain maximum brightness gains by deinking, either by washing, flotation or a combination of both. In our study, deinking was by flotation, and brightnesses are reported as brightness

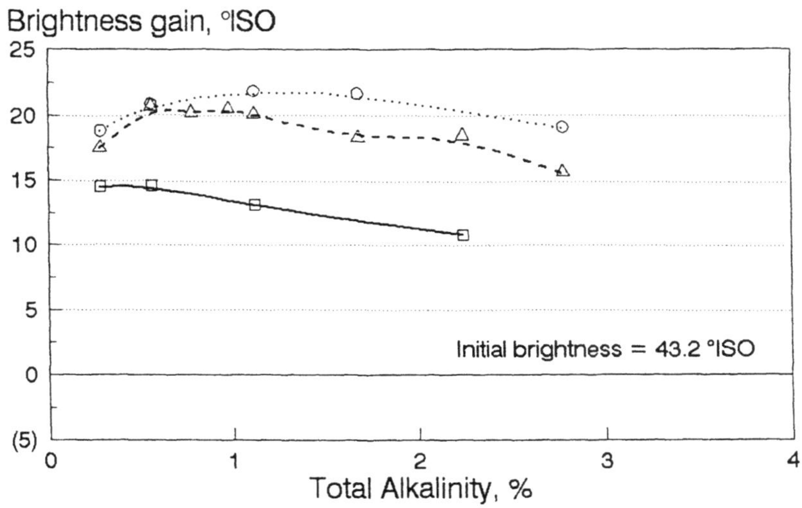

Figure 5.1 Total alkalinity optimisation in the pulper. □, 0% H_2O_2; △, 1% H_2O_2; ○, 2% H_2O_2.

gain, which is the brightness after flotation compared to the initial brightness after pulping without any chemicals being present.

The need for alkalinity to facilitate ink release by swelling the fibres, and perhaps by saponification of the ink binders, is well known [2]. In general, caustic soda (NaOH) is the main source of alkalinity in deinking.

Sodium silicate, also frequently used in deinking, contributes to the total alkalinity. The total alkalinity (t.a.) of a pulping liquor can be calculated as follows:

$$\text{t.a. (\% odp)} = \text{NaOH (\% odp)} + 0.112 \times \text{silicate (\% odp)}$$

with NaOH as 100% and sodium silicate as Waterglass 38 Bé.

For hydrogen peroxide and caustic soda, the chemical addition levels are given as a weight percentage on oven dry pulp (% odp) of 100% reagent. Sodium silicate is used on a solution basis (Waterglass 38 Bé), with its addition level being expressed as a percentage on odp of the product as received.

5.3.1 Total alkalinity optimisation

The optimisation of the total alkalinity level is illustrated in Figure 5.1, with different levels of hydrogen peroxide and 2.5% sodium silicate being used. Without hydrogen peroxide addition to the pulper, a maximum brightness gain of 14.8 degrees ISO was obtained by flotation deinking. It was shown that in the absence of hydrogen peroxide the brightness after flotation drops as soon as the total alkalinity level exceeds 0.6% odp. In order to understand this brightness drop, the results of pulping and flotation trials on

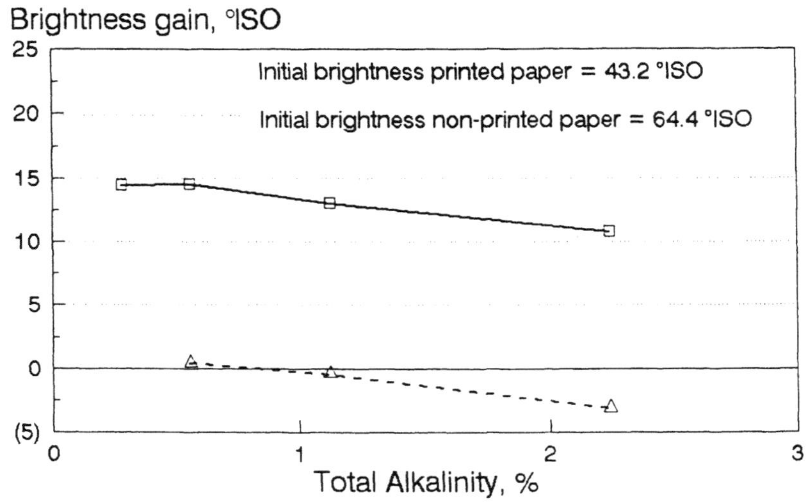

Figure 5.2 Effect of excess total alkalinity in the pulper. □, printed; △, non-printed.

non-printed furnish are presented in Figure 5.2. It can be seen that increased total alkalinity causes a brightness drop with the non-printed furnish. This brightness drop is known as alkaline darkening. As a result of the alkalinity, lignin is partially solubilised, which produces coloured lignin derivatives.

Also in Figure 5.1, it is shown that with hydrogen peroxide addition in the pulper, the optimum total alkalinity level is slightly higher, i.e. 0.8% total alkalinity on odp for 1% H_2O_2 and 1% on odp when using 2% H_2O_2. Part of the total alkalinity is consumed by hydrogen peroxide to produce the active perhydroxyl anions. From examination of the shape of the brightness curves it can be seen that the brightness was more stable with changing total alkalinities when hydrogen peroxide was added to the pulper. This means that even when the optimum total alkalinity demand changes due to variables in the furnish (ONP/OMG ratio, age, contaminants, etc.), or in the process (temperature, retention time, pulp consistency, etc.), a higher and more stable brightness after flotation can be ensured by using some peroxide in the pulper.

A fairly easy control for the total alkalinity is the pH measurement. As shown in Figure 5.3, once the optimisations have been carried out and the optimum total alkalinity levels are known, the pH curves developed can be used as controls on a daily basis.

5.3.2 Sodium silicate optimisation

The role of sodium silicate in deinking and in peroxide bleaching has been the subject of many papers [3–5]. In pulp bleaching, sodium silicate is

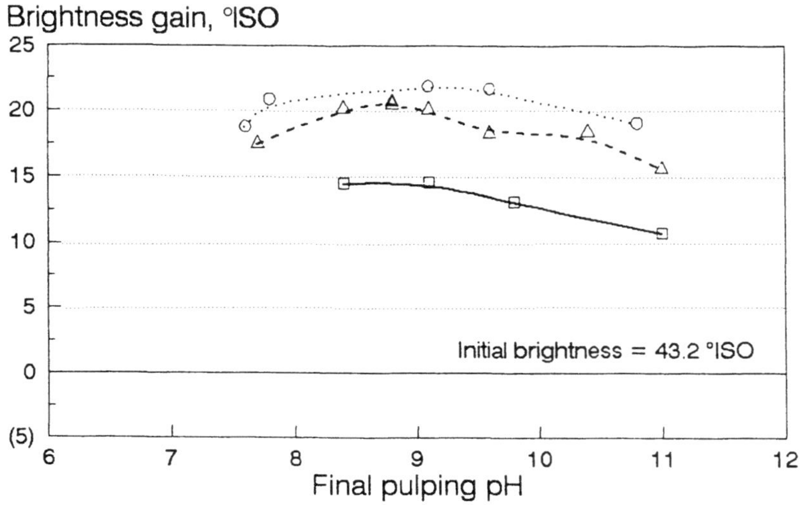

Figure 5.3 Process control in the pulper by pH measurements. □, 0% H_2O_2; △, 1% H_2O_2; ○, 2% H_2O_2.

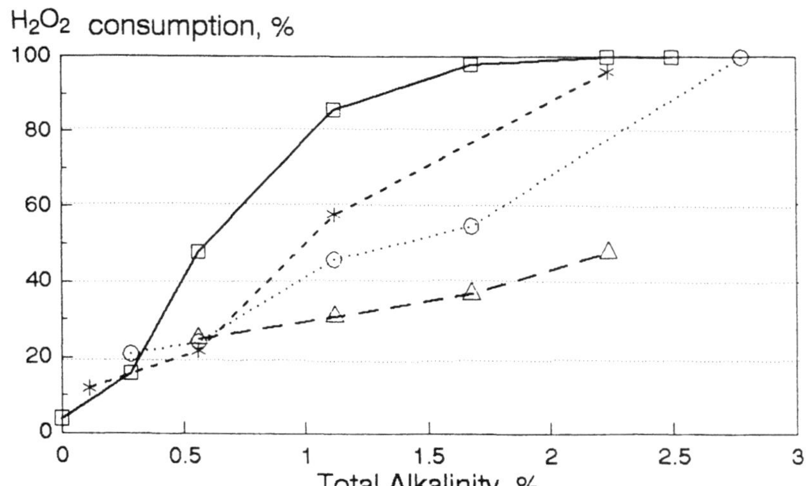

Figure 5.4 Stabilisation of hydrogen peroxide in pulper applications. □, 0% Si; *, 1% Si; ○, 2.5% Si; △, 5% Si; 2% initial H_2O_2.

associated with the stabilisation of hydrogen peroxide, and also in pulper applications, sodium silicate stabilises hydrogen peroxide, as shown by Figure 5.4. For a given total alkalinity, 1.1% on odp for example, the peroxide consumption decreases from 87%, without sodium silicate application in the pulper, to 30% with 5% sodium silicate present. These are the percentages of the initial amount of peroxide found in the pulping liquor after pulping. Figure 5.5 shows that stabilisation is not the only role of

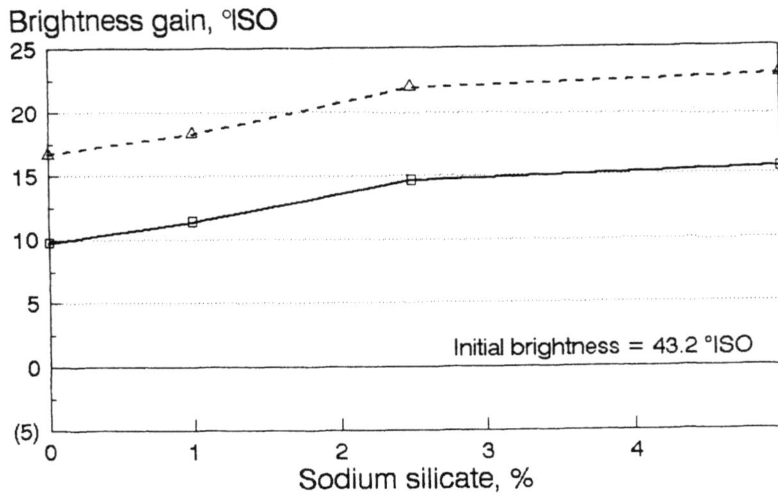

Figure 5.5 Optimisation of sodium silicate in the pulper. □, 0% H_2O_2; 0.56% t.a.; △, 2% H_2O_2, 1.12% t.a.

sodium silicate; it is found that a significant brightness gain is obtained by flotation deinking through the use of sodium silicate in the pulper. Since the two curves, with and without peroxide are parallel, the brightness gain due to the application of sodium silicate is independent of the brightness gain due to peroxide bleaching.

To understand the role of sodium silicate in deinking, image analysis was used. Ink particle histograms (by size) and the percentage area covered by ink particles were measured after pulping. The role of sodium silicate as an ink collector is shown in Figure 5.6. Sodium silicate prevents ink from redepositing on the fibres and agglomerates small ink particles to form larger ones ($>10\,\mu m$), so that they have a more suitable size for flotation. Figure 5.6 shows that more ink particles were found in the larger-size classes by increasing the sodium silicate level, and moreover, the surface covered by ink particles larger than $10\,\mu m$ was significantly increased. The percentage of the measured area covered by these ink particles after pulping is given by the curve on the graph, with the axis on the right.

5.3.3 Bleaching and deinking

A distinction between bleaching of recycled fibres and deinking, which is the elimination of ink particles, was achieved by comparison of the results obtained by processing printed wastepaper with the results from non-printed paper under exactly the same process conditions. For the non-printed furnish, the only possible way to increase brightness is by bleaching. For the printed furnish, brightness gains can be due to either bleaching or

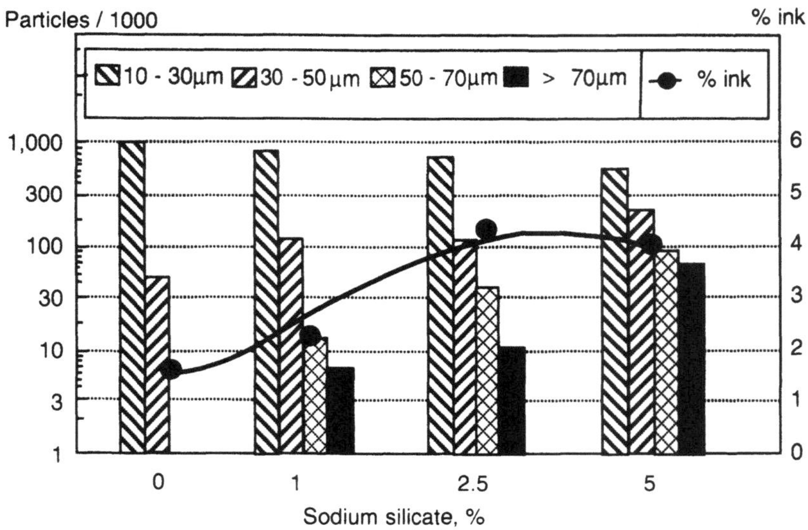

Figure 5.6 An illustration of the role of sodium silicate as an ink collector, showing particle size and % ink coverage as a function of sodium silicate content; pulping 2% H_2O_2, 1% t.a.

deinking. Figure 5.7 shows the results obtained for different hydrogen peroxide addition levels in the pulper with printed or with non-printed furnish. For each hydrogen peroxide level, the optimised quantities of sodium silicate and total alkalinity were applied. The brightness gain obtained by deinking is shown by the difference between the curve for printed paper and the curve obtained with non-printed paper. The brightness gains obtained by the hydrogen peroxide addition are similar for the printed and non-printed furnishes, and are therefore due to bleaching of the fibres.

5.4 Hydrogen peroxide in tower bleaching

When high brightness deinked pulps are required, hydrogen peroxide bleaching following deinking is often considered. Optimisation of the chemicals has again to be performed, in order to ensure the best hydrogen peroxide efficiency, and also to avoid undesirable decomposition reactions and alkaline darkening.

The results obtained from tower bleaching described hereafter were obtained with an equivalent mixture of newsprint and magazines. Pulper conditions were kept constant as follows: 50°C, 30 min, 11% consistency, 1% fatty acid soap solution, 10°dH water hardness, 0.56% total alkalinity and 2.5% sodium silicate. The pulp was then diluted to 0.8% consistency with water at 10°dH water hardness, and deinking by flotation was

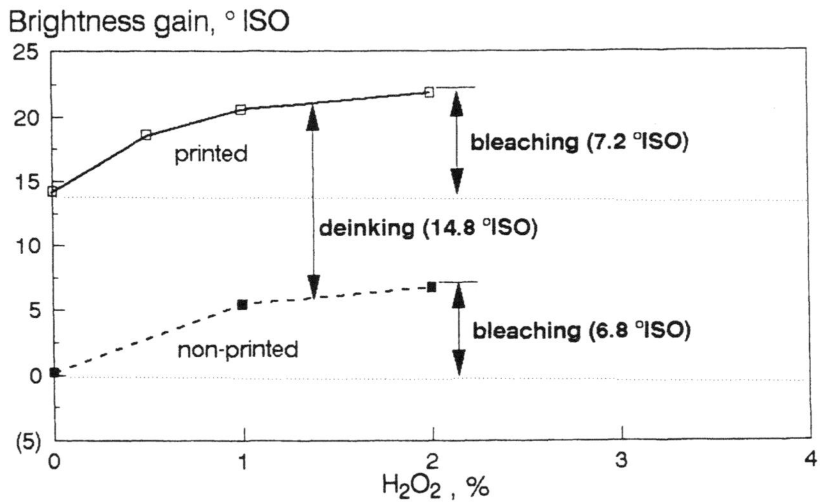

Figure 5.7 Bleaching and deinking in the pulper with printed (initial brightness = 43.2°ISO) and non-printed (initial brightness = 64.4°ISO) furnish.

then carried out at 40°C for 20 min. Tower bleaching was performed in polyethylene bags placed in a warm water bath at 70°C for 60 min.

5.4.1 Pulp consistency

As is the case with mechanical pulp bleaching using hydrogen peroxide, high consistency bleaching leads to higher brightnesses than with medium consistency treatment [6]. This is illustrated in Figure 5.8, where the consistency was varied between 10 and 30%. With 1% hydrogen peroxide and optimised levels of total alkalinity and sodium silicate, a brightness increase of 3 degrees was gained by increasing the consistency from 10 to 25%.

5.4.2 Optimisation of total alkalinity and sodium silicate

The optimisation of total alkalinity and the addition levels of sodium silicate must be carried out with due consideration being given to the furnish composition and physical process variables such as temperature, retention time, etc. Figure 5.9 illustrates this optimisation for an equivalent mixture of newsprint and magazines, bleached in a tower with 2% hydrogen peroxide, at 25% consistency, 70°C and a retention time of 60 min. As before, post-deinking bleaching results are presented as brightness gains compared to the initial brightness of the pulp after pulping without chemicals. The hydrogen peroxide consumptions are presented in the graphs, together with

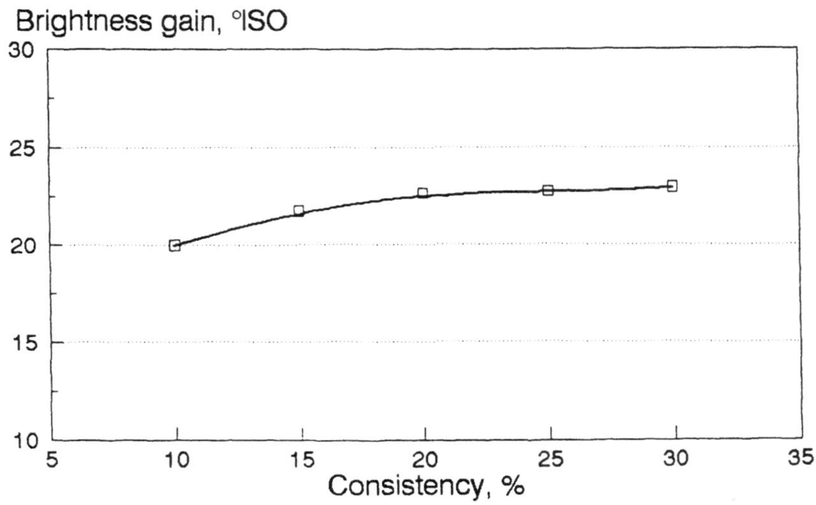

Figure 5.8 Effect of pulp consistency on brightness in tower bleaching: initial brightness = 44.1°ISO; pulping 0% H_2O_2, 0.56% t.a., 2.5% silicate.

the brightness gains. It appears necessary to utilise a dosage of 2.5% silicate (Figure 5.9a) and a total alkalinity of 1.2% on odp (Figure 5.9b) to reach the optimum brightness development after flotation. It is also shown that in order to ensure a high bleaching efficiency, some residual hydrogen peroxide must be found at the end of the bleaching. In a similar way to bleaching in the pulper, this is to protect the pulp from alkaline darkening.

Figure 5.10 shows the optimisations of the total alkalinity level for different hydrogen peroxide dosages. It is again necessary to increase the total alkalinity level when the hydrogen peroxide level is increased.

5.5 Pulper bleaching versus tower bleaching

In Figure 5.11, a comparison is made between the results obtained by post-deinking bleaches and the brightness gains obtained by bleaching the recycled fibres in the pulper, with the latter being the most common application in the paper recycling industry. The highest peroxide efficiencies, or in other words, the highest brightness gains, were observed when hydrogen peroxide was applied in post-deinking bleaching. This is related to the more appropriate conditions for hydrogen peroxide bleaching, namely a higher temperature, a higher consistency and a longer retention time. Nevertheless, these results cannot be generalised to cover all industrial installations, since the physical conditions in the pulper and bleaching tower can be very different from those examined here.

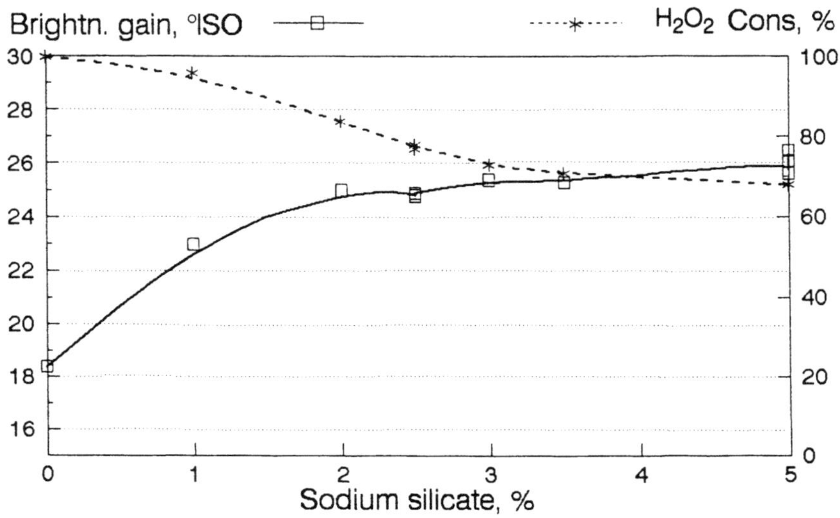

(a)

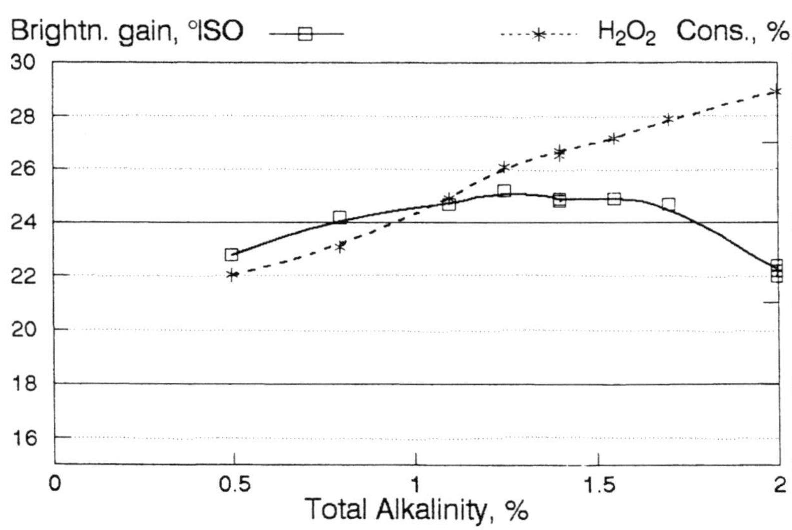

(b)

Figure 5.9 Optimisation in tower bleaching: (a) sodium silicate (2% H_2O_2, 1.4% t.a.); (b) total alkalinity (2% H_2O_2, 2.5% silicate). Initial brightness = 44.1°ISO; pulping 0% H_2O_2, 0.56% t.a., 2.5% silicate.

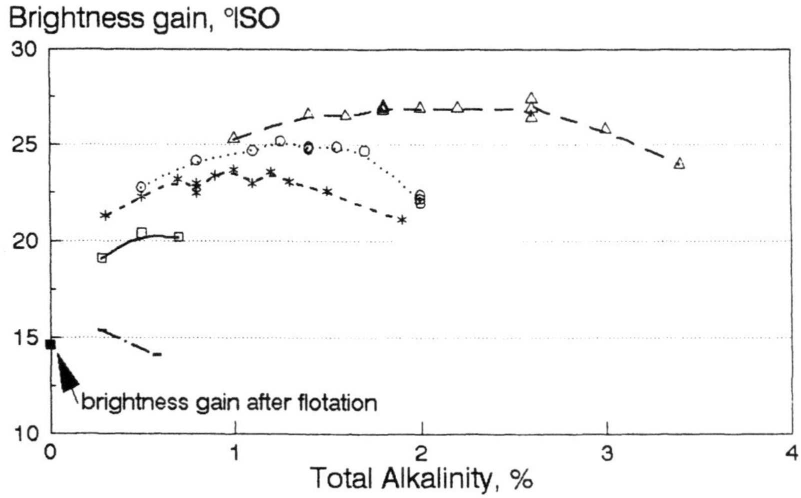

Figure 5.10 Total alkalinity optimisations for different levels of peroxide: initial brightness = 44.1° ISO; pulping 0% H_2O_2, 0.56% t.a., 2.5% silicate. Dashed line (no symbols), 0% H_2O_2; □, 0.5% H_2O_2; *, 1% H_2O_2; ○, 2% H_2O_2; △, 4% H_2O_2.

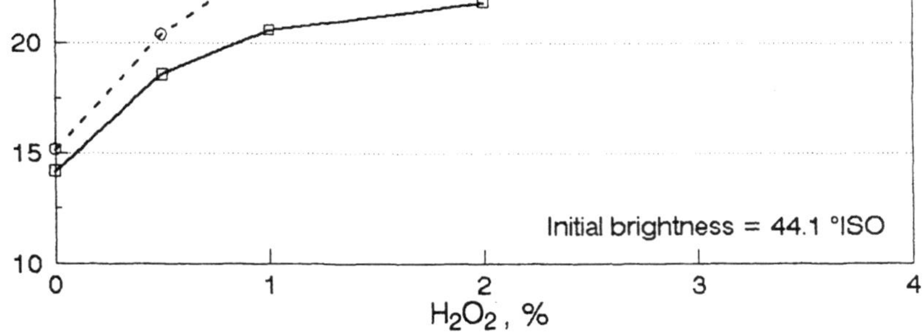

Figure 5.11 Comparison of the effect of peroxide level on brightness for pulper (□) and tower (○) bleaching.

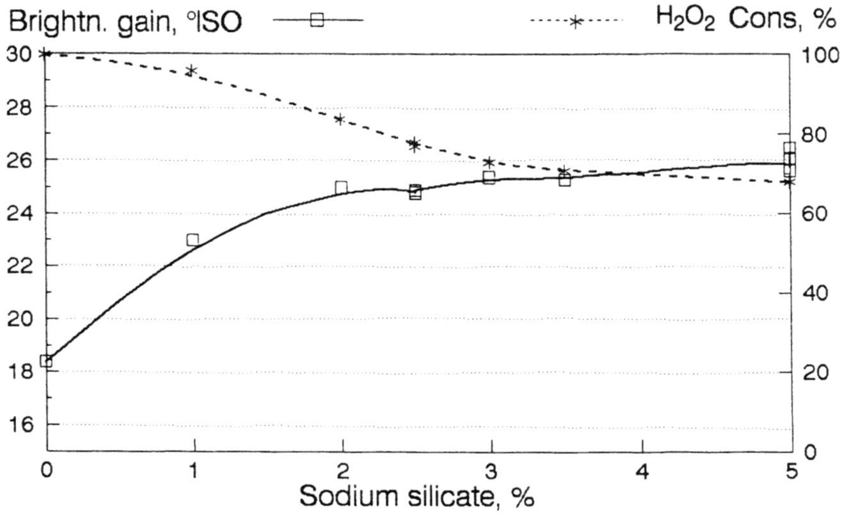

(a)

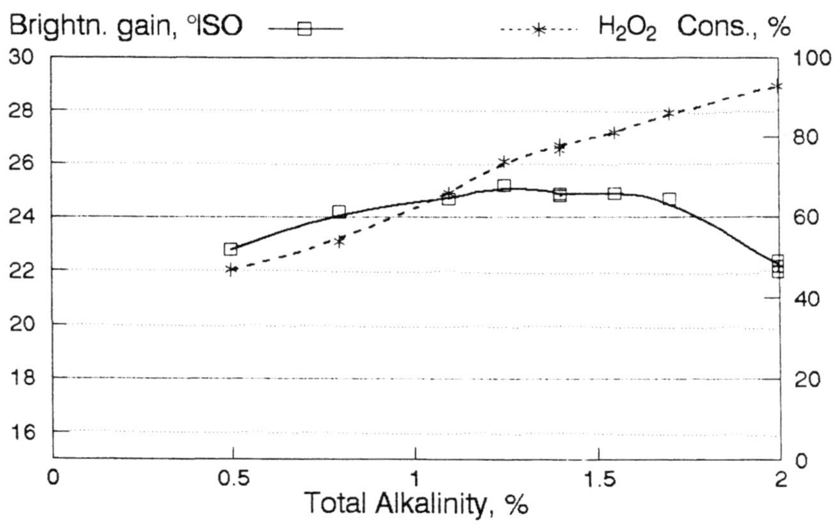

(b)

Figure 5.9 Optimisation in tower bleaching: (a) sodium silicate (2% H_2O_2, 1.4% t.a.); (b) total alkalinity (2% H_2O_2, 2.5% silicate). Initial brightness = 44.1°ISO; pulping 0% H_2O_2, 0.56% t.a., 2.5% silicate.

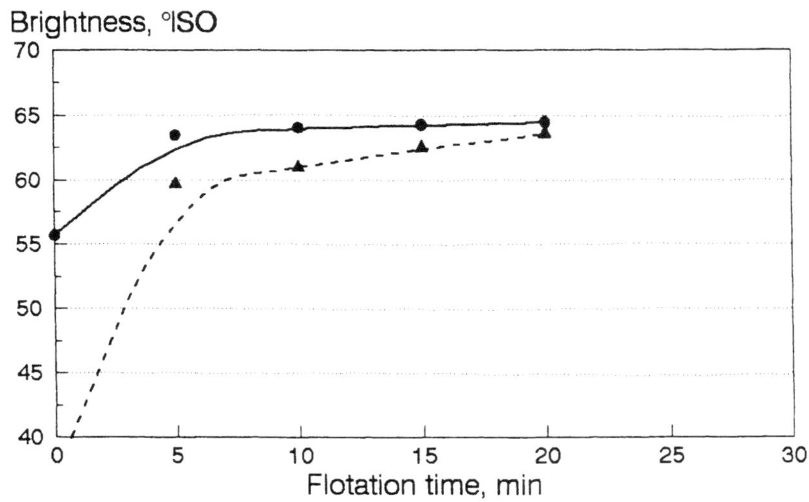

Figure 5.13 Brightness before (●) and after (▲) dispersion as a function of flotation time for the first deinking step (prior to dispersion).

positive effect of this practice on cleanliness is very clear, ink dispersion has a negative effect on the brightness. As a dispersion unit is usually operated at high consistency and high temperature, it is a suitable location for hydrogen peroxide bleaching. This section describes how the brightness loss due to ink dispersion can be compensated for by using the disperser for hydrogen peroxide bleaching, as well as for ink dispersion [7].

5.7.1 Deinking prior to dispersion

The hypothetical deinking line consists of a pulper followed by a first flotation deinking stage. The third step is ink dispersion and the fourth step (optional) is a post-dispersion deinking by flotation.

The efficiency of the first deinking step is an important variable in determining the brightness loss in the disperser. This first deinking efficiency was simulated in the laboratory by varying the flotation time. It is well known that for an industrial deinking line, parameters such as pulp consistency, wastepaper furnish, air-bubble size, etc. will also influence the deinking efficiency. As shown in Figure 5.13, the brightness after a flotation time of 5 min is less than 2°ISO lower than the brightness after 20 min flotation. Nevertheless, the brightness drop observed when these pulps are treated in a dispersion unit (without chemicals) is very different, i.e. 1°ISO brightness loss for the cleanest pulp (20 min flotation) and almost 4°ISO brightness loss for the pulp with a flotation of only 5 min. This leads to a difference of almost 5°ISO between both pulps after the dispersion unit.

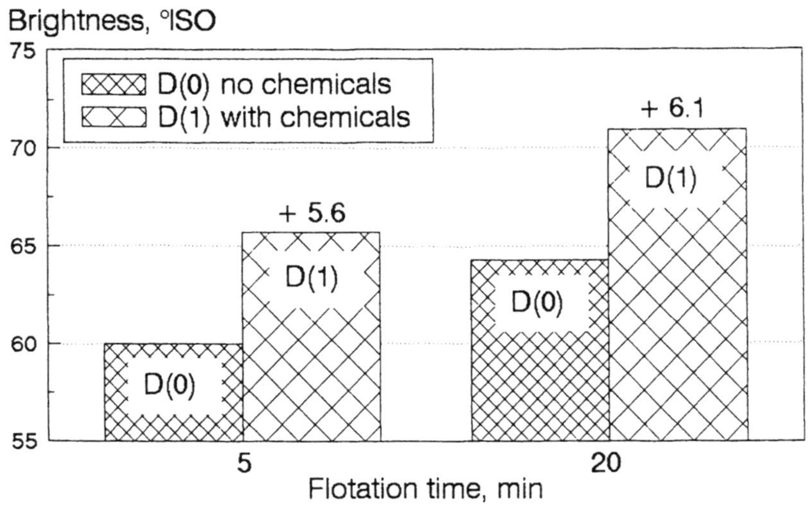

Figure 5.14 Disperser bleaching with 1% H_2O_2, for systems without (D(0)) and with (D(1)) chemicals.

5.7.2 Bleaching results

Although the brightness loss due to ink dispersion depends on the pulp cleanliness prior to the disperser, the brightness gain obtained with hydrogen peroxide bleaching is not related to this brightness loss. This is shown by the results given in Figure 5.14, where both a clean and a less clean pulp were treated in the disperser without (D(0)) and with (D(1)) chemicals. For both pulps, a similar brightness gain was obtained. The disperser trials with chemicals were carried out with 1% hydrogen peroxide and optimised total alkalinity and sodium silicate. The disperser temperature was 70°C with a consistency of 25%; the pulp was kept for 30 min at this temperature and consistency, which was found necessary to fully bleach the pulp.

Addition levels of total alkalinity and sodium silicate for disperser bleaching need to be optimised in order to have the highest bleaching efficiency. It was shown that the required charge of sodium silicate to have a maximum brightness gain is ~3% on odp; the optimum total alkalinity level is 1% odp.

5.7.3 Post-dispersion flotation efficiency

For a disperser installed between two deinking steps, the influence of the chemicals in the disperser on the post-deinking efficiency was studied. The results of laboratory trials are presented in Figure 5.15, with the results of pilot plant trials shown in Figure 5.16; the latter were carried out in the

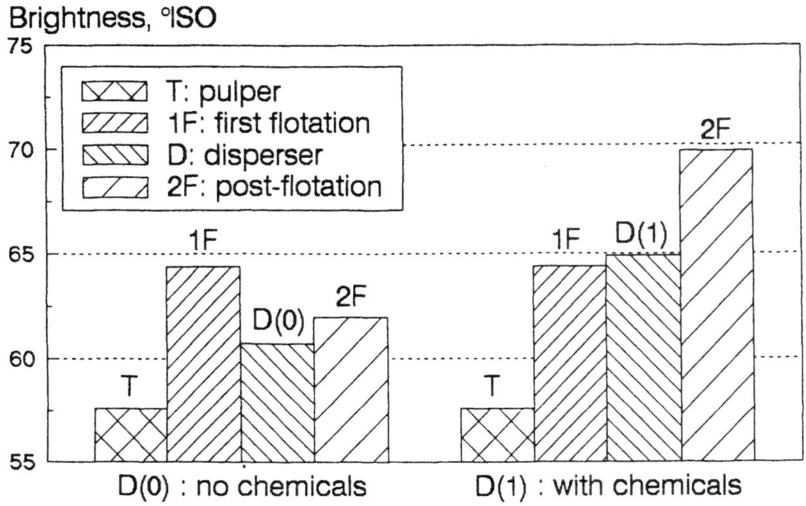

Figure 5.15 Influence of the presence of chemicals in the disperser on post-deinking – results of laboratory trials.

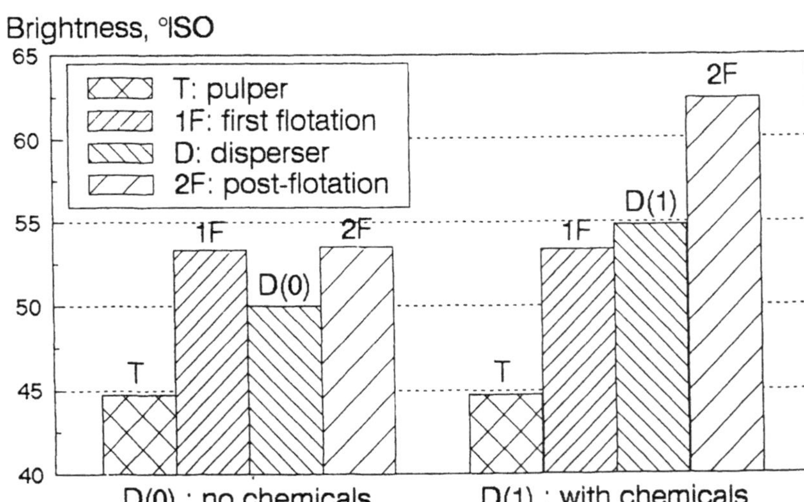

Figure 5.16 Influence of the presence of chemicals in the disperser on post-deinking efficiency – results of pilot plant trials.

LAMORT Research and Development Centre in France [8]. Development of brightness in a deinking line consisting of a pulper, a first flotation, a disperser and a post-dispersion flotation is shown in this figure. The brightness after the disperser and the brightness after the post-dispersion flotation are presented, both for a disperser with, and without, chemicals. It is shown that the use of 1% hydrogen peroxide in the disperser eliminates

Table 5.1 Comparison of brightness gains (°ISO) obtained with disperser bleaching

	No chemicals in disperser	With chemicals in disperser
Disperser outlet	−3	+1
After post-dispersion flotation	+1	+8

the brightness drop of ~3°ISO due to ink dispersion in the disperser and increases the brightness above that following the first flotation stage. A second advantage of the use of chemicals in the disperser is the increased post-flotation efficiency. A moderate brightness gain was obtained by flotation after a disperser without any chemicals, compared to the brightness gain obtained with flotation after the disperser with the optimised peroxide-based chemical formulation. The laboratory post-flotation efficiency was increased from 2°ISO without chemicals to 5°ISO with chemicals. In the pilot plant 7°ISO was obtained with chemicals in the disperser, compared to 3°ISO without chemicals.

Table 5.1 summarises the brightness gains obtained with disperser bleaching. Brightnesses at the disperser outlet and after post-dispersion flotation are compared to the brightness at the disperser inlet.

5.7.4 Disperser energy

The energy of the dispersion system is also an important parameter. Without chemicals in the disperser, the brightness loss increases with increasing disperser energy (Figure 5.17). This can be explained by a simple increase in the dispersion efficiency with higher energies. However, the brightness gain obtained by post-dispersion flotation is similar, whatever the applied energy.

The use of chemicals in the disperser makes it possible to achieve higher brightnesses at the disperser outlet. Figure 5.18 shows an optimum for the brightness at the outlet of the disperser at a specific energy of 45 $\mathrm{kWh\,t^{-1}}$. Between 45 and 70 $\mathrm{kWh\,t^{-1}}$, the brightness at the outlet of the disperser decreases by about 2 points. A post-flotation makes the pulp more uniform: no difference can be found after post-flotation for a pulp dispersed with an energy input between 45 and 70 $\mathrm{kWh\,t^{-1}}$.

5.8 Parameters to control hydrogen peroxide efficiency

Although bleaching with hydrogen peroxide is less efficient for recycled fibres than for virgin fibres, careful control of some process parameters can help to maximise the brightness gain.

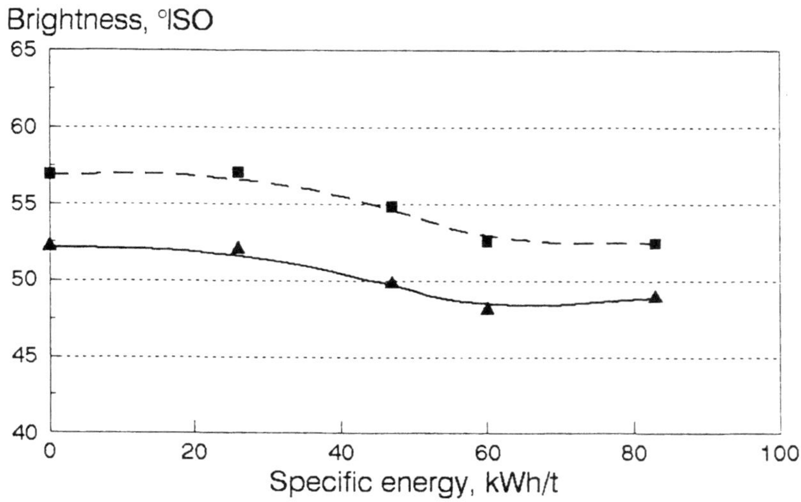

Figure 5.17 Disperser energy optimisation without chemicals in the disperser; 0 kWh t^{-1} = disperser inlet. ▲, disperser outlet; ■, after post-flotation

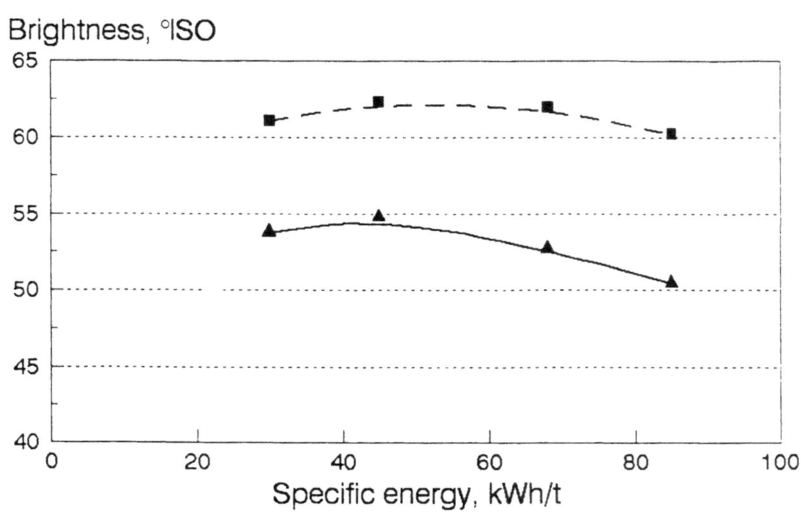

Figure 5.18 Disperser energy optimisation, with 1% H_2O_2, 1% t.a., and 3% silicate in the disperser; brightness at disperser inlet = 52.2°ISO. ▲, disperser outlet; ■, after post-flotation.

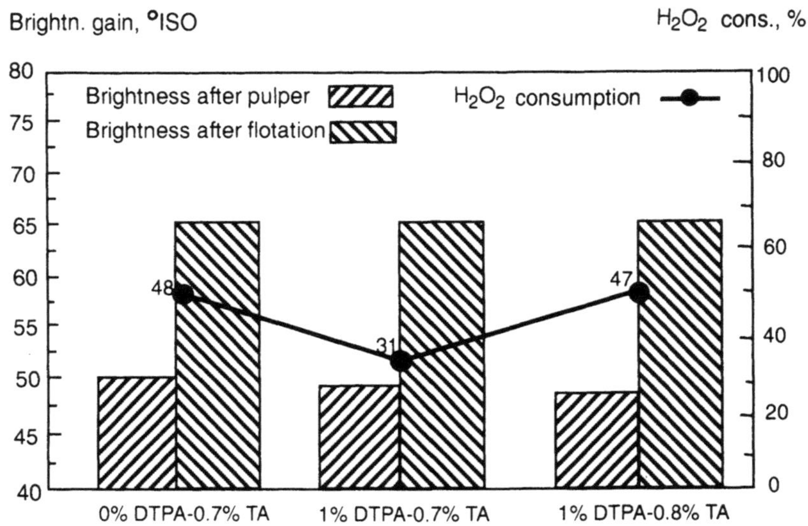

Figure 5.19 Effect of addition of DTPA to the pulper on brightness gain; pulping 0.5% H_2O_2, 2.5% silicate.

The easiest way to detect poor hydrogen peroxide efficiency is to determine, in parallel with the brightness, the hydrogen peroxide consumption. Generally, a total peroxide consumption means a significant loss of the peroxide efficiency. Some of the important parameters causing this peroxide decomposition are reviewed.

5.8.1 Metal ions

Catalytic decomposition of hydrogen peroxide by metal ions is well known, as is the stabilisation of hydrogen peroxide with sodium silicate and/or sequestering agents. Two sequestering agents were tested, namely pentasodium diethylenetriaminopentaacetate (DTPA) and a sodium salt solution of diethylenetriaminopentamethylene phosphonic acid (DTPMP). When hydrogen peroxide is stabilised with 2.5% sodium silicate, the addition of DTPA, in addition to the sodium silicate, does not increase the brightness after flotation, as shown in Figure 5.19 [9].

In post-deinking bleaching, where the temperature, pulp consistency and retention time are more suitable for bleaching the fibres, stabilisation of the hydrogen peroxide must be carefully controlled. Figure 5.9a shows the optimisation of the sodium silicate level for a given peroxide and total alkalinity level. In addition, for post-deinking bleaching, 2.5% sodium silicate was found to be optimal. To increase the brightness by means of a better stabilisation, two sequestering agents were examined, i.e. DTPA and DTPMP. The addition levels are given for the product as received;

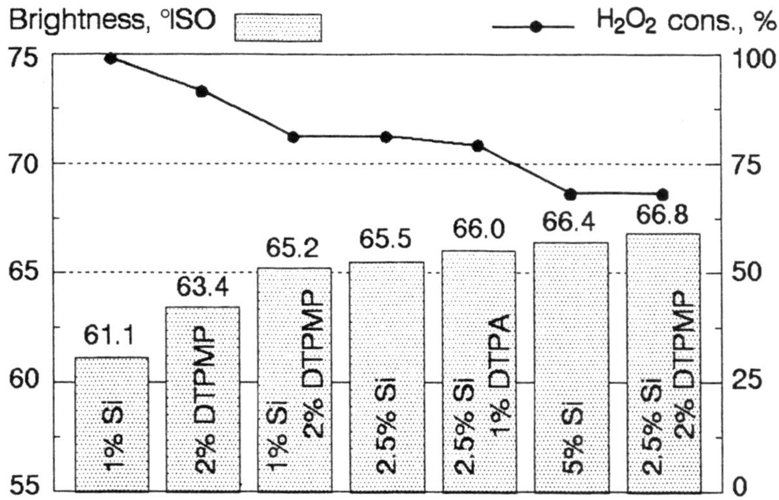

Figure 5.20 Results obtained for post-deinking bleaching trials carried out with sodium silicate only, or in combination with DTPMP or DTPA; tower bleach, 70°C, 60 min, 20% consistency, 1% t.a., 1% H_2O_2.

namely 25 and 40% active material for DTPMP and DTPA, respectively. Figure 5.20 presents the brightness and peroxide consumption for post-deinking bleaching trials, carried out with sodium silicate only, or in combination with these sequestering agents. Post-deinking bleaching conditions were held fixed at 70°C, 20% consistency and 60 min retention time, and 1% hydrogen peroxide and 1% total alkalinity were used. For the pulp examined in the laboratory, a significant brightness gain (1.3°ISO) could be obtained by the use of DTPMP in addition to 2.5% sodium silicate. However, the use of DTPA could improve the final brightness of the pulp by only 0.5°ISO. An increase in the level of sodium silicate to 5% on odp led to a brightness increase of 0.9°ISO. A fairly good result was also obtained with 2% DTPMP and 1% sodium silicate, which was 0.3°ISO below the brightness when 2.5% silicate was used. If the level of DTPMP is optimised (lower than 2%, because of economics), this could be an interesting approach to overcome the scaling problems which result from high silicate levels.

For the formulations where a significant decrease in hydrogen peroxide consumption was observed, some process variables were changed in order to activate the hydrogen peroxide to achieve a higher bleaching efficiency (increase in alkalinity, temperature or retention time). These results are shown in Figure 5.21. Activation of the less decomposed peroxide with an additional amount of caustic soda was not successful in the case of DTPA or DTPMP. In addition, an increase in the bleaching temperature alone,

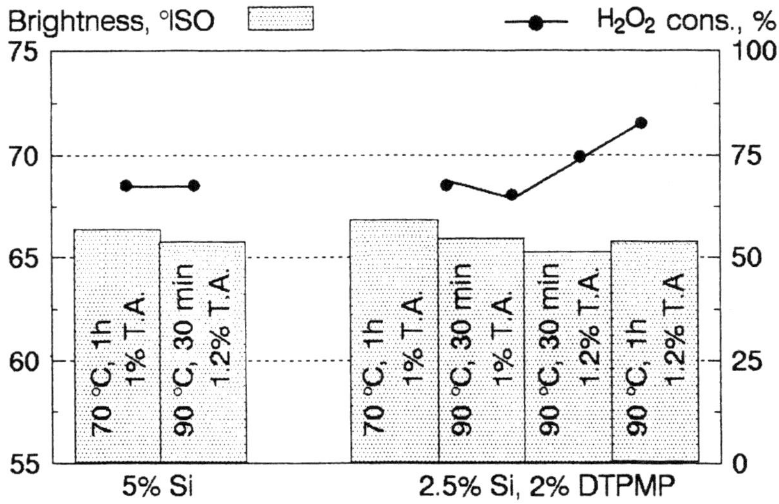

Figure 5.21 Effects of changes in process variables for post-deinking bleaching trials carried out with sodium silicate only, or in combination with DTPMP; tower bleach, 20% consistency, 1% H_2O_2.

or in combination with a higher alkalinity level, could not improve the final brightness.

The stabilisation of hydrogen peroxide with sequestering agents depends on the metal contamination of the pulp and the water circuit in the pulp mill. Therefore, each pulp has to be evaluated separately to obtain the highest hydrogen peroxide efficiency by optimising total alkalinity, sodium silicate level and, in some cases, a sequestering agent. While carrying out this optimisation, both hydrogen peroxide consumption and brightness have to be examined.

5.8.2 *Catalase enzyme*

An uncontrolled decomposition of hydrogen peroxide is initiated by catalase, an enzyme produced by micro-organisms to destroy the hydrogen peroxide formed during metabolism in some aerobic cells. This hydrogen peroxide is decomposed because it has the capacity to destroy cellular structures.

The main known sources of catalase-forming micro-organisms are the wastepapers and the slime in the waste water from the paper machine which is sent back to the deinking plant.

The presence of catalase can easily be detected: the addition of a known concentration of hydrogen peroxide to the filtrate containing the catalase will result in a quick decomposition of the peroxide with the formation of oxygen bubbles, with total decomposition being reached after 15 min.

The destruction of catalase by a thermal treatment (> 70°C) has been described. Although efficient, this treatment is not suitable for all stages in a deinking line. In the pulper, for example, the temperature is limited to 50–55°C, above which stickies start to melt and cause serious problems. An alternative solution is the application of a biocide, such as a peroxygen compound, peracetic acid. The catalase-forming micro-organisms are killed by the application of peracetic acid, and catalase formation is also inhibited. Peracetic acid is injected into the white-water circuit. As with all biocides, peracetic acid has to be applied in 'shock doses', due to its powerful nature and its lack of residual activity. Within 24 h, and with four different shock doses of peracetic acid, the water circuit of a deinking plant can be disinfected. The peroxide stability in the deinking circuits can thus be restored and normal brightness gains are observed.

5.8.3 Residuals of reductive bleaching agents

For several reasons, predominantly colour removal, sodium hydrosulphite is sometimes used in combination with hydrogen peroxide in a two-stage bleaching of deinked pulp. Hydrosulphite can also be used to bleach high-yield pulp to be mixed with deinked pulp for the production of paper containing a proportion of recycled fibres.

Residual amounts of hydrosulphite in the water circuit can cause a loss of hydrogen peroxide efficiency. Moreover, hydrosulphite by-products, such as sodium bisulphite and sodium thiosulphate, have both a peroxide and caustic soda demand. This means that not only must the initial level of hydrogen peroxide be increased to destroy residual hydrosulphite but also that the caustic soda demand can vary in a deinking or bleaching line when applying both reductive and oxidative bleaching chemicals. This has to be considered in the optimisation of both caustic soda and hydrogen peroxide levels.

5.9 Oxygen and ozone bleaching for recycled fibres

Recently several technical reports have been published on the bleaching of recycled fibres with oxygen [10, 11] or ozone [12–14]. These strong oxidative bleaching agents can only be used for recycled fibre bleaching if no, or very little, mechanical pulp is present in the wastepaper furnish.

Oxygen alone can be used to increase the brightness of high-quality office waste. For these furnishes, a higher efficiency was found when peroxide was added together with oxygen [11]. For mixed office waste, containing a considerable amount of mechanical pulp, the bleaching sequence E_{op} P was found to be more efficient than the traditional hypochlorite final bleaching stage, for furnishes with a kappa number greater than 5 [11].

The use of ozone in wastepaper bleaching is fairly recent, compared to other commonly used techniques [12]. In general, recycled fibre bleaching with ozone is limited to high-quality office waste with very little mechanical pulp. A recent report on the ozone bleaching of mixed office waste showed limited results (1–2°ISO) by the use of ozone alone [14]. To obtain high-brightness pulps, a two-stage bleach, i.e. ozone followed by hydrogen peroxide, was suggested.

5.10 Reductive bleaching agents

The most common reductive bleaching agent for wastepaper is, as for mechanical pulp, sodium hydrosulphite. The best bleaching conditions for hydrosulphite are as follows: low pulp consistency (3–5%); pH 6; 1–2 h at 60°C [15]. It was also reported that for high brightness gains the use of hydrogen peroxide, or a combination of hydrogen peroxide and hydrosulphite bleaching, is most efficient. Sodium hydrosulphite has a positive effect on colour destruction in the wastepaper furnishes and is therefore often used in combination with hydrogen peroxide [16].

5.11 Sodium hypochlorite

Sodium hypochlorite is an effective colour stripping agent with woodfree wastepaper furnishes, but is ineffective when more than ~10–15% high-yield pulp fibres are present. Optimal conditions are as follows: pH ~10.0; consistency 10–15%; temperature 40°C; retention time 2 h. It is necessary to have an active residual at the end of the bleach stage to prevent lignin derivatives from colouring the pulp. Sodium hypochlorite is also an effective bacterial control agent.

Concerns regarding the production of chlorinated derivatives during hypochlorite bleaching have led to a reduction in use. Chloroform production is dependent on the addition rate of hypochlorite. In one study in four mills, using a single hypochlorite stage with an addition rate of between 2 and 3% hypochlorite produced an average of 0.34 kg chloroform per tonne of pulp, whereas with addition rates less than 1%, the average from two mills was 0.13 kg per tonne [17]. It is probable that the production of chlorinated derivatives released in wastepaper follows similar trends, with levels of less than 1 kg per tonne found in effluents from secondary fibre mills using low additions of hypochlorite [18].

References

1. Renders, A. (1993) *Tappi J.* **78**(11), 155.
2. Carmichael, D.L. (1990) *Pulp Pap. Can.* **91**(10), T365.
3. Ali, T., McLellan, F., Adiwinata, J., May, M. and Evans, T. Functional and performance characteristics of soluble silicates in deinking. Part I: alkaline deinking of newsprint/ magazines. *CPPA First Recycling Forum, Proceedings*, Vol. 2, 1991.
4. Ali, T., McArthur, D., Stott, D., Fairbank, M. and Whiting, P. (1986) *J. Pulp Pap. Sci.* **12**(6), J166.
5. Burton, J.T. (1986) *J. Pulp Pap. Sci.* **12**(4), J95.
6. Meyrant, P. and Dodson, M. (1989) *TAPPI Pulping Conf.* Book 2, 669.
7. Renders, A. and Hoyos, M. (1993) *TAPPI Pulping Conf.* Book 3, 1017.
8. Serres, A., Colin, P. and Lascar, A. (1992) *Rev. ATIP* **46**(7), 201.
9. Dionne, P.Y. and Renders, A. In mill optimisation of deinking chemicals by laboratory experiment. *PITA Annual Conference 1993*, Bolton, UK.
10. Naddeo, R., Magnotta, V., Kulikowski, T., Ayala, V. and Jezerc, G. (1992) *Pulp Pap.* **66**(11), 71.
11. Heimburger, S.A. and Meng, T.Y. (1992) *Pulp Pap.* **66**(2), 117.
12. Gangolli, J. (1982) *Pap. Technol. Ind.* **26**(5), 152.
13. Kogan, J. and Muguet, M. (1992) *Prog. Pap. Recycling*, 37, November.
14. Forsberg, P., Genco, J.M., Ballanttyne, W., DiNovo, S.T. (1993) *Prog. Pap. Recycling* 53, November.
15. Putz, H.J. and Göttsching, L. *21st Eucepa Conference*, Torremolinos, 1984, p. 297.
16. Hache, M. and Joachimides, T. (1992) *Tappi J.* **75**(7), 187.
17. Miner, R. *et al.* (1993) in *Secondary Fiber Recycling*, ed. Spangenberg, R.J., TAPPI Press.
18. McKinney, R.W.J. Private Communication.

6 The effects of recycling on pulp quality

R.C. HOWARD

6.1 Introduction

Recycling is not a new technology. It might be said to have become a commercial proposition when Mathias Koops established Neckinger Mill, Bermondsey, UK, in 1800. By 1886, recycling was sufficiently important in the USA that a standard textbook on papermaking devoted a section to recycling [1]. In the UK, Strachan published a book in 1918, which was entirely devoted to the recycling of papers [2]. Since that time, paper and board recycling has been widely practised where it made sense in terms of raw material cost and end product quality. The new dimension today, however, is the growing perception of the buying public that recycled fibre is 'more environmentally friendly' and the resulting legislation designed to promote recycling. One example of such legislation is the current requirement in twelve US states that newsprint should contain a specified amount of recycled fibre [3]. Another example is the German 'Green Dot' Program which legislates for the collection and recycling of packaging materials [4]. It is evident that all paper makers must now consider recycled fibre as a potential component of their furnish, and in most cases this needs to be without concession to, or compromise of, end product quality requirements. This situation has given rise to concern over the impact of recycled fibre on pulp, paper and board properties.

Despite the ~200 year history of recycling, there were very few investigations of recycled pulp properties until the late 1960s. Pfaler [5], and Brecht [6, 7] were ahead of their time in this regard. From then on, a considerable amount of work was carried out to identify the effects of recycling on pulp properties, and the causes of these effects. Establishing the cause and effect relationships has naturally led to procedures for ameliorating some of the less desirable results of recycling. A review of the relevant literature can most usefully be summarized under the following headings:

1. The general effect of recycling.
2. Factors affecting recycle potential.
3. Methods of modifying recycled pulp properties.

6.2 The general effect of recycling

Considerable numbers of workers have examined what has been called the fundamental problem in recycling, i.e. how fibres are altered by recycling procedures, and what resulting effects are seen in paper made from those fibres. Methods of investigation have been many and varied. Furnishes range from unbleached chemical pulps to mechanical pulps, including blends. Recycling procedures have used the British Standard handsheet machine, other sheetmaking procedures, pilot machines, or combinations of these. Even the recycling of shredded, dried pulp has been examined. On occasion, the recycled pulp has been beaten before each re-making, in which case the beating might be to a specific freeness, or to a specific paper property such as apparent density or breaking length. Drying of the handsheets has been by the standard method or by a variety of procedures. Fines may or may not have been recirculated during sheetmaking. Despite these considerable differences in experimental detail, some consistent trends emerge.

6.2.1 Refined chemical pulps

For refined chemical pulps, there is general agreement that recycling causes a major reduction in breaking length, burst, and fold with a lesser reduction in apparent density and stretch. Increases in tear, stiffness, scattering coefficient, opacity and air permeability are usually observed. The first recycle causes the greatest change in any property, and this appears to be true regardless of whether the virgin pulp was originally dry or moist. Figure 6.1, based on the work of McKee [8], is representative of the general experience. With extended recycling, i.e. in excess of four recycles, most physical properties have stabilized, although there is some evidence that tear, in particular, may go through a maximum before dropping back a little [9–12]. Stock freeness generally shows a slight increase, provided that the pulp has not been re-beaten to restore physical properties, in which case a considerable decrease in freeness is observed.

Some properties have been examined only rarely. In the case of brightness, one group of workers found a small increase (~5%) for both bleached and unbleached softwood and hardwood kraft pulps [11]. Elastic modulus was measured in three different investigations, and found to fall [8, 10, 13]. Surface properties such as roughness, pick, etc. have also been rarely examined. Van Wyk and Gerischer noted a fall in sheet surface pick strength (as measured by the IGT test instrument) [9], and McKee observed a fall in *z*-direction strength [8].

The main thrust of the research work in the 1960s and the 1970s was to understand the cause of the observed general trends for chemical pulps. It has been shown that the principal cause of the changed sheet properties is

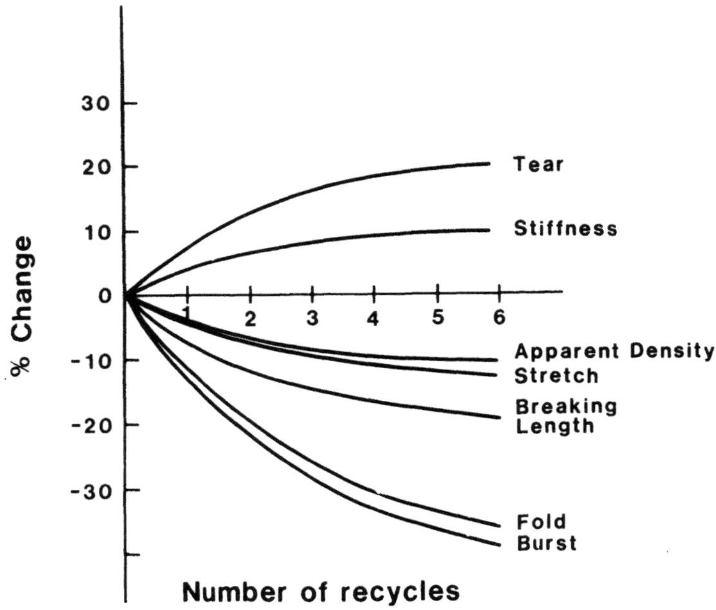

Figure 6.1 The general effect of recycling on refined chemical pulps (data for unbleached softwood kraft taken from ref. 8).

the reduced bonding ability of the fibres. Cildir and Howarth [14] showed that it was possible to quantify that reduction by measuring the wide span and zero span breaking length of handsheets during the recycling of a beaten, bleached sulphite pulp, and computing the bonding contribution to sheet strength at each recycling step by using a simplified form of Page's tensile strength equation [15], as follows:

$$\frac{1}{T} = \frac{1}{Z} + \frac{1}{B} \tag{6.1}$$

where T is the sheet breaking length at a 10 cm span, Z is the sheet breaking length at zero-span, and represents the fibre strength, and B is the bonding contribution.

This simplification assumes no change in the fibre morphology during recycling.

McKee [8] suggested that at least a partial cause of the loss of bonding potential could be due to loss of fibre wall swelling. Using the water retention value (WRV) as a measure of internal fibre swelling, McKee observed a reduction in the WRV during his experiments, and in particular noted that the most rapid decrease occurred in the first two cycles. Lundberg and de Ruvo, recycling a 50/50 bleached birch kraft stone groundwood (SGW) mixture, found a direct relationship between WRV and tensile strength [16],

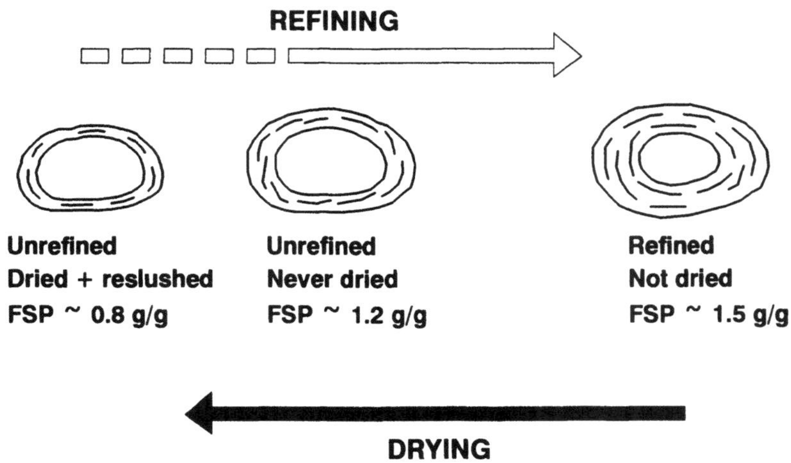

Figure 6.2 Diagramatic representation of fibre wall swelling during refining and drying; the degree of swelling in each case is expressed in grams of water per gram dry fibre.

and Fellers and coworkers commented that recycled fibres beaten to the same degree of swelling as the virgin fibres produced paper of the same strength [17].

Szwarcsztajn and Przybysz [18, 19] confirmed the importance of swelling and its relationship to recycle potential, but they further pointed out that in the case of the unbleached kraft which they recycled, both the fibre fraction and the fines fraction showed a reduced WRV.

An alternative, and perhaps more direct, method of measuring the swelling of the fibre wall is to determine the fibre saturation point (FSP) using the procedures developed by Stone and Scallan (20). Howard and Bichard used this approach during a laboratory recycling experiment, and found that for refined chemical pulps, the FSP fell markedly during recycling, with the greatest change occurring, as for most other properties, at the first cycle [21].

The mechanism by which fibre wall swelling affects bonding potential is conceptually simple (Figure 6.2). In the swollen state, the wet fibre wall is delaminated, thus making the fibre surface conformable and promoting intimate fibre–fibre contact during the processes of sheet formation and consolidation. On drying, extensive interfibre bonding takes place. However, during drying (even air-drying), the internal surfaces of the delaminated fibre walls themselves are brought together by surface tension forces, as well as applied external forces, and the adjacent wall surfaces bond internally. It may be, as Lundberg and de Ruvo and their coworkers have suggested [22, 23], that reorientation of microfibrils and better alignment of cellulose chains take place, with the whole resulting in a more intensely bonded wall structure. Scallan and Tigerström have recently shown that this *intra*-fibre hydrogen bonding stiffens the fibre wall to the extent of doubling the elastic modulus of the wall from 2 to 4 MPa [24]. It may also correspond to the observations of Burgajer [25] and Yamagishi and Oye [11] that multiple recycling leads to an increase in cellulose crystallinity. Overall, the phenomenon is frequently termed 'hornification' or 'irreversible hornification' after Jayme, who coined the term to describe the effect of drying chemical pulps [26]. This subject has recently been fully reviewed by Laivins and Scallan [27]. Mere reslushing is insufficient to fully reverse the effect, and therefore a swollen chemical pulp (such as a refined pulp or a 'never-dried' pulp) will always be weaker after drying or recycling.

The hypothesis that the loss of bonding ability of recycled fibres is entirely due to loss of swelling has been challenged by other workers. Essentially, they suggest that the phenomenon may be the result of two effects: changes occurring principally to the surface of the fibre, and changes occurring principally to the bulk of the fibre. Eastwood and Clarke [28], for example, when recycling once-dried, semi-bleached kraft on a pilot paper machine and with rosin/alum sizing, noted that the relationship between the apparent density and the breaking length at each level of recycling was not linear, as might have been expected if swelling controlled the bonding by virtue of its effect on flexibility. Pycraft and Howarth [29, 30] measured two reaction constants for the reaction between cellulysin and pulp, and found a very strong correlation between the recycle potential of the pulp and both of these constants. They suggested that one constant relates to the rate of degradation of the bulk of the fibre, whilst the other represents the accessibility of the fibre surface to enzyme attack, i.e. the extent to which the fibre surface is 'open' or 'sealed'. Support for this line of reasoning may be found in the work of Ogiwara and Arai [31] and Oltus *et al.* [32]. Structural changes occurring specifically to the surface of fibres during recycling have been demonstrated by using transmission electron microscopy (TEM) cross-sections (Okayama *et al.* [33]) and by scanning electron microscopy (SEM) examination (Sachs [34, 35]), but while these gross morphological changes are dramatic, it remains unresolved as to how important they are.

Therefore, at the present time, it is believed that the main cause of strength loss in recycled chemical pulp is the loss of fibre wall swelling. Although minor differences in recycling response have been observed between different chemical pulp types, such as unbleached and bleached sulphite and unbleached kraft [36] or kraft pulps of different yield [37], this overall principle is believed to apply for all refined chemical pulps.

6.2.2 *Mechanical pulps*

Few evaluations have been made of the recycle potential of mechanical pulps, and until recently, those evaluations invariably examined stone

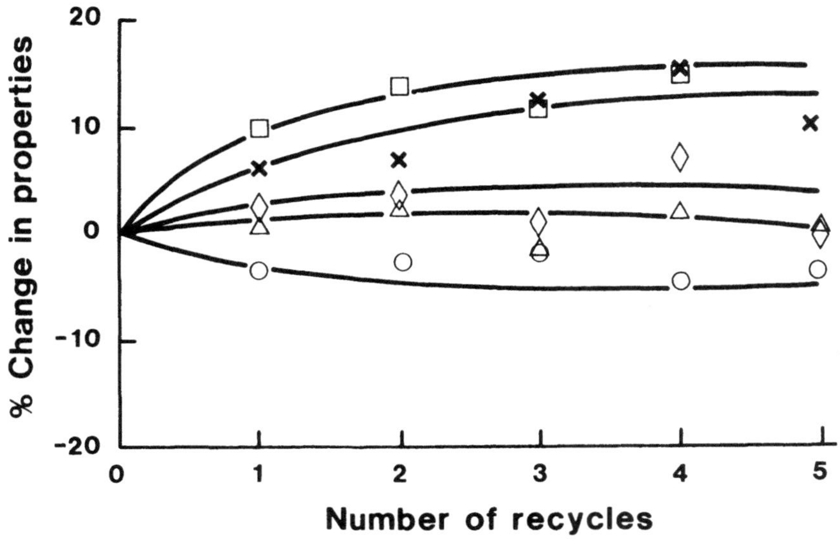

Figure 6.3 The general effect of recycling for mechanical pulps: (O) scattering coefficient; (△) tear; (◇) burst; (X) breaking length; (□) density (data for stone groundwood taken from ref. 21).

groundwood, and compared its performance with chemical pulps. Usually, the comment was made that the effects of recycling stone groundwood were essentially the same as for chemical pulps, but less evident [8, 36–38]. Recent work, however, has shown that, in fact, mechanical pulps recycle in an entirely different way from chemical pulps [21, 39]. Whereas refined chemical pulps lose density and tensile strength, mechanical pulps show small gains in strength and density (Figure 6.3). There are two reasons for this different behaviour. The first is that, unlike chemical pulps, the walls of mechanical pulp fibres are not extensively delaminated in the wet state. Hornification during drying is, therefore, limited and will have little, if any, effect on inter-fibre bonding. The reason for the *increase* in strength and density is believed to be because of progressive flattening and flexibilizing of the stiff, uncollapsed fibres during each successive papermaking and reslushing cycle. The flatter, more flexible fibres bond better and give a thinner, denser sheet.

It might be thought that fines would have a role in property development during mechanical pulp recycling, but there is no evidence of that in the literature. In fact, in the data of Howard and Bichard, while a slight increase in fines is evident, the scattering coefficient fell during recycling, whereas a rise would be expected if an increase in mechanical pulp fines was responsible for the strength/density increase [21].

In discussions of the general effect of recycling, fibre fragility and loss of fibre strength are frequently mentioned as causes of reduced recycled pulp strength, perhaps because it seems intuitively obvious that 'used' fibres should be weaker. However, the scientific evidence does not support this assumption. Strength losses have been observed by some workers [8, 10, 14], while others have observed no change [9, 21], or even a gain [40]. Howard and Bichard [21] have reviewed these diverse results in the light of their own data, which were obtained from laboratory recycling experiments on eleven pulps. They concluded that there was no fundamental reason why a drying and reslushing cycle should change the fibre strength. Where such an effect has been observed in recycling experiments, it may have resulted from some incidental aspect of the paper making–recycling operation, such as fines loss, or fibre damage due to calendering. Alternatively, it could be an artefact of the fibre strength test, arising from a change in inter-fibre bonding or fibre curliness.

6.3 Factors affecting the recycle potential

Essentially, recycled fibres are contaminated, used fibres. Recycled pulp quality is, therefore, directly affected by the 'history' of the fibres, i.e. by the origins, processes and treatments which these fibres have experienced. This history can be split into five periods:

1. The fibre furnish and pulping history.
2. The paper making process history.
3. The printing and converting history.
4. The consumer and collection history.
5. The recycling process history.

Within each of these periods, certain factors affecting recycled pulp quality have been identified in the literature, and possible explanations have been advanced for their influence. These will now be reviewed.

6.3.1 The influence of fibre furnish and pulping history

Of all of the aspects of recycled fibre history, this one is the most important because, as described in the previous section, different pulp types respond in different ways to recycling. The differences between refined chemical pulps, and mechanical pulps such as stone groundwood and thermomechanical pulp, have been discussed. But what about other high-yield pulps, unrefined chemical pulps and blends of pulps? And what of hardwoods and non-wood pulps? Although there is little in the literature, the general principles already outlined can guide our expectations.

Unrefined chemical pulps do not necessarily behave in the same way as refined chemical pulps, with the actual behaviour depending on whether or

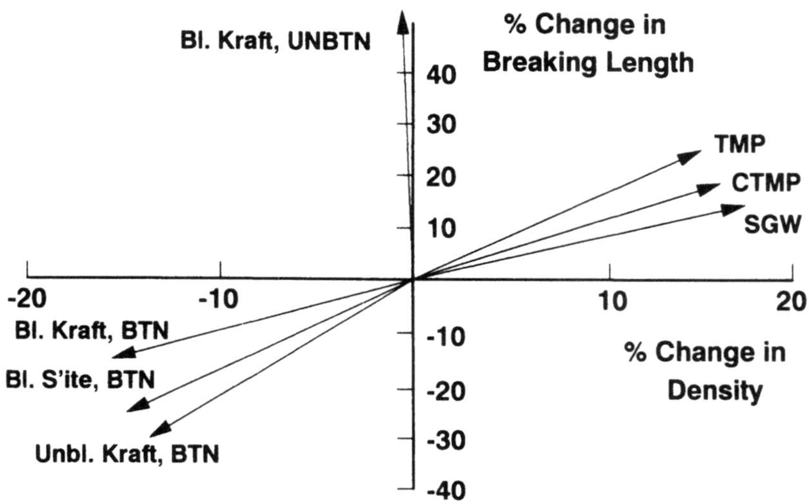

Figure 6.4 The recycle potential of different pulp types; arrows point in the direction of increasing numbers of cycles [21].

not the pulps have been dried (Figure 6.2). Never-dried chemical pulps are quite swollen even in the unrefined state and will make a strong virgin sheet. However, these pulps will lose bonding potential due to hornification during drying and, therefore, after repulping they will form a weaker, bulkier sheet. Dry pulps, on the other hand, have already been hornified by the pulp drying process. Further drying during virgin paper-making does not reduce to any great extent the already low level of swelling, and the recycled pulp would therefore not be expected to lose much bonding potential. However, the fibres of dry pulps (and especially bleached pulps) are frequently curly, and such curl reduces the pulp strength potential [41,42]. In laboratory experiments, it has been found that the curl is partially removed by recycling and the de-curled fibres then form a stronger sheet [21]. If this applies in a mill environment, it is possible that some unrefined, dry pulps may become stronger after recycling.

Chemithermomechanical pulp (once-dried) has been found to behave like stone groundwood and thermo mechanical pulp (TMP), but with the detailed difference that almost all of the pulp property development occurred at the first cycle, whereas for the latter two pulps, the pulp properties changed progressively over several cycles [21]. This more rapid response is attributed to a reduction in fibre wall rigidity arising from the chemical pretreatment.

The recycling performance of the pulps discussed so far, can be summarized in terms of their tensile strength/density relationships. This is shown in Figure 6.4. Given the very different recycling responses, the question naturally arises as to whether a furnish exists which would show a zero property change during recycling. Evidence from Chatterjee *et al.* [39] suggests that such a situation might exist with semi-chemical pulps or never-dried, high-yield pulps. In that case, hornification of the partially delignified fibre wall could be offset by increased fibre flattening and flexibility. The response of a 50/50 blend of TMP and refined kraft has also been investigated [21]. Recycling of this furnish did not give zero change because the rates of response of the two pulp components are different. Hornification of the kraft occurred mainly at the first recycle, and the resulting strength loss was not immediately offset by improved bonding of the mechanical pulp component.

Hardwood pulps have rarely been used in recycling research, and only a very limited amount of data exists to show their behaviour. What is available suggests that hardwoods will behave in exactly the same manner as softwoods [11, 43]. Indeed, there is no reason to suppose otherwise.

The same principles apply to *non-wood pulps*, as demonstrated by Mansito *et al.* [44]. They showed that a bleached bagasse soda pulp lost bonding capacity due to swelling reductions. Hornification was less pronounced in an unbleached bagasse chemithermomechanical pulp (CMP), a fact which the authors attributed to the lignin content.

6.3.2 The influence of paper making history

Paper making process history may be defined as the sequence of events which occurs from the pulp storage chests to the paper machine reel. The paper making history therefore includes stock preparation (including chemical treatments), pressing, drying, calendering and after-treatments such as coating.

6.3.2.1 Effect of virgin pulp refining. There is general agreement that the greater the initial degree of chemical pulp refining, then the greater the change of pulp quality on recycling. This has been demonstrated, for example, by Higgins and McKenzie [45], McKee [8] and Lundberg and de Ruvo [22] (Figure 6.5). The current understanding of this loss of recycle potential is that it corresponds to the loss of fibre wall swelling described earlier. Refined pulp is internally delaminated to a greater extent than unrefined pulp, and these delaminations close up during the drying process. The greater the extent of initial swelling, then the greater the change in pulp properties after drying. Furthermore, Stone and Scallan have shown that not only do the walls not fully reswell when reslushed, but it may be impossible to regain the initial high levels of swelling, even after re-refining [46].

Stock refining always creates fines, and unless these fines are lost from the system during paper making or subsequent recycling, they will be

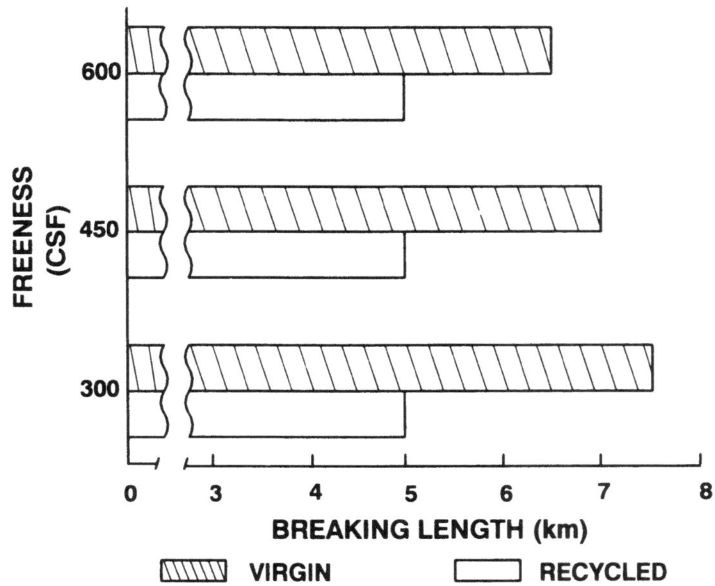

Figure 6.5 Breaking length, before and after recycling, as a function of the initial degree of refining: (□) virgin pulp; (□) recycled pulp (data for bleached softwood kraft taken from ref. 8).

present when the recycled fibres are re-used. Many workers have noted that recycled pulp freenesses tend to be lower than for the equivalent virgin pulp and they have attributed this to fines [36, 47]. Szwarcsztajn and Przybysz have investigated the issue further. They suggested that recycled fines serve only to fill the sheet, and play no part in improving the bonding [48]. However, Hawes and Doshi disagreed to some extent. They found that recycled fines could make a modest contribution to the recycled pulp strength, but that this contribution was much less than could be obtained from fines created by subsequent refining [49].

Stock preparation inevitably reduces the fibre length as well as creating fines. A recycled pulp will therefore always have a shorter average fibre length than the original virgin pulp, as well as a higher fines content, unless the shorter material is lost during recycling.

Taken overall, a recycled refined chemical pulp will show a marked loss of swelling, a reduction in fibre length and an increase in fines, all of which will contribute to its reduced recycle potential.

6.3.2.2 The effect of wet pressing. Little has been published concerning the possible effect of the degree of wet pressing on recycle potential. However, Carlsson and Lindström [50] conducted a detailed study with a

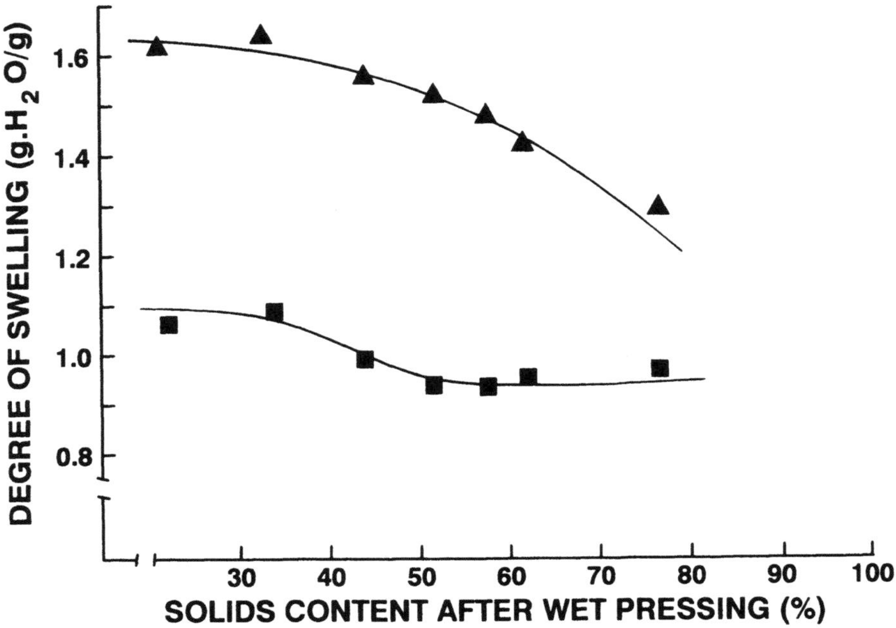

Figure 6.6 The effect of handsheet wet pressing pressure on the degree of swelling after reslushing of handsheets made from a refined, bleached kraft pulp: (▲) reslushed without drying; (■) dried at 105°C and reslushed (adapted from ref. 50).

variety of pulps. They found that for chemical pulps at moderate levels of refining, heavier pressing caused a loss of swelling of both the reslushed, never-dried paper (i.e. equivalent to wet press broke) and the reslushed paper after drying. The greater the initial level of refining, then the more marked was the effect, but in most cases the effect only became significant at ex-press solids contents exceeding ~35% (Figure 6.6).

These results were confirmed by Pycraft and Howarth [30]. They used an enzyme degradation technique to examine the quality of recycled fibre obtained from pilot machine-made paper. It was found that the recycle potential of the more highly wet pressed paper was lower than that of the less highly pressed material, but they also observed that the difference manifested itself only after drying; samples taken after the presses showed no detectable difference in recycle potential, in contrast to the results of Carlsson and Lindström [50]. However, the solids content after the presses would not have been more than 35% in their case, which was too low to show a significant effect in the undried sheet.

Loss of recycle potential in chemical pulps as a result of pressing, whether it can be detected before drying in the case of wet pressing to a high solids content, or after drying for a moister sheet, is due to the same process of intra-fibre bonding of the swollen and delaminated fibre wall as has been proposed to explain irreversible hornification. Pressing assists bringing together delaminations in the fibre wall, thus resulting in a more hornified fibre after drying. The point at which hornification takes place depends on the initial degree of swelling. De Ruvo and Htun [47] have shown that for a more swollen fibre, hornification begins at a higher moisture content. In terms of everyday paper making, the effect of pressing, although it will affect the recyclability of dry paper, is unlikely to affect the recyclability of wet broke until high ex-press solids contents are reached.

Two mechanical pulps were included in Carlsson and Lindström's experiments. The WRVs of both pulps fell by less than 5% even when pressed to solids contents as high as 73%. These results are not surprising. The fibre walls of mechanical pulps, even if mildly refined at low consistency, do not delaminate extensively, and pressing would therefore not be expected to cause a significant reduction in swelling, or, consequently, in recycle potential. Indeed, as shown in laboratory experiments [21, 39], pressing might be expected to flatten mechanical pulp fibres and make them more flexible, thus increasing the recycled pulp strength rather than reducing it.

6.3.2.3 The effect of drying. Many experiments have been conducted to examine the effect of drying conditions on recycle potential, since this is relevant not only to the recycling of paper, but also to the use of once-dried market pulps, which in some respects can be considered as being recycled when used. Jayme [26], for example, investigated the effect of pulp drying and showed that for an unbleached kraft pulp (rather than paper) a drop in the WRV can be observed after either air-drying or drying at 70°C. Treiber and Abrahamson [51] noted a considerable drop in fibre saturation point when drying a dissolving pulp at 120°C compared to 25°C. Okayama *et al.* [52] dried both shredded pulp and standard handsheets using a variety of techniques, including freeze drying and microwave drying, and found that each technique produced its own distinct effect on the WRV of the recycled pulp.

The effect of different methods of drying paper has also been examined. Pycraft and Howarth [30] demonstrated that the drying cylinder temperatures on a pilot paper machine affected the recycle potential. Lundberg and de Ruvo, and their coworkers [22, 23] noted that increased drying temperatures during the drying of Formette-Dynamique prepared sheets gave rise to reduced swelling, and restrained drying reduced the swelling a little more. They used a commercial bleached kraft pulp (Table 6.1). Lundberg noted that pulp prepared from paper dried at the higher temperature could not regain the WRV of the initial pulp, even after prolonged beating.

In all of the above cases, chemical pulps were used, and in all cases hornification of the fibre wall accounts for the basic phenomenon. The

Table 6.1 The effect of different drying regimes on the degree of swelling after reslushing of a refined bleached kraft pulp (data taken from ref. 22)

Drying conditions	Degree of swelling ($g\ H_2O\ g^{-1}$)
Never-dried	1.90
Dried unrestrained at 20°C and reslushed	1.50
Dried unrestrained at 120°C and reslushed	1.42
Restraint-dried at 120°C and reslushed	1.38

question of the importance of the method of drying has recently been reviewed by Laivins and Scallan [27]. They have concluded that the method is of secondary importance, as it merely increases or decreases the extent of hornification at the drying step.

Drying procedures have been shown to affect more than just bonding. It has been observed that in comparison to restrained sheets, freely dried sheets (or pulp), when remade into paper, exhibit increased stretch at break [22, 52]. No explanation has been advanced, but it seems certain that this is the result of the shrinkage putting microcompressions into the fibres, as originally described by Page and Tydeman [53]. An important ramification of this observation is that where recycling assessments have been made using standard handsheets, the observed sheet properties will be different from what might be expected from recycling machine-made paper where the sheet has been allowed to shrink in the cross direction.

Once again, it must be remembered that mechanical pulps will behave differently from chemical pulps, because the wet fibre wall is not swollen. Normal drying procedures will not reduce the recycle potential.

6.3.2.4 The effect of calendering. The effect of calendering on the recycle potential of both groundwood-containing and woodfree paper has been examined by Göttsching and Stürmer [54], and Gratton [55]. They showed that the heavier the calendering, then the greater the loss in the breaking length of handsheets made from the calendered paper. Tear strength also suffered, and the drainage rate of the stock worsened. These effects corresponded to both a loss in WRV (in the case of woodfree paper) and a reduction in fibre length and strength. Steel-on-steel nips were compared with steel-on-paper nips (as in a supercalender) and with temperature-gradient calendering; the hardest nip was found to cause the greatest loss in recycle potential (e.g. see Table 6.2). The cause of these phenomena is attributed simply to the mechanical forces in the nip compressing and damaging the fibres. Therefore, the modern trend away from hard nips is highly desirable as the level of recycling increases.

6.3.2.5 The effect of chemical additives. Commercial paper making almost invariably involves the use of chemical additives of one sort or

Table 6.2 The properties of handsheets made after reslushing virgin newsprint which had been calendered in different ways [55]

	Reslushed pulp properties		
	Tensile index ($N\,m\,g^{-1}$)	Tear index ($mN\,m^2\,g^{-1}$)	Average fibre length (mm)
Uncalendered	36	6.8	1.51
Temperature-gradient calendered	34	5.4	1.42
Conventionally calendered	29	4.7	1.31

another. Horn [10], Eastwood and Clarke [56], and Guest and Voss [57] found that rosin/alum sizing in the original paper caused a further loss of recycle potential which was over and above what would have been expected anyway. No mechanism has been demonstrated, although Guest and Voss suggested that the sized fibres retain their hydrophobic surfaces and thus to some extent inhibit bonding in the recycled sheet [57]. Forester [58], in a limited experiment, found that neutral sized paper (using an alkyl ketene dimer size) retained more of its original burst strength, but less of its breaking length. Guest and Voss, on the other hand found the reverse [57], and also found that the recycle potential of alkyl ketene dinner (AKD)-sized paper was considerably worse than for rosin/alum sized paper. No explanation for either set of results was offered, and no consideration was given to the influence of the ionic state of the pulp before it was made into paper for the first time.

This latter point could be critical. Lindström and Carlsson [13] have demonstrated the extent to which the chemical environment affects the fibre swelling of virgin chemical pulp, and also the strength of paper made from it. They also discussed how the chemical environment during the first sheet-making controlled the swelling of the pulp after drying and recycling. Unbleached pulps dried under acid conditions in the first making showed considerably less swelling (and consequently strength) than pulps initially dried under alkaline conditions (Figure 6.7). On the basis of their very carefully controlled conditions, the authors were able to identify the importance of the acid group content of the pulp (and its ionic form) to the initial swelling. Bleached pulps are unaffected by initial pH because of their low acid group content. All of this work was carried out on chemical pulps, while mechanical pulps still remain to be studied.

Few other systematic studies have been reported, whereas one might have expected to find work identifying the effect of calcium carbonate loadings in the original paper compared to clays, the effect of starch in and on the 'to-be-recycled' paper, and the effect of coating. Those studies which have been carried out have tended to concentrate on adverse papermaking reactions involving incompatibility, e.g. chalk-loaded paper, both recycled and in an acid paper making system.

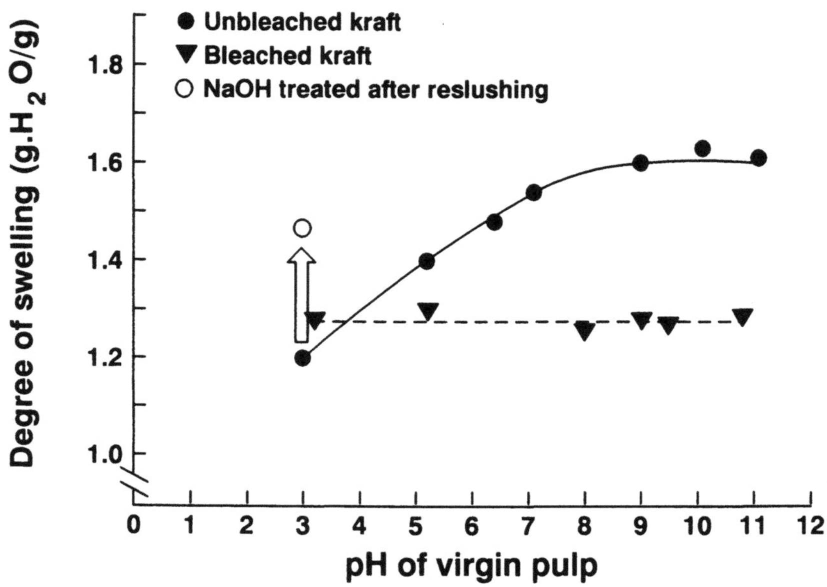

Figure 6.7. The impact of stock pH during paper making on the swelling behaviour after reslushing, of a bleached (▼) and an unbleached (●) kraft pulp; the arrow indicates the increase in swelling caused by NaOH treatment (O), after reslushing unbleached paper initially made under acid conditions [13].

6.3.3 The influence of converting history

The two main converting operations which affect recycled pulp properties are casemaking and printing. Little attention has been paid to the former, but important work by Inoue *et al.* has shown that hydrophobic materials applied during casemaking can be retained in the pulp suspension after repulping, and deposit on the surface of the fibres, thus lowering the surface free energy of the recycled product [59]. It seems reasonable to suppose that these same materials will inhibit fibre bonding in a similar manner to that suggested for sizes.

Printing operations affect primarily the optical properties of recycled pulp. It is now clear that it is not just the amount of residual ink after the deinking stage which controls the recycled pulp brightness, but also the particle size distribution. Ink particles smaller than ~50 μm, cannot be resolved individually by the human eye, but their presence lowers the brightness and lends a grey tone to the pulp. Above 50 μm, ink particles appear as specks. Recent developments in measuring equipment allow both the particle size and the total amount of ink to be determined [60, 61], and provide the recycler with the tools necessary to produce brighter, cleaner pulp.

6.3.4 The influence of the consumer and collector

Today, there is much emphasis on the use of 'post-consumer' waste, rather than 'post-commercial' material, but there has been no systematic study reported of the additional effects that the consumer might have on recycle potential. Apart from adding further contamination to an already contaminated fibre, a very significant post-consumer effect could be ageing. For example, Andrews has shown that the longer newsprint is stored prior to recycling, then the lower the recycled pulp strength [62]. The cause of this potentially serious problem has not yet been explained.

The consumer and collector also influence the quality of recycled pulp in another important way, and that is by mixing together totally different grades of paper such as those containing mechanical pulp with woodfree grades. The quality of the recycled pulp then falls somewhere between the quality of the best fibre in the wastepaper, and the worst. The important implication here is that over and above any loss of quality arising *per se* from the reuse of a particular paper grade, there will be a further change due to 'dilution' by other paper grades. Defoe has shown this to be an important consideration in the quality of recycled linerboard [63].

6.3.5 The influence of recycling processes

It will be apparent that the quality of recycled pulp is heavily influenced by events that take place *before* the recycling process. This is an important point because one frequently reads statements such as '. . . recycling produces a deterioration in the strength of the paper produced from the fibres . . .', implying that the *process* of recycling reduces pulp quality. In fact, while the recycling process itself includes a variety of unit operations, each of which may involve mechanical, chemical or thermal treatment, no systematic study of the effects of the processes on fibre properties has been reported. However, some general comments and predictions can be made. These effects will be reviewed under three headings, namely consistency, fines removal, and chemical treatment.

6.3.5.1 Consistency effects. From extensive work with virgin pulps, it is now well known that mechanical action at consistencies above, e.g. 8%, will impart significant curl and microcompression into fibres. This effect increases with consistency, resulting in a weaker, bulkier, more stretchy pulp [64]. In recycling operations, high-consistency mechanical action can occur at the pulper (pulping consistencies are often between 10 and 15%, and may be as high as 35%), in dispersion units (25–35%) and in bleaching (10–25%). The permanence of the curlating effect will depend on the fibre furnish, as well as on the mechanical treatments and temperatures which follow the high-consistency stage and which may tend to remove the imposed curl [65].

6.3.5.2 Fines removal. In all recycling operations, fines, together with fillers, are preferentially removed during deinking, cleaning and thickening. For some purposes, such as in preparing recycled pulp for absorbent grades (e.g. tissue), this is quite deliberate. The longer fibres are retained giving a porous, free-draining pulp with good tear strength. However, for other grades (e.g. newsprint) this is an incidental effect, the impact of which on recycled pulp properties has not yet been established.

6.3.5.3 Chemical treatment. Sodium hydroxide has been used in conjunction with wastepaper since 1800, either to help ink removal or assist the disintegration of heavily sized papers. No doubt, its beneficial effects on strength were noted then. In more recent times, the effect of sodium hydroxide has been investigated for mixed wastepapers [56], for recycled kraft pulp [36], for newsprint [66] and for sized handsheets [11], and today a 'caustic soak' is sometimes included in newsprint and old corrugated container (OCC) recycling operations [67, 68]. Resulting breaking length improvements have been noted. Typically, less than 1.0% sodium hydroxide solutions (on bone-dry fibre) were used. While none of the above workers measured the fibre swelling properties, it seems certain that sodium hydroxide is swelling the fibre in the same manner as that discussed by Katz *et al.* [69] and Lindström and Carlsson [13], resulting in a more flexible, better bonding fibre. Lindström and Carlsson, in fact, showed that from an understanding of swelling phenomena, one would not expect the beneficial effect of sodium hydroxide to be very marked on bleached fibre, even though a clear benefit could be obtained on unbleached pulp (Figure 6.7) [13]. This is exactly what Horn found [10]. Where lignified fibres are concerned, some beneficial change to the surface chemistry of the fibre is also conceivable, resulting from the removal of hydrophobic materials such as resins. All these changes are achieved without major reductions in drainage rate, but there may be other undesirable effects, such as loss of brightness.

Wastepaper processing frequently involves the use of other chemicals. For example, during the repulping of old newspapers and magazines, typical additives at the pulper will include sodium hydroxide, hydrogen peroxide, sodium silicate, surfactants (such as fatty acid soaps), and chelating agents. While sodium hydroxide may promote pulp strength, an opposite result could arise from the presence of surfactants. If these are not washed out, but instead carry over into the paper machine system, they may lower the surface tension, and reduce bond strength and paper strength in the manner observed by Springer *et al.* [70]. The effect of other recycling chemicals, however, remains unknown at the present time.

6.4 Methods of controlling the paper making potential

In considering the quality changes that are inherent in recycling, it is important to remember that the impact on quality is not necessarily negative. For example, newsprint is frequently made using high levels of recycled fibre, often comprising a 70/30 mix of old newspapers and old magazines. The recycled mechanical pulp fibres will be flattened and flexibilized, and this will assist the bonding. The presence of chemical pulp from the magazines will also help furnish strength, and, together with fillers, will assist brightness. Taken overall, a strong, bright sheet of newsprint is possible. In another example, tissue manufacturers use significant proportions of recycled fine papers, and computer printout. The hornified chemical pulp fibres help to give a bulky, absorbent product. Finally, in making printing and writing grades, recycled fibres can help bulk, opacity and formation. Nothwithstanding this, recycled pulp quality may need some modification before use, and there are six possible approaches.

6.4.1 Refining

The paper making potential of recycled chemical pulps can be changed by refining, just as it can be for virgin pulps, and the purpose of refining is essentially the same, i.e. to improve fibre bonding potential by regaining the fibre wall swelling which was lost when the fibre was dried (Figure 6.2). This approach has been successful both in laboratory refining [37, 71, 72], and in pilot scale and full scale refining [73–77]. However, as already discussed, recycled pulps start with two significant disadvantages: they tend to have a shorter fibre length and they are usually of lower freeness, due to the presence of fines. Furthermore, Szwarcsztajn and Przybysz have shown that the swelling level of the fines cannot be recovered by refining, unlike that of the recycled fibres [48]. Thus, although Hawes and Doshi found that the new fines created by refining are beneficial [49], the overall result is that refined recycled pulps are still weaker than virgin pulps at the same drainage rate. Similarly, if refined to the same strength as the virgin pulp, the recycled pulp will have significantly slower drainage which may impair the paper machine output.

It should be noted that both the fines increase and the fibre length reduction are inevitable results of re-refining pulp. They are not caused by increased fibre fragility, as is sometimes suggested [78]. Indeed, there is no direct evidence that recycled fibres are more fragile than virgin fibres or that they are weaker [21]. However, if, in a commercial operation, the wastepaper supply contains mechanical pulp, the situation is worsened because the fibres of mechanical pulps contribute little to chemical pulp strength and are very easily broken up by refining.

Low intensity refining strategies which preserve length and minimize fines production, while maximizing fibre swelling, are generally recommended [73]. Unfortunately, these methods are less efficient in their use of power.

High-consistency refining has been examined as a method of developing pulp properties without undue drainage rate reduction, and the potential of this approach has been demonstrated by Fellers *et al.* [17]. Energy consumption is much higher, however, and the combination of pulp properties developed by this approach is different from those developed by low-consistency treatments. Dispersion treatments which utilize high consistencies can also give beneficial strength improvements [79]. Such processes may see more development in the future.

6.4.2 *Prevention of hornification*

Refining is essentially a method of reversing the effects of hornification. The mechanical action of the refiner delaminates again the fibre wall which hydrogen-bonded together during the previous drying process. The question arises as to whether hornification could be avoided in the first place by a suitable chemical pretreatment. This question has been addressed in the past [13, 45] and the subject has recently been reviewed and new results given [27].

One approach to preventing hornification involves modifying the cellulose hydroxyl groups by partial substitution, so that the adjacent microfibrils in the fibre wall do not bond. Methylation is a suitable procedure, and it has been shown that substituting as few as 2.2% of the hydroxyl groups is enough to disrupt internal bonding sufficiently to prevent hornification. On a commercial scale, however, this method would probably be too expensive.

Rather than substitution, an alternative approach could involve the treatment of pulp with a water soluble material which is capable of bonding to the hydroxyl groups. Such a material would then obstruct any intimate contact of cellulose microfibrils during drying. Sucrose and glycerol have been found effective in laboratory studies, but solution concentrations as high as 20% are required, and their effectiveness is limited to one cycle, because, being water soluble, they will be washed out during the next repulping operation.

In summary, therefore, while the requirements for suitable pretreatments have been established, no commercially viable process is yet available.

6.4.3 *Blending with virgin pulp*

Recycled pulp potential can be maximized by blending with a virgin pulp of appropriate properties. Szwarcsztajn and Przybysz observed that virgin pulp had a disproportionately large effect in improving recycled pulp properties [19]. They attributed this to the fact that a virgin pulp contains not only more 'active' fibres, but also more 'active' fines. They found that adding a beaten virgin pulp produced an even greater beneficial effect; this work was carried out on chemical pulps.

In concept, the use of a stronger 'reinforcing' pulp with relatively weak, short-fibred pulp is not new. It has been common practice in newsprint manufacture, where a proportion of strong, long-fibred softwood pulp carries a mechanical pulp component. Recent work has shown particular advantages in using thin-walled, northern softwood pulps [80]. In comparison to coarser, thick-walled pulps, there are more fibres per gram and the fibres are more flexible. Wet and dry tensile properties, as well as opacity and formation, are improved to a greater extent by fine pulps. High-quality virgin fibre helps recycled pulp in another way, i.e. when blended with recycled fibre, it has the desirable consequence of upgrading the quality of the material in the 'recycling loop' when the paper is itself recycled.

In the future, careful selection of furnish components may be the key to utilizing high levels of recycled pulps.

6.4.4 *Chemical additives*

Cationic starch is probably the most common additive for improving recycled pulp strength. It is frequently added at the wet end of the paper machine, for example, in recycled linerboard production. It probably works by improving the bond strength per unit bonded area of the sheet, in the same manner that has been demonstrated for virgin chemical pulps [81]. Other methods of chemically upgrading recycled fibre have been investigated, although none has gained wide acceptance. These include the use of dry strength resins such as anionic copolyacrylamides, which are claimed to promote bonding [82]. Ammonium zirconium carbonate has also been used [83], and has been found to restore considerably the strength lost at the first recycle, although no mechanism has yet been established.

For completeness, it should be remembered that sodium hydroxide is often used during repulping, and while its primary use may be to assist defibering and ink detachment, it can also swell unbleached fibres and promote bonding potential.

6.4.5 *Fractionation*

Fractionation is an option which can be considered when the waste raw material contains large proportions of different fibre types. It involves separating the recycled pulp into both long-fibre and short-fibre fractions. For example, it is reported that in several German mills which use a waste-based, kraft-containing furnish to produce test liner and corrugating medium, the stock is separated by screening and the long-fibre fraction is beaten separately. It is then re-used according to one of the following schemes: (a) utilization of the two fractions in the production of two different grades on separate machines; (b) utilization of the two fractions in two or more plies in the same sheet of paper; (c) remixing the beaten long-fibre fraction with the unbeaten short fraction. In this study [84] it was found that the first two methods were the most effective in terms of sheet physical properties, and the efficient use of energy. In a different study [85], an Italian board mill reported its use of the above scheme (b), with excellent product quality being claimed from low-grade raw material.

Fractionation is quite common in Europe, but is rare in North America, probably because of the better availability (at the present time) of high-quality waste.

6.4.6 Chemical reprocessing

Rather than merely returning a recycled pulp to the quality level of its equivalent virgin pulp, an alternative approach is to upgrade the quality by chemical reprocessing. In the approach suggested by de Ruvo *et al.* [86], recycled pulp prepared from a range of waste papers was given an oxygen delignification treatment. In all cases, very significant improvements in pulp strength properties were observed, leading to the conclusion that the amount and type of wastepaper used in recycled linerboard could be substantially increased. An additional bonus was that contaminants of a waxy or sticky nature were also reduced. Kangas and Lindsay have proposed a mild ozone treatment as a method for improving the strength of old newspapers (ONP) pulps [87]. A 25% strength gain was obtained from a 2% ozone charge applied to the pulp at high consistency. The authors believe that ozone is attacking the lignin and therefore exposing more cellulose. However, they suggest that the fibre surface may also be undergoing a chemical modification in some way. Although interesting, this approach is currently too expensive for commercial operation. Recently, a different approach has been proposed. Nguyen and coworkers have patented a process for recycling old corrugated containers to give a bleached pulp suitable for use in fine papers. The process involves a short kraft pulping operation (including chemical and heat recovery), followed by conventional bleaching [88]. Two mill installations are planned.

6.5 Conclusions

In this review we have seen how recycled pulp quality is influenced first by the pulp type, then by the processes of papermaking, converting, use and storage, and finally by the recycling operations. The principal effects can be summarized as follows:

1. Different pulp types recycle in fundamentally different ways. Refined or never-dried chemical pulps lose swelling capacity during drying. This 'hornification' stiffens the fibre wall, thus lowering the bonding potential. Mechanical pulp fibres do not hornify, but instead become flattened and more flexible as a result of successive paper making operations. They bond better. The performance of other pulps, and blends, will depend on the relative rate of change of these and other fibre properties. There is no evidence of fibre strength loss or fragility.
2. Most aspects of the manufacturing process affect recyclability, i.e. initial refining, wet pressing, drying and calendering. Generally speaking, the more 'severe' the treatment, then the greater the loss of recycle potential. The chemical condition of the stock during the first making also affects some pulps. In addition, the converter and the consumer have an impact on the recycled pulp quality.
3. Recycled pulp properties can be modified by a variety of processes. These have traditionally included stock refining (for recycled chemical pulps), chemical additives, fractionation, and blending with virgin fibre. Very recently, chemical reprocessing of wastepaper to upgrade pulp quality has become viable. Meanwhile, it is generally true that the better the virgin fibre quality, the better the quality of the recycled pulp and the blends that can be made with that pulp.

Finally, while research work over the decades and in various countries has enabled us to identify many of the factors involved in recycled pulp quality, there are still gaps in our knowledge, and more research is required if we are to understand the implications of using higher levels of recycled fibres, and to devise appropriate treatment strategies to enable us to do this.

References

1. Davis, C.B. (1886) *The Manufacture of Paper*, Philadelphia.
2. Strachan, J. (1918) *The Recovery and Re-manufacture of Wastepaper*, The Albany Press, Aberdeen.
3. Anon. (1992) *Pap. Recycler* **3**(11), 8.
4. Wessel, D. (1993) *Paper* **218**(3), 32.
5. Pfaler, E. (1933) *Pap. Zeit.* **48**(76), 282.
6. Brecht, W. (1947) *Das Pap.* **1**(1/2), 16.
7. Brecht, W. (1947) *Das Pap.* **1**(3/4), 60.
8. McKee, R.C. (1971) *Pap. Trade J.* **155**(21), 34.
9. Van Wyk, W. and Gerischer, G. (1982) *Paperi Ja Puu* **64**(9), 526.
10. Horn, R.A. (1975) *Pap. Trade J.* **159**(7/8), 78.
11. Yamagishi, Y. and Oye, R. (1981) *Jpn Tappi* **35**(9), 33.
12. Chatterjee, A., Kortschot, M., Roy, D.N. and Whiting, P. (1993) *Tappi J.* **76**(7), 109.
13. Lindström, T. and Carlsson, G. (1982) *Sven. Papperstidn.* **85**(15), 146.
14. Cildir, H. and Howarth, P. (1972) *Pap. Technol.* **13**(5), 333.
15. Page, D.H. (1969) *TAPPI* **52**(4), 674.
16. Lundberg, R. and de Ruvo, A. (1978) *Sven. Papperstidn.* **81**(12), 383.

17. Fellers, C., Htun, M., Kolman, M. and de Ruvo, A. (1978) *Sven. Papperstidn.* **81**(14), 443.
18. Szwarcsztajn, E. and Przybysz, K. (1976) *Cell. Chem. Technol.* **10**(6), 737.
19. Szwarcsztajn, E. and Przybysz, K. (1974) *Zellst. Pap. (Berlin)* **23**(7), 203.
20. Stone, J.E. and Scallan, A.M. (1968) *Cell. Chem. Technol.* **2**(3), 343.
21. Howard, R.C. and Bichard, W. (1993) *J. Pulp Pap. Sci.* **19**(2), J57; **18**(4), J151, 1992.
22. Lundberg, R., and de Ruvo, A. (1978) *Sven. Papperstidn.* **81**(11), 355.
23. de Ruvo, A., Htun, M., Ehrnroot, E., Lundberg, R., and Kolman, M. (1980) *Industria Della Carta* **18**(6), 287.
24. Scallan, A.M. and Tigerström, A.C. (1992) *J. Pulp Pap. Sci.* **18**(5), J188.
25. Bugajer, S. (1976) *Paperl* **37**(12), 108.
26. Jayme, G. (1944) *Wochenbl. Papierfabr.* **6**, 187.
27. Laivins, G. and Scallan, A.M. in *Products of Papermaking*, ed. C. Baker, Fundamental Research Committee, London, 1993, pp. 1235–1260.
28. Eastwood, F.G. and Clarke, B. in *Fibre-Water Interactions in Papermaking*, Vol. I, British Paper and Board Industry Federation, London, 1978, pp. 835–848.
29. Pycraft, C.J.H. and Howarth, P. (1980) *Paper Technol. Ind.* **21**(9), 283.
30. Pycraft, C.J.H. and Howarth, P. (1980) *Paper Technol. Ind.* **21**(10), 321.
31. Ogiwara, Y. and Arai, K. (1968) *Text. Res. J.* **38**(9), 885.
32. Oltus, E., Mato, J., Bauer, S. and Farkas, V. (1987) *Cell. Chem. Technol.* **21**(6), 663.
33. Okayama, T., Kitayama, T. and Oye, R. (1981) *Jpn Tappi* **35**(12), 27.
34. Sachs, I.B. (1985) *Paper Technol. Ind.* **26**(1), 38.
35. Sachs, I.B. (1988) *Wood Fibre Sci.* **20**(3), 336.
36. Bovin, A., Hartler, N. and Teder, A. (1973) *Paper Technol.* **14**(5), 261.
37. Kolman, M., Fellers, C., Lundberg, R. and de Ruvo, A. (1976) *Preprints EUCEPA Conf., Secondary Fibres and Their Utilization in the Paper Industry*, Bratislava.
38. Guest, D. and Weston, J. (1986) *Preprints Tappi Pulping Conf.*, Atlanta.
39. Chatterjee, A., Roy, D.N. and Whiting, P. (1992) *Proc. of 78th CPPA Annual Meeting*, Montreal, Canada, p. A277.
40. Bobalek, J.F. and Chatervedi, M. (1986) *Tappi J.* **72**(6), 123.
41. Page, D.H. (1985) *Sven. Papperstidn.* **88**(3), R30.
42. Mohlin, U-B. and Alfredsson, C. (1990) *Nord. Pulp Pap. Res. J.* **4**, 172.
43. Klye, R.C. (1961) *Appita* **14**(6), xxi.
44. Mansito, O., Agnero, C. and Sosa, M.E. (1992) *Papel* **29**(June/July), 69.
45. Higgins, A.G. and MacKenzie, A.W. (1963) *Appita* **16**(5), 145
46. Stone, J.E. and Scallan, A.M. in *Consolidation of the Paper Web*, Vol. 1, British Paper and Board Makers Association London, 1966, p. 145.
47. de Ruvo, A. and Htun, M. in *The Role of Fundamental Research in Papermaking*, Vol. 1, British Paper and Board Industry Federation, London, 1981, p. 195.
48. Szwarcsztajn, E. and Przybysz, K. in *Fibre-Water Interactions in Papermaking*, Vol. II, British Paper and Board Industry Federation, London, 1978, p. 857.
49. Hawes, J.M. and Doshi, M. (1986) *Preprints Tappi Pulping Conference*, p. 613, Atlanta.
50. Carlsson, G. and Lindström, T. (1984) *Sven. Papperstidn.* **87**(15), 119.
51. Treiber, E. and Abrahamson, B. (1972) *Holzforsch. Holzverwert.* **24**(3), 54.
52. Okayama, T., Okada, Y. and Oye, R. (1982) *Jpn Tappi* **36**(3), 42.
53. Page, D.H. and Tydeman, P.A. in *The Formation and Structure of Paper*, British Paper and Board Makers Association, London, 1962, pp. 397–413.
54. Göttsching, L. and Stürmer, L. in *Fibre-Water Interactions in Papermaking*, Vol. II, British Paper and Board Industry Federation, London, 1978, pp. 877–896.
55. Gratton, M.F. (1992) *J. Pulp Pap. Sci.* **18**(6), J206.
56. Eastwood, F.G. and Clarke, B. (1977) *Paper Technol. Ind.* **18**(5), 155.
57. Guest, D.A. and Voss, S.P. (1983) *Paper Technol. Ind.* **24**(7), 256.
58. Forester, W.K. (1985) *TAPPI Pulping Conf.* Book 1, 141.
59. Inoue, M., Gurnagul, N. and Aroca, P. (1990) *Tappi J.* **73**(12), 81.
60. Jordan, B., Nguyen, N. and Trepannier, R. (1993) *Proc. Tappi Recycling Symposium*, New Orleans, 28 Feb.–4 Mar., p. 377.
61. Jordan, B.D. and Popson, S.J. (1993) *Preprints 2nd Research Forum on Recycling*, St. Adele, Québec, 5–7 October, p. 153, CPPA, Montreal.

62. Andrews, W.C. (1990) *Pulp Pap.* **64**(9), 125.
63. DeFoe, R.J. (1993) *Tappi J.* **76**(2), 157.
64. De Grâce, J.H. and Page, D.H. (1976) *Tappi J.* **59**(7), 98.
65. Page, D.H., Barbe, M.C., Seth, R.S. and Jordan, B.D. (1984) *J. Pulp Pap. Sci.* **10**(5), J74.
66. Hayashi, Y. (1979) *Abstracts Special Research Symp. in Environmental Sci.* **2**, pp. 20–23, Japan.
67. Heimburger, S.A. and Meng, T.Y. (1992) *Pulp Pap.* **66**(1), 79.
68. Healey, E. (1990) *Pulp Pap.* **64**(9), 138.
69. Katz, S., Liebergott, N. and Scallan, A.M. (1981) *Tappi J.* **64**(7), 97.
70. Springer, A.M., Dullforce, J.P. and Wegner, T.H. (1986) *Tappi J.* **69**(4), 106.
71. Kitayama, T., Okayama, T. and Oye, R. (1987) *J. Soc. Fibre Sci. Technol. Jpn* **43**(9), 456.
72. Göttsching, L. and Stürmer, L. (1976) *Preprints EUCEPA Conf. Secondary Fibres and Their Utilization in the Paper Industry*, Bratislava.
73. Levlin, J.-E. (1976) *Preprints EUCEPA Conf. Secondary Fibres and Their Utilization in the Paper Industry*, Bratislava.
74. Danforth, D.W. (1971) *Pap. Trade J.* **155**(37), 76.
75. Robinson, D.H., Defoe, R.J. and Fredriksson, B. (1985) *Pulp Pap.* **59**(5), 69.
76. Canon, J.G. and Defoe, R.J. (1984) *Proc. TAPPI Papermaker's Conf.*, pp. 41–45.
77. Borsch, B.M.H. (1986) *Tappi Papermakers Conf.* pp. 78–86.
78. Attwood, D. (1990) *Pap. Technol. Ind.* **31**(11), 3.
79. Rangamannar, G. and Silveri, L. (1989) *TAPPI Pulping Conf.* Book 1, 381.
80. Seth, R.S. and Kingsland, M.A. (1990) *Pulp Pap. Can.* **91**(7), T273.
81. Howard, R.C. and Jowsey, C.J. (1989) *J. Pulp Pap. Sci.* **15**(6), J225.
82. Chan, L-L. (1976) *Pulp Pap. Can.* **77**(6), T93.
83. Howarth, P. in *Fibre-Water Interactions in Papermaking*, Vol. II, British Paper and Board Industries Federation, London, 1978, pp. 823–831.
84. Göttsching, L., Török, I., and Putz, H.-J. (1989) *Paper Technol. Ind.* **30**(6), 14.
85. Culicchi, P. (1989) *Proc. Eucepa Symposium*, Ljubljana, pp. 187–207.
86. de Ruvo, A., Farnstrand, P.-A., Hagen, H. and Haglund, N. (1986) *Tappi J.* **69**(6), 100.
87. Kangas, M.F. and Lindsay, J.D. (1993) *Preprints TAPPI Engineering Conference*, p. 407.
88. Nguyen, X.T., Shariff, A., Earl, P.F. and Eamer, R.J. (1993) *Prog. Pap. Recycling* **2**(3), 25.

7 Water and waste water treatment in recycling mills

R.W.J. McKINNEY

7.1 Introduction

Water is the matrix of life - all biological reactions occur in water. Early life forms evolved in the oceans and eventually colonised land, leading to astonishing biodiversity. Without water the earth would be barren. It should therefore be no surprise that water pollution is a major environmental concern and since paper making is a water intensive process, its use of water should be closely scrutinised. Most industrialised nations have imposed standards on waste water discharges and the intent of these is to ensure that society should not bear any costs arising from industrial water use, that is, that water discharged after treatment should be returned fit for almost any purpose. As standards become more stringent, more emphasis is on the total closure of paper-mill water systems, that is, zero liquid effluent.

In wastepaper recycling, the goal of zero water use has been extensively researched, for example, by the Kimberly Clark Corporation in the mid 1980s [1], and reached the stage of a pilot plant facility. However, the quality of fibre produced did not reach the standards achieved by conventional wet processing, so their dry system was abandoned in favour of traditional wet processing, used in their most recent mills. Dry processing of wastepaper is likely to remain an elusive goal, although if air were used as a transport medium for cleaning operations, fibre cleanliness would probably be superior to that achieved when water is used as a transport medium. The main problem in dry processing is getting fibre separation without fibre damage. Paper is a network of fibres and its unusual strength is due largely to hydrogen bonds formed as the paper web is dried. A good illustration of the importance of hydrogen bonds is the difference between the wet and dry strengths of a sheet of paper. Consequently, the separation of fibres, which is required for cleaning operations during reprocessing as well as subsequent paper making, involves less fibre damage when hydrogen bonding is weakened.

In the dry state, cellulose fibres are in a compressed state, held by hydrogen bonds between hydroxyl groups on cellulose surfaces. Wetting a fibre (with water, a polar solvent) weakens these compressive forces. As the hydroxyl groups on cellulose surfaces are wetted, hydrogen bonds between cellulose surfaces are replaced by bonds with water molecules, making it much easier to separate wet fibres. Dry separation invariably results in fibre damage as fibres separate by breaking rather than by pulling away from neighbouring and overlapping fibres in the matrix [2]. Hence water will remain the preferred medium for paper and board recycling, despite producing large volumes of waste water.

In conventional paper making, effluent loads are a consequence of retention on the paper machine being $<100\%$, that is, if retention on the paper making machine were 100% there would not be a need for treatment of excess paper-machine water. This is not the case in recycling, where the need is to separate fibre from non-fibrous materials. Since water is used as a transport medium to carry fibres through various cleaning and screening stages, inevitably a water stream carries unwanted materials which have been separated from fibres and these can create an effluent load. Despite a large proportion of these originating from the paper being recycled, the complexity of a paper product prevents the reuse of all of the original components, when paper is made from recycled fibres. Paper products are highly engineered materials so that, for example, a coating layer of clay and binder, when broken up by pulping coated papers, cannot be reformed as a coating layer during paper making; it is mixed throughout the fibre suspension and is undesirable. Other unwanted materials are those added during paper converting operations, for example, inks, glues, etc., as well as materials inadvertently included during wastepaper collection and delivery, for example, grit, food debris, etc. The presence of these and similar materials would adversely affect recycled paper making operations and recycled paper quality, hence removal is necessary and ultimately a large proportion of non-fibrous materials removed are either suspended or dissolved in water. To avoid problems arising from pollution, these materials must be removed before water is discharged from the recycling mill, hence the need for waste water treatment.

7.2 Freshwater use in recycling

If fresh water were used for a single pass through a mill it would result in very high water consumption, with consequent high treatment costs, both capital and operating; a need for very large volumes of water; and possible criticism of a policy of high water abstraction. To avoid this, virtually all recycling mills derive most of their process water from recycled, clarified water. Several mills have closed their water systems completely, so that there is no liquid effluent discharge [3]. The actual volume of fresh water used is very dependent on a number of factors, including:

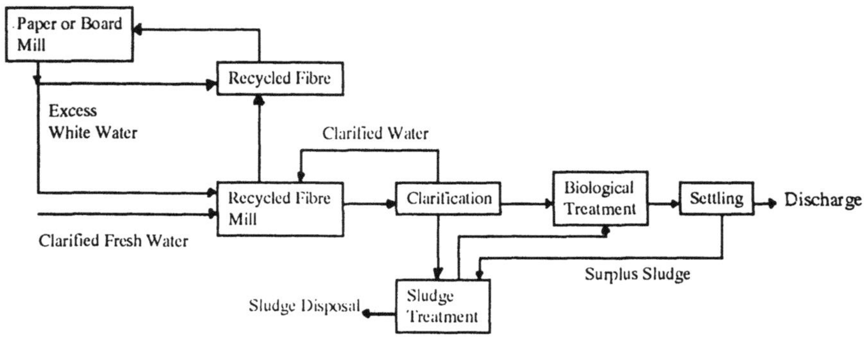

Figure 7.1 Schematic illustration of a recycling mill's water circuit.

- type of wastepaper being recycled;
- product being produced;
- integration of recycling process with paper or board mill;
- water availability and cost;
- water losses in product, sludge, evaporation, etc;
- sophistication of water clarification systems.

Freshwater use can be limited to that required for boiler makeup, cooling, sealing and some chemical makeup uses. These could total as little as $2\,m^3\,t^{-1}$ of recycled product, for a very simple recycling process, but sealing water alone could consume $4\,m^3\,t^{-1}$ in a complex recycling mill, if sealing water is not collected, filtered and reused. A typical, simple water circuit for a recycling mill is illustrated in Figure 7.1

Boiler feed water requires treatment to a high standard and is frequently provided by treating a potable water supply. Other mill water needs can be provided by fresh water, from a variety of sources - surface water, well water, etc. Ideal quality requirements for cooling and sealing waters are given in Table 7.1, with typical standards of recycled mill clarified water, using conventional dissolved air flotation for clarification.

It is clear from Table 7.1 that clarified water is not usually of the quality required for sealing or cooling water. Chemical makeup water quality requirements are dependent on the chemical; for example, in the makeup of a solid (powder) polymer for use in sludge treatment or primary clarification, it is better to use clean, fresh water.

To minimise freshwater use, where possible it should be collected and reused by diversion to the freshwater tank, for example, cooling water from compressors or sealing water from vacuum pumps. If it is not possible to reuse in a freshwater duty (due to high temperature) it should be used in the application with the next highest quality demand. Using a cascade system, the cleanest duty cascading to the lowest quality demand allows

Table 7.1 Sealing water quality requirements and typical clarified water properties

	Sealing water	Clarified water
pH	6.8 to 7.3	6 to 9
Dissolved solids (mg l^{-1})	<250	≫250
Suspended solids (mg l^{-1})	<10	>50
BOD (mg/l)	<2	>250
Scale formation	Negligible	Moderate to high
Corrosion properties	Negligible	Low to high

freshwater use to be minimised. In areas with low freshwater availability, sealing and cooling water can be collected, filtered, cooled by heat exchangers and reused. Losses can be made up with fresh water. Conductivity measurements warn of contamination from leaking seals, etc.

If a high-quality recycled pulp, with an emphasis on brightness (such as a deinked market pulp) is produced, to achieve maximum chemical cleanliness the final stage dilution and wash water should be high-quality water, colourless, with negligible biological activity and low ($<0.2\,mg\,l^{-1}$) iron and manganese content, with other properties similar to those listed for sealing water.

Current best practice for specific water consumption in European mills, with integrated recycling plants, (other than those that are completely closed) are typically in the ranges given below.

- Packaging board: 6–8 $m^3\,t^{-1}$ of board,
- Newsprint: 12–15 $m^3\,t^{-1}$ of newsprint,
- Tissue: 12–15 $m^3\,t^{-1}$ of tissue,
- Printing and writing: $>20\,m^3\,t^{-1}$ of paper.

Water systems in North America tend to be more open.

7.2.1 Freshwater treatment

The degree of treatment required depends on raw water-quality and its end use. It may be possible to use fresh water without any treatment, for example, well water. If colour removal is necessary, for high-quality products, chemical coagulation with, for example, aluminium sulphate and a secondary polyelectrolyte may be required. If suspended solids removal is necessary, then storage, perhaps followed by filtration and sterilisation is adequate. Various treatment stages are illustrated in Figure 7.2, which would take a low-quality freshwater input to a high-quality level, suitable for cooling and sealing duties, but not boiler makeup.

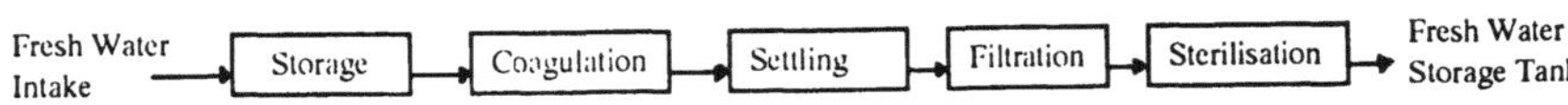

Figure 7.2 Schematic illustration of freshwater treatment.

Table 7.2 Total specific systems loads from wastepaper (kg $tonne^{-1}$ wastepaper input)

	Wastepaper grade		
	Fine papers	News and magazines	Packaging
BOD_5	15-30	15-25	15-40
Suspended solids	150-350	100-200	50-150

7.3 Effluent loads

Suspended solids in the effluent stream from a recycling mill are a mixture of fibres, fibre fines, crill, mineral loadings, ink particles, colloidal inorganic and organic materials, etc. Proportions of these vary, according to the wastepaper being recycled; for example, packaging grades tend to have relatively low specific solids loads, due to their relatively low mineral filler content, whereas coated papers give rise to a high solids load, with total losses of up to 40% during the wastepaper recycling process. Mineral fillers include calcium carbonate, clay, talc and titanium dioxide.

Organic constituents also vary, according to wastepaper type, but the most significant are carbohydrates, either from cellulose degradation or from starch. Hence effluent COD:BOD (chemical oxygen demand:biological oxygen demand) ratios are low, typically in the range 2 to 4:1, indicating an effluent which is relatively easy to treat biologically. If present, lignin derivatives can give rise to a strong colour. Other organic components include proteins, adhesives, coating binders, food residues, etc – a very complex organic 'soup' is created during wastepaper pulping. The strength of the 'soup' is dependent on the types of wastepaper being treated. Typical effluent loads for different wastepapers are given in Table 7.2. Actual loads are dependent on a number of factors, including:

- geographic region (for example, more starch is used in Europe than North America);
- actual wastepaper mix (for example, magazines have a higher specific BOD than newsprint);
- internal wastepaper treatment systems (for example, internal water clarification loops remove some BOD and most of the suspended solids from effluent streams);
- specific water consumption.

The level of water closure has a marked impact on the organic strength

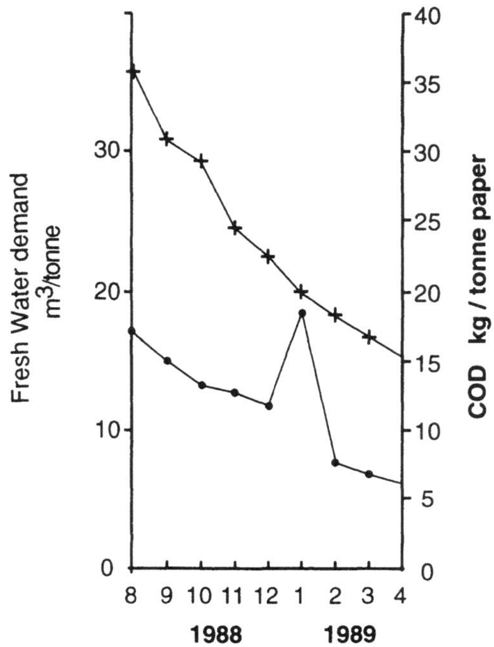

Figure 7.3 Specific COD load falls with lower freshwater demand [4]. +, COD kg t^{-1} paper; ●, freshwater demand.

of effluents. As less freshwater is used, the specific organic load falls [4] (see Figure 7.3), a phenomenon which is discussed in section 7.8.

Individual processing stages create organic loads; for example, bleaching with hydrogen peroxide dissolves natural resin and fatty acids. Resin and fatty acids, or active chlorine residues contribute a major portion of acute toxicity of mill effluents, but they are easy to remove, the latter by treatment with reducing agents and the former by biological effluent treatment, so they pose no substantial environmental hazard or treatment difficulty.

7.3.1 Specific chemical constituents of effluents

Heavy metal content is probably the issue most discussed and this has been extensively researched [5–8]. Most results show that in recycling mill effluents heavy metals are present in small quantities and most of the heavy metals are found in sludges, in which concentrations are lower than in municipal sewage sludge [5]. Land spreading of sludge is accepted, provided normal heavy metal monitoring programmes are adopted. This is the subject of ongoing research, for example, by the Environmental Protection Agency in the USA (USEPA). In Figure 7.4, two deinking mill sludges

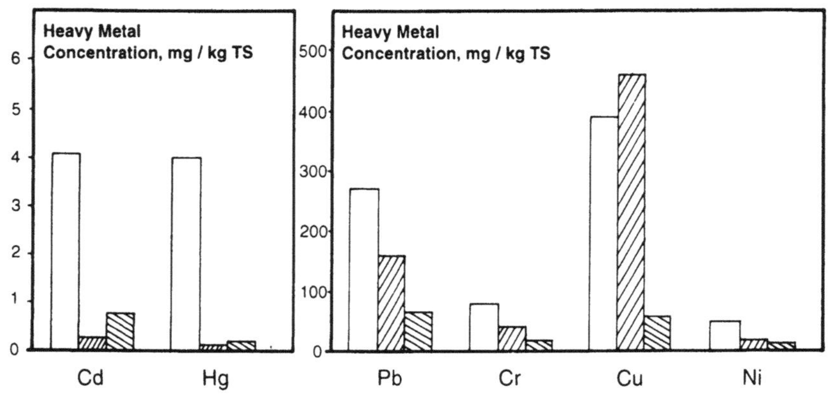

Figure 7.4 Heavy metal concentration in deinking sludge (□), municipal sewage sludge (□) and activated sludge (deinking mill) (▧) [5].

(primary and from an activated sludge treatment plant) are compared with a municipal sewage sludge [5].

Concentrations of specific potentially toxic chemicals have been measured by several regulatory authorities [8–10]. In the USEPA study [10] treated and raw mill effluents were collected and were screened for the presence of 130 chemicals. After this the presence of 64 chemicals was examined in 19 secondary-fibre mill effluents. Not all of these were found in recycled mill effluents. Chemicals found during this study are given in Table 7.3. Following the study, the USAEPA elected to implement limitations for only tri- and penta-chlorophenol contained in slimicides. Results from another three paper board mills surveyed are given as an average in Table 7.3 [8,9]. The source of chloroform was a bleach stage using hypochlorite; loads are up to $16\,g\,t^{-1}$ [8]. Polychlorinated biphenyls (PCBs) have ceased to be a cause for concern since their use in paper making was discontinued in the 1970s, for example, in 1971 in the USA.

Nutrient levels are low, especially phosphorus. Typical ratios for BOD:N:P are 100:1.5:0.04, so nutrients must be added to sustain biological treatment processes. However, most added nutrient is incorporated into biomass produced during the biological treatment process and so levels in effluents can be very low, even following biological treatment. Nutrient loads on receiving waters are thus low, much lower than the nutrient load from municipal effluents.

As advances in analytical equipment and techniques occur, the components of deinking effluents will be placed under more intense scrutiny. If results from increased scrutiny indicate the presence of trace amounts of other chemicals this will not necessarily indicate increased pollution, but will merely reflect increased sensitivity of tests.

Table 7.3 Chemicals identified in recycled mill effluents [8-10]

Category	Chemical	No. of samples	Average influent, ($\mu g\ l^{-1}$)	Average effluent, ($\mu g\ l^{-1}$)
Deink fine [10]	Chloroform	3	4190	145
	Naphthalene	3	142	nd
	Pentachlorophenol	3	15	12
	Tetrachloroethylene	3	95	nd
	Toluene	3	58	nd
	Trichloroethylene	3	493	7
	PCB 1242	3	3	nd
	Lead	3	149	28
Deink newsprint [10]	Butyl benzyl phthalate	3	5	na
	Cyanide	3	1560	na
Deink tissue [10]	Trichlorophenol	6	48	41
	Chloroform	6	1367	55
	Naphthalene	6	48	nd
	Pentachlorophenol	6	38	34
	Phenol	6	119	nd
	PCB 1254	6	1	nd
	PBB 1260	6	1	nd
Paper board from recycled paper [10]	Trichlorophenol	18	360	430
	Bromoform	18	40	21
	Pentachlorophenol	18	356	400
	Phenol	18	204	144
	Butyl benzyl phthalate	18	61	21
	Di-n-butyl phthalate	18	32	44
	Diethyl phthalate	18	183	138
	PCB 1248	18	9	nd
	Lead	18	443	51
	Zinc	18	1811	469
Paper board (recycled) [8, 9]	Chloroform	-	-	0.3
	Methylene chloride	-	-	26.3
	Toluene	-	-	3.3

Three samples taken per facility. nd - not detected. na - not analysed.

7.4 Effluent treatment standards

Recycled paper and board mill effluents, if untreated, can have a substantial impact on receiving waters in the discharge area, with the impact thereafter dependent on dilution. Immediate effects are due primarily to the deposition of suspended solids with later effects due to oxygen depletion. Due to high suspended solids and organic strengths these impacts can be considerable; suspended solids are predominantly inert fillers, such as calcium carbonate, clay and organic fibres. These settle rapidly, usually near the point of discharge, provided the velocity is low; otherwise solids are carried by the flow and deposited in slow moving waters some distance from the discharge point. Banks of sludge form, which blanket plants and if not removed, kills them, so depriving of food the food chain leading to fish. Organic components of the sludge are broken down anaerobically, especially in warm weather and so strip the water of some dissolved oxygen. Bubbles of gas released from the anoxic sludge rise to the surface, carrying some sludge, so that an unpleasant scum forms, as well as the production of a foul odour.

High organic strength of effluent can lead to an even more severe impact. In a healthy water course, dissolved oxygen is high, but as organic material is degraded, oxygen is removed from solution, possibly faster than it is replaced by natural reoxygenation. If the organic strength of an effluent is high compared to the volume of the receiving water, that is, there is a low dilution factor, the oxygen concentration may fall to zero, leading to anaerobic decomposition of the organics. The effect on receiving waters of this is severe: as oxygen levels fall, various life forms cannot survive and are eliminated in that section of the water course. For example, game fish require a minimum of 4–6 $mg\ l^{-1}$ oxygen and coarse fish 2–4 $mg\ l^{-1}$ oxygen. If septic conditions develop, virtually no higher life forms, plant or animal survive and the river is clearly heavily polluted. In addition, foul smells are produced. These conditions may develop some distance away from the discharge point.

These effects of untreated discharges are not restricted to recycled paper mills; any discharge which contains suspended solids and dissolved organics would have similar effects, with the severity dependent on dilution and the health of the receiving water; for example, the impact of municipal effluents is discussed in standard effluent treatment texts [11]. However, unlike municipal effluent discharges, there are no severe public health hazards with recycled mill effluents since in the latter case no pathogenic organisms are present.

In most countries regulations were developed in the late 1960s and 1970s to restrict discharges of suspended solids and effluents with a high oxygen demand, usually measured as biological oxygen demand (BOD). More stringent restrictions are now being proposed or implemented. As other effluent properties – acute and chronic toxicities, bioaccumulation of potentially toxic substances by fish, etc. – have come under scrutiny for all pulp and paper mills, so have those from recycled-fibre mills. Pollution taxes may be imposed, for example, on phosphorus discharges in Germany, to encourage reduction below a given maximum discharge. Some national standards are summarised in Table 7.4.

To meet regulatory standards most mills have some form of primary clarification and secondary biological treatment, or discharge their untreated or partially treated effluent to sewer, for treatment in a municipal sewage treatment works. As more stringent standards are introduced these may impact on recycling processes and chemicals used, so the installation of some form of tertiary treatment will probably become necessary.

Table 7.4 Effluent discharge standards in various countries

	BOD_5 (mg l^{-1})	COD kg/AD[d]	Suspended solids (mg l^{-1})	Nitrogen (mg l^{-1})	Phosphorus (mg l^{-1})	Other kg/ADMT[e]
Germany[a]	25	5	-	10	1.5	AOX[f] <0.012
UK[b]	20	-	30	-	-	Various
USA[c] (kg $tonne^{-1}$)	3.1	-	4.6	-	-	pH 5.0-9.0

[a]WHG Paragraph 7a, Administrative Ordinance, Annex 19, Part B, dated 21 September 1990. [b]UK standards vary according to dilution and quality of the receiving water. Values given are typical, but could be much lower. [c]Currently (1994) under review by the USEPA. It is possible that additional limits in specific chemicals from recycling mills will be imposed in addition to those on tri-and penta-chlorophenols. Values given are for fine paper, NSPS 30-day average standards. [d]air dry. [e]air dry metric tonnes. [f]adsorbable organic chlorine.

Table 7.5 Testing schedule for recycled mill effluent [9]

Daily	Thrice weekly	Weekly	Monthly	Semi-annual
COD/TOC[a]	BOD_5	Nitrogen	Resin acids	Chlorinated organics
TSS[b]		Phosphorus	Fatty acids	Dioxins and Dibenzofurans
pH		Volatile SS[c]	Toxicity	Extractables
Conductance		Aluminium	Total metals	
		Zinc	Mercury	

[a]TOC = total organic carbon. [b]TSS = total suspended solids. [c]SS = suspended solids.

As well as limits on specific substances, various detailed studies [9, 10] recommend routine monitoring of a variety of substances, which may lead to limits on some of these being imposed. Regular analysis of a wide variety of materials is mandatory in Ontario, Canada [9] and based on this, Table 7.5 outlines analyses and frequency relevant to recycling mill operations.

7.5 Physiochemical treatment processes - waste water clarification

Waste water streams carry with them the materials removed during wastepaper processing. If this water is to be reused, which is normal practice, then these must be removed, otherwise concentrations will build up to the extent that the water loop functions as a reservoir of contaminants, merely evening out varying concentrations of contaminants arising from the input of different types of wastepaper. Thus water clarification has a major role in the success or otherwise of a recycling operation, particularly in the case of deinking mills. In some deinking systems, two stage or even three stage water loops are used, and in a few of these different pH conditions apply to the first and second process loops [12]. Segregation of water loops ensures minimal recontamination of fibre by dissolved or suspended materials. Alkali and acid loops are used to optimise the removal of all

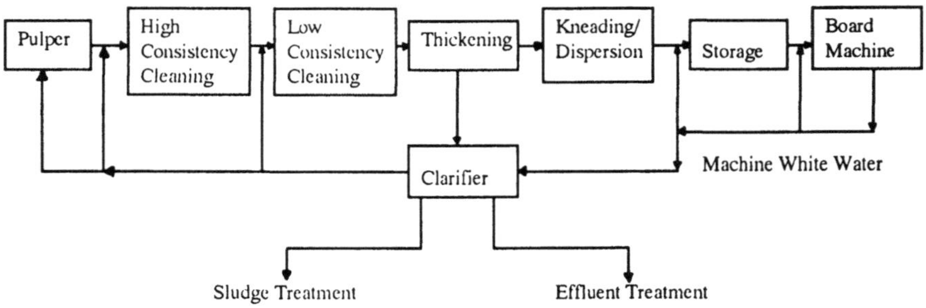

Figure 7.5 Schematic illustration of a simple water loop, packaging grade.

organic compounds [12–14] including dissolved and suspended organics and are most relevant when the paper or board machine operates under acid conditions and wastepaper processing is under alkaline conditions (cf. chapter 3).

Packaging wastepaper processing lines generally have lower quality demands and so frequently have a single water loop, illustrated schematically in Figure 7.5. In this case it may be possible to operate a waste processing system with a zero effluent discharge [3, 15], as illustrated, with clarification only.

Few recycling mills have reached this level of water closure, in which water losses are confined to evaporation, moisture content of sludges and other solid wastes. Where complete closure has been achieved, much more complex water treatment systems are used, including spill collection, closed loop sealing water, the use of cooling towers, etc. and an extensive effluent treatment system, designed to maximise the removal of COD. This type of system is illustrated in Figure 7.6, and includes several clarification stages.

More open water systems are preferred, due to problems which arise as water systems are closed up. These include:

- increased corrosion;
- increased deposition, both inorganic (scale) and organic (slime);
- increased microbial action, due to high organic strength and temperature;
- effects on wet end chemistry of paper or board machines;
- increased temperature (can be positive, since drainage rate increases with temperature).

As closure proceeds there is a reduction in specific organic material dissolution from wastepaper, illustrated in Figure 7.3 [4] but the system COD will increase, as will dissolved and suspended solids. It is increased dissolved solids and organic strengths which are responsible for most of these problems, since suspended solids are removed by clarification. To

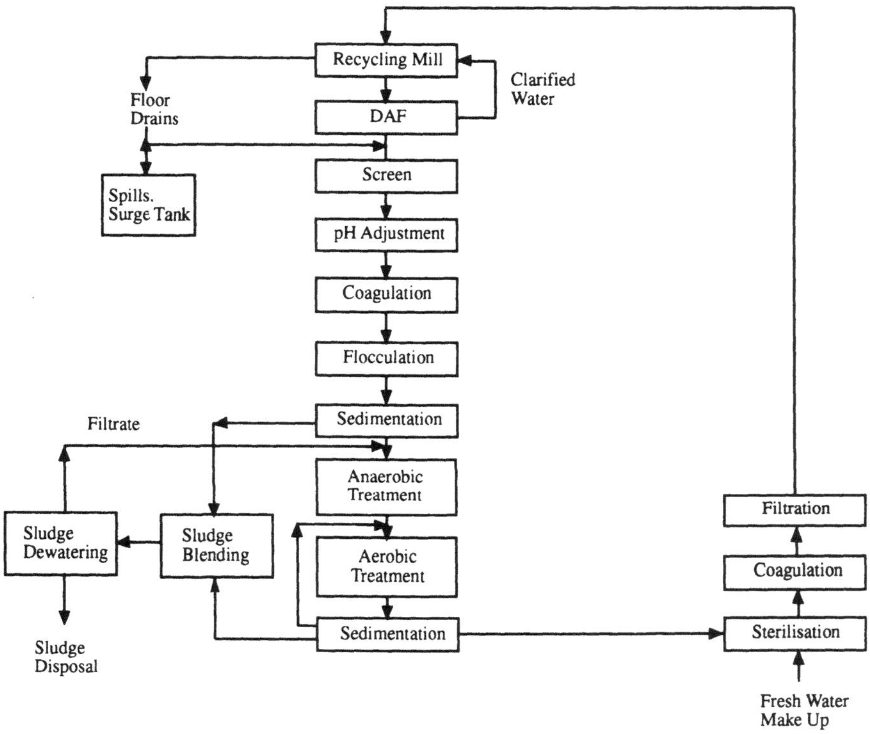

Figure 7.6 Schematic illustration of a closed water system. DAF = dissolved air flotation.

minimise problems, corrosion resistant materials should be used and chemical programmes to control scale deposition and slime introduced.

There are two reasons for water clarification: the first is to reduce the organic and solids load on a biological treatment plant, as is normal in effluent treatment, whereas the second is to produce water of a quality which can be reused in the wastepaper recycling process. In this way the freshwater consumption of recycled-fibre mills is reduced. Clarification processes are similar to those used as primary treatment stages in conventional waste water treatment and include sedimentation, dissolved air flotation (DAF) and filtration. Of these, dissolved air flotation is the most common. It is a high-rate solids removal process and so requires a smaller area for installation. Dissolved air flotation is often used for internal water clarification, but sedimentation is used in external effluent treatment, prior to a biological treatment system. In some instances it is possible to combine these functions, so that only one clarification unit is installed.

Design considerations are similar to those of any waste water treatment process, detailed in a standard text [11]. The review below is to highlight differences from standard design and specific operational differences in recycling mill effluent treatment.

7.5.1 Pre-treatment

There can be several stages of pre-treatment, including:

- surge control, to prevent hydraulic overloads on clarification systems;
- pH adjustment, either to meet the needs of biological stages or chemicals used in clarification;
- coarse screening, to remove polythene fragments and other processes screening rejects.

In paper recycling, surge control can be very important, since the quality of recycled clarified water has a major impact on final fibre quality. If surges are not restricted to within the hydraulic range of the clarification unit then recycled clarified water quality will fall, and recycled fibre quality will follow. Since high flows are common, especially in integrated mills, some surge control is essential.

pH control is usually more important in deinking mills under alkaline conditions, when it can be necessary to reduce pH to a range suitable for a biological treatment stage. Chemicals used for clarification may become less effective at extremes of pH and so some pH control may be necessary to control chemical addition rates. Suspended solids usually have a negative charge, which moves closer to zero as the pH falls. Hence a neutral pH is usually optimal for chemical addition. If clarified water is reused, pH adjustment may not be necessary.

All recycling mills have various rejects such as polythene fragments which inevitably find their way into effluent systems, where they can block sludge pumps, etc. Removal is relatively easy, using a continuous screen.

Foam production can be an unsightly problem and steps may have to be taken to control foam, especially when clarification is by sedimentation. Droplet spray systems should be used in preference to chemical control, since the performance of DAF, flotation deinking and activated sludge systems can be reduced by the use of defoamers.

7.5.2 Sedimentation

In paper recycling, sedimentation has not been widely used as an internal process stage for water clarification prior to re-use, though it is commonly used to reduce the solids load on a biological process in effluent treatment. Its major advantage is that, unlike dissolved air flotation, there is no reliance on chemicals to achieve solids removal, which is easiest with effluents from recycling of packaging grades. In the cases of waste waters from newsprint and woodfree paper recycling, dispersed fine mineral filler particles, which carry a strong negative charge, are very stable and are not removed by conventional sedimentation without chemical use.

When sedimentation is used to treat effluents from deinking mills as a primary stage before biological treatment, allowance must be made for high solids content of primary sludges. Sludge thickens quickly and allowance in scraper design, pump type, etc. must be made, especially if the clarifier is receiving backwater from a washing stage in a deinking mill. If there is an internal water clarification stage, then the solids load on the primary clarifier is very considerably reduced. If long sludge retention times are allowed, sludge will become anaerobic and apart from creating the nuisance of foul odours, sludge will float to the surface and increase the suspended solids in the overflow. If this persists, a thick sludge mat can form on the surface. This is often a problem with warm effluents or during summer months. Sludge removal has to be rapid, even at the cost of reduced solids content of the sludge. Nor should long volumetric retention times be allowed in sedimentation tanks, since the bulk liquid can also become anaerobic. In the case of warm effluents, clarifiers should be designed with low retention times, with fairly high overflow and underflow rates and with shallow depths. Low retention times and aeration are two of the major advantages of dissolved air flotation units.

Sedimentation tanks can be rectangular or (more often) circular. There is no conclusive evidence that one form of tank is superior, provided the total volume and surface areas are equivalent. Cost is the usual deciding factor. Typical surface overflow rates are in the range 0.9–1.3 $m^3\ m^{-2}\ h^{-1}$ (m^3 waste water/m^2 surface area/h). When mineral fillers are present and suitable chemicals are not used, solids removal efficiency is poor, in the range 70–90%. BOD removal efficiency is usually in the range 15–20%. Chemical addition will increase efficiencies, with suspended solids removal up to 98%, dependent on chemical dose, which can be very high. Maximum BOD removal is typically 20–30%. Lower pH helps to reduce chemical dosage rates, since the negative charge on suspended solids falls as pH falls. Inert solids carry-over to a biological plant is usually quite high, especially in the case of a deinking effluent and this must be reflected in the return activated sludge recirculation rates and secondary sludge handling capacity in the secondary settlement tank.

7.5.3 Dissolved air flotation (DAF)

Solids removal by dissolved air flotation differs from flotation deinking in several import respects. In DAF the objective is to remove all suspended solids from the liquid phase, so producing a clarified stream and a concentrated solids (sludge) stream. Bubble size is <0.1 mm and these are created by dissolving air under pressure in water; usually some of the clarified water is recycled, pressurised and semi-saturated with air, which is mixed with the influent stream (to be clarified) just before the DAF unit. Air comes out of solution, throughout the liquid volume, in minute bubbles, which attach themselves to suspended solids and so increase their buoyancy, enabling the

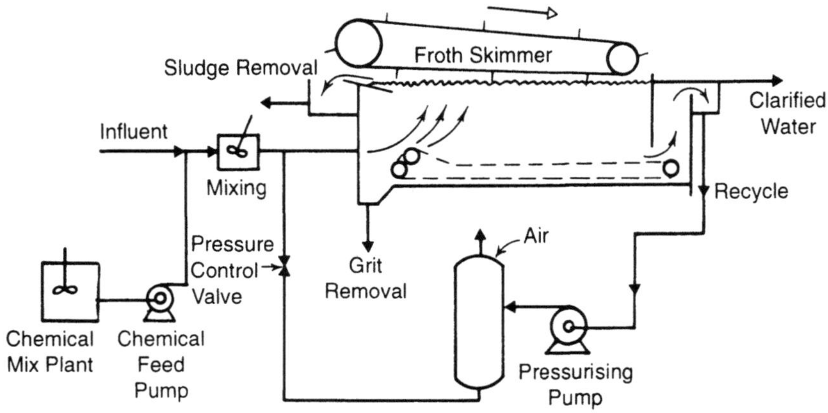

Figure 7.7 Schematic illustration of a dissolved air flotation clarifier.

particles to rise to the surface, even when particles have a higher density than water. A thick froth forms on the surface, which is removed by a skimming mechanism or rotating scoop. Grit and other heavy particles settle at the bottom and can be raked to a central sump for removal. A typical DAF clarifier is illustrated in Figure 7.7.

Flotation deinking is a solids fractionation process, reliant on surface chemical phenomena to separate ink from cellulose fibres. Air bubbles in the size range 0.1–1.0 mm are created by air injectors; air is introduced directly into the liquid phase, through an impeller or by a Venturi effect. Surface chemistry phenomena enable ink particles to be attached to a relatively large air bubble, which rises to the surface, forming an inky foam, which either overflows a weir, or is removed by a vacuum system.

In DAF solids removal, efficiency is dependent on a range of factors, including the types of suspended solids, chemical addition rates, addition point and ratio of air to solids.

The most difficult effluents to clarify are from deinking mills using coated and heavily filled wastepapers, so that the waste water includes inorganic and organic suspended, colloidal and dissolved materials. With these, chemical use to assist flocculation is essential and performance depends largely on chemical types, addition point and air-to-solids ratio. A dual chemical system is frequently used with the first added to reduce the repulsive charges between particles and with a second added to pull particles together. This is discussed in section 7.8.

Air-to-solids ratios for deinking mill clarification are in the range 2 to 4:1, though this varies with DAF unit type. Very high solids loadings rates can be used, up to 1200 kg m^{-3} day^{-1}. Overflow rates are typically 2.0–2.5 $m^3 m^{-2} h^{-1}$. The Krofta DAF unit, which uses a rotating inlet to achieve zero velocity, has a very high overflow rate, up to 7.5 $m^3 m^{-2} h^{-1}$.

Solids removal efficiency during DAF can be high; up to 99+% removal, though this inevitably involves chemical addition. Requirements for clarified water duties should set appropriate clarified water standards, since high operating costs can be incurred through using high chemical addition rates to achieve an unnecessary clarity. Chemical overdosing is a frequent problem.

Operating problems include instability due to either changes in solids or hydraulic loads. Some units, such as the Krofta, use a recycle loop to maintain a fixed solids loading, possible only when solids levels are higher than design and flow stabilisation permits dilution.

7.5.4 Filtration

Drum filters have been used successfully as a primary treatment for deinking mill effluent [16]. Performance of this type of filter is dependent on the formation of a fibre mat on the surface of the filter, which can be easily removed as sludge. If a fairly high proportion of solids present are fibre and fibre fines this allows good performance levels to be reached, especially when chemical flocculants are added. With low fibre content effluents, the units can work only if a stable floc is formed by chemical treatment, otherwise the filter cloth will blind, particularly if concentrations of mineral fillers are high. Advantages of this type of unit include its small footprint and a considerable tolerance with respect to flow and load changes.

Vibrating filters are also used to remove residual solids after other forms of clarification, to achieve a high-quality clarified water suitable for use in applications which need low suspended solids, such as high-pressure showers. In-line static filters can also be used for this duty.

Drum filters with very fine mesh cloths can be used as a tertiary treatment, for example, to remove fine suspended solids after biological treatment. Solids removal efficiency tends to be low, 40–60%, and blinding of the mesh by microbial growth can be an operating problem. As effluent quality standards increase, it is probable that microfiltration will become more important in meeting these higher standards.

7.6 Biological treatment systems

Biological treatment processes are used to remove dissolved organic materials which could cause oxygen depletion in receiving waters. This is achieved by providing an ideal environment for rapid microbial (mainly bacterial) use of carbonaceous materials, either for carbon assimilation into new cells, or for respiration. This involves supplying the required volume of oxygen and any nutrients which are growth limiting factors. Cellular material produced by microbial growth must be removed from suspension, usually by sedimentation. Colloidal solids can also be removed through

Table 7.6 Comparison of aerobic and anaerobic systems

Characteristic	Aerobic (activated sludge)	Anaerobic (USAB)[a]
Land requirement	Moderate to high	Low
Costs		
- capital	High	High
- operating	High	Low
Sludge production	High	Low
Energy requirements	High	Low to zero
Sensitivity		
- toxic shock	Moderate	High
- pH	Moderate	High
- temperature	Low	Moderate
Nutrient requirement	High	Low
Start-up	Fast	Slow
Effluent types	All	Restricted

[a]USAB = upflow sludge anaerobic blanket.

their incorporation into agglomerates produced by bioflocculation. Efficient sedimentation plays a major role in determining the success or otherwise of a waste water treatment system, since much of the residual BOD, measured after biological treatment, is associated with suspended solids.

Many different systems are in use treating effluent from recycling mills, divided into either aerobic or anaerobic treatment processes. In some cases these can be combined into one system to provide maximum removal of carbonaceous material. Effluent characteristics frequently are the deciding factor when choosing between these systems; some are not suitable for anaerobic treatment. Operating cost advantages are reported for anaerobic systems [4], though anaerobic systems alone will not usually meet discharge standards, so the choice is usually between a combined anaerobic and aerobic or aerobic only. A comparison of systems is given in Table 7.6.

The most important advantage of the anaerobic system is that organic loads are converted primarily into useful biogas, rather than into sludge and this conversion occurs without the need for high energy consumption, hence giving lower operating costs. As carbon taxation policies become more widespread, biogas production will become more of an advantage, which will lead to a greater use of anaerobic processes, since biogas production is sustainable and reduces the reliance on fossil fuel combustion.

7.6.1 Anaerobic treatment processes

Most common in recycling mills is the upflow sludge anaerobic blanket (USAB) reactor, which has been used in a number of recycling mills, primarily packaging, but also one recycled-tissue mill [17]. Other anaerobic systems are in use in the pulp and paper industry [18] but the USAB reactor is the most important in recycling. Others are not suitable due to the strength of most recycled mill effluents, although anaerobic filters are used in Japan for effluent treatment in some deinking mills.

USAB reactors are used when effluent strengths are quite high, typically $BOD_5 > 1000\,mg\,l^{-1}$ and $COD > 2000\,mg\,l^{-1}$. Loading rates are high, typically 5–10 kg BOD m^{-3} day^{-1}. Due to this and problems that can occur when sulphate concentrations are high they have been used mainly in the treatment of effluents from packaging recycling mills, such as waste-based liner, corrugating, box board, etc., which produce high-strength effluents. In at least one case, an anaerobic system is used to reduce recirculating COD in a totally closed mill [17]. Effluents with much lower effluent strengths have been successfully treated [17] and in one instance, with an inlet BOD_5 of about 500 mg l^{-1} BOD removal efficiency of 80% was achieved, averaged over one year [19].

When sulphate concentrations are high, reduction to sulphide produces high concentrations of hydrogen sulphide, which can inhibit methane production, as well as causing odour and corrosion problems. Although the ratio of COD to sulphate should not be < 5 [4] (or COD:S < 15) there are several reports of successful treatment at ratios below this; in one case [19] of treatment at a relatively low temperature and a COD:sulphate ratio of 3.8, BOD removal averaged 80%, although when the ratio fell to 2:1 [20] BOD_5 removal fell to $< 70\%$. Relatively high sulphate levels also appear to lower biogas production [19, 20]. Techniques to control sulphide concentrations can be used [21]. Odour problems can be solved with closed containers and plant design, for example, the inclusion of a subsequent aerobic stage. Low dissolved oxygen levels in the final settling tank, after the aerobic stage, have been shown to have a deleterious impact on final effluent quality. After the dissolved oxygen concentration was increased there was a considerable improvement in final effluent quality [22].

7.6.1.1 Anaerobic process microbiology. There are numerous steps involved in anaerobic digestion, but two different microbial populations carry out different but basic reactions. The two groups are known as 'acid formers' and 'methane formers'. Their roles are illustrated in the anaerobic cycle, illustrated in Figure 7.8.

Acid formers (non-methogenic bacteria, for example *Clostridium* spp.) are responsible for the hydrolysis of complex organic compounds to simple organic acids, such as acetic and propionic acid. These are very common in paper mill water systems; high levels of organic acids are common in paper mill water systems [23]. Of considerable importance is the potential for hydrogen gas generation (under anaerobic conditions) by acid formers. This has led to several explosions and fatalities in paper mills in situations

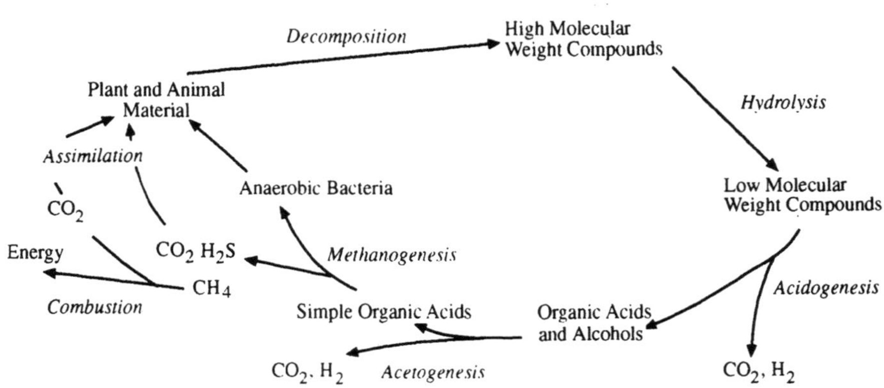

Figure 7.8 The anaerobic carbon cycle.

where anaerobic conditions developed, not necessarily related to effluent treatment, but within the mill [24]. Safety must be an important design consideration [25].

The second group of bacteria, 'methogenic', are strict anaerobes, for example, *Methanobacterium*, which converts simple organic acids to methane and carbon dioxide, similar to those bacteria found in the stomachs of ruminant animals. Of these, the most important are those which utilise acetic and propionic acids, which have very slow growth rates, usually considered to be the rate limiting step. Reactions involved are complex and are reviewed elsewhere [10, 23].

High levels of BOD removal from recycling plant effluents are achieved, typically 80–85% removal of BOD_5, 70% COD removal [17]. Biogas production is in the range 380–500 m^3 per tonne COD removed [19, 26]. Biogas usually has a methane content of about 80%, with a heat value of about 28.6 MJ m^{-3}. When combined with an aerobic system, overall BOD_5 removal can be very high, up to 99.7% [26] with COD removal of up to 86% [15].

7.6.2 Aerobic treatment processes

Most deinking mills use an aerobic treatment process. There are two types of process, either suspended growth (activated sludge) or fixed growth (biological filters). The most commonly used is activated sludge; for example, in the UK only one deinking mill has a filter process, whereas seven others use variants of the activated sludge process. A comparison between the processes is given in Table 7.7 and the aerobic carbon cycle illustrated in Figure 7.9.

Table 7.7 Comparison between and filter and activated sludge processes

	Trickling filter	Activated sludge
Land requirement	Low	Moderate to large
BOD removal	Moderate to high	High
Energy requirement	Low	High
Sludge production	Moderate to high	High

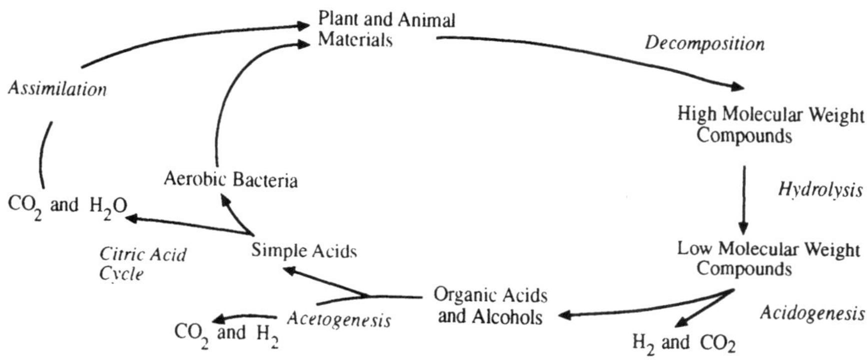

Figure 7.9 The aerobic carbon cycle.

7.6.2.1 Filter processes. Trickling filters were the first waste water treatment systems, used first in England in 1893, since when the process has been developed into a number of variants. Due to the high strength of recycling mill effluents, high-rate filters are used, which generally have synthetic media. Recirculation of filter or final effluent allows high organic loadings (up to 2 kg BOD m^{-3} day^{-1}) to be used and improves final effluent quality, as well as helping to minimise ponding and to reduce problems from filter flies and odours.

Process microbiology. Organic materials in the effluent are degraded by a population of micro-organisms within a film attached to the filter media, illustrated in Figure 7.10. In the outer layers of the slime film aerobic micro-organisms predominate, which use the organic materials which are adsorbed onto the surface layer and penetrate by diffusion. As the slime film grows and increases in depth, oxygen cannot diffuse throughout the film and so an anaerobic environment develops at the media/slime interface. Also with this increase in depth, external carbon sources do not reach micro-organisms at the filter media interface and so endogenous metabolism predominates; the slime film is washed off the media and a new

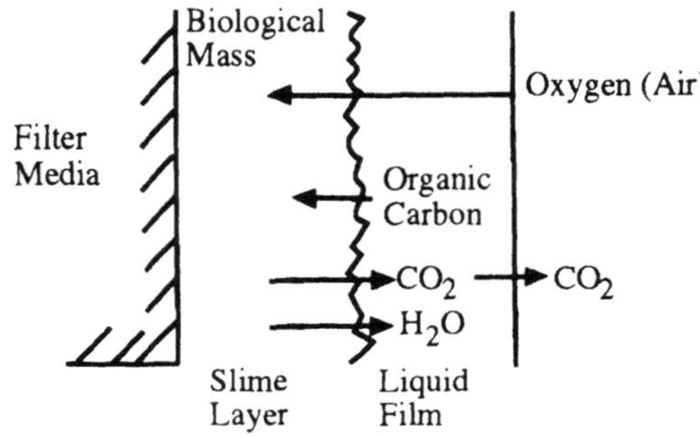

Figure 7.10 Schematic illustration of slime layers on filter media.

slime film starts to build up. At this point, aerobic bacteria predominate at the interface but when a relatively deep slime layer is formed again, metabolism may be anaerobic. Other micro-organisms such as protozoa, fungi and algae can be present and under low-load conditions, higher animals, such as worms, insect larvae and snails are also present.

The development of anaerobic conditions is primarily responsible for odour problems with filters, which can be exacerbated by high levels of sulphate, resulting in the production of sulphides, including hydrogen sulphide. Since sulphate pulping predominates, most paper mill effluents have fairly high sulphate concentrations; newsprint mills have relatively low sulphate concentrations. A UK deinking mill (deinking primarily woodfree wastepapers) has a high-rate filter installation and has experienced extended periods with odour problems. If filters are considered for effluent treatment of a recycled mill-effluent, odour control must be given a high priority.

Suspended solids (biofilm washed off the media) are removed by sedimentation. Filter biological growth usually has a higher density and better settling properties than activated sludge, recycling rates are lower and so smaller settling tanks can be used, compared with activated sludge.

Process variants. These are reviewed in standard effluent treatment texts [11] but one relatively recent variant is used in one recycling mill in the USA [27]. This is the submerged aerated biofilter reactor, which uses an open gravel (usually silica) bed which allows two types of biomass to develop; biofilm on gravel surfaces and activated sludge solids retained in gravel bed voids. This allows high loading rates to be used, 4.9 kg BOD m^{-3} day^{-1}, with BOD removal at 78% [27]. However, the final effluent quality described in this case would not be acceptable in many situations.

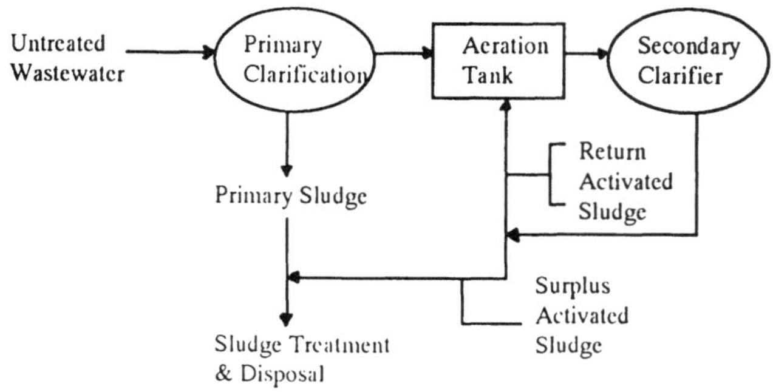

Figure 7.11 Schematic illustration of the activated sludge process.

7.6.3 Activated sludge

The activated sludge process was developed in Manchester, England and was first used in 1914. It is now the most common effluent treatment system in use. It has been developed in many adaptations for a wide variety of processes, but all share the common feature of the return of a sludge stream to be mixed in a reactor with the effluent undergoing treatment. Since the sludge stream being returned has been produced by reaction with the effluent, it is termed 'activated'. The process is illustrated in Figure 7.11

In general terms, the process is used to treat mid-strength effluents, with BOD_5 concentration in the range 100–500 mg l^{-1}. Higher rate processes, treating higher strength effluents, such as the deep shaft process, have been used as a first stage aerobic treatment process [28]. Activated sludge variations used in treating secondary-fibre mill effluents include:

- oxidation ditch [29]
- surface aeration
- deep shaft process [28]
- pure oxygen processes [30]
- jet aeration [31]

It is beyond the scope of this chapter to review these processes in depth, but points specific to recycled-fibre mill effluent plant design are reviewed.

Process microbiology. Bacteria degrade the organic components in the waste water, using a proportion of this as food, to produce energy so that they can use the remaining organic carbon to produce new cellular material. A wide variety of predominantly Gram-negative bacteria are involved, including members of the genera *Pseudomonas*, *Zoogloea*, etc. as well as filiamentous forms such as *Sphaerotilus*. Other micro-organisms are important; protozoa and rotifers help to 'polish' the effluent by grazing on dispersed bacteria which have not been incorporated into biological flocs.

Protozoa can indicate the health of the sludge: dominance by ciliated protozoa such as *Vorticella* and *Carchesium* indicates a healthy sludge [32]. In poor sludges, flagellated protozoa dominate and more filiamentous bacteria are present. 'Bulking' is the term given when activated sludge has poor settling properties, usually due to excessive growth of filiamentous bacteria. This results in poor sludge compaction and can lead to high solids overflow. Causes of bulking include nutrient deficiency (either micro or macro) and variable temperatures, especially when temperatures increase to 30°C and beyond. Steps which can be taken to minimise bulking include:

- use of a low capacity selector mixing tank to give a high loading, for a short period, before recycled activated sludge and effluent are mixed;
- installation of chemical dosing facilities to control filiamentous bacteria, for example, by chlorine dioxide;
- use of low overflow rates for the secondary clarifier, for example a maximum of $0.7\,m^3\ m^{-2}\ h^{-1}$;
- instrumentation to indicate failure of nutrient dosing pumps, high sludge blanket levels, etc.

Other common problems in secondary-fibre activated sludge effluent treatment include foam generation and high inert solids load.

Foam production is due to the presence of surface active agents, particularly when alkali lignins and lignosulphonates are present [33]. These may be created by alkali pulping or other alkaline stages. Other surface active agents usually present include fatty acids, rosin acids, etc. Where fillers are present they can concentrate around each bubble, contributing to foam stability. In the recycling process problems due to foam include: reduced washer capacity and efficiency, a reduction in pumping capacity, reduction in screen capacity and screening efficiency, inconsistent ink removal during flotation, housekeeping and safety, etc. In an effluent plant, foam contributes to solids carry-over and can be unsightly. Even after biological treatment residual surfactants can create foam on the surface of receiving waters, which is associated with industrial pollution, hence it should be avoided. Attention should be given to the design of the out-fall, avoiding turbulence and where possible, submerging the out-fall beneath the receiving water surface.

A high inert suspended solids load on the activated sludge plant is inevitable when treating deinking mill effluents, due to the carry-over of inert colloidal mineral filler particles from the clarification stage. This solids load should be considered when considering mixed liquor suspended solids concentrations (use volatile solids only) and especially in the design of the secondary settling facilities, sludge recycling rates, etc. Some chemical assistance to flocculation may be necessary, with chemicals added immediately before secondary settling, to remove these inert solids. However, they are normally incorporated into biological flocs and give the sludge relatively good settling properties.

7.6.4 *Aerated lagoons (aerated stabilisation basins)*

These have been popular in the North American pulp and paper industry, due to relatively low costs. Their microbiology is similar to the activated sludge process, but without sludge recycle. Retention times are much higher, usually more than 10 days. A major disadvantage is their land requirements and, in general, lower levels of organics removal than is achieved with the activated sludge process [7]. However, other capital and operating costs are lower. Algal growth during summer months can create a high BOD in overflows (associated with the algae) and various methods are used to control algae. Odour can also be a problem in summer months. In winter, due to severe climatic conditions, organic removal rates can fall, as temperatures fall but other problems (algae, odour) are eliminated.

Following treatment in aerated lagoons conventional sedimentation tanks can be used to remove suspended solids, but it is more usual to use one or more large settling basins. Periodically these have to be shut down and solids deposited in the basin removed. As discharge standards become more stringent it is probable that the use of aerated lagoons will fall.

7.7 Sludge treatment and disposal

Regulations controlling the disposal of solid wastes continue to become more stringent and as old landfills close, relatively fewer new sites are licensed, so overall landfill capacity has fallen; the result is very substantially increased landfill charges. Since paper and board recycling generates large amounts of solid wastes this has a major impact on recycling economics. In some paperboard mills, it is possible to reuse primary sludges in the manufacturing process, but the high biological activity of secondary sludges prevents their use. In most cases sludge must be disposed of by external means, usually following dewatering, to reduce volume and weight.

Within most recycling mills there are several sources of solid wastes, including pulper screen rejects, other screen and cleaner rejects, as well as a variety of sludges from, for example, DAF clarification, flotation deinking units, surplus activated sludge, etc. Relatively dry reject streams, for example, rejects from pulping, may be kept separate from other waste streams, since ultimate disposal of rejects and sludge can be by different methods; for example, this type of waste is not suitable for land spreading, disposal is either by landfill or incineration. Magnetic separation can be used to remove ferrous materials. If PVC is included in rejected plastics, then incineration should be by an incinerator which provides controlled combustion conditions to avoid the production of noxious fumes and prevent severe air pollution. PVC in plastics in rejects is much less commonly used than polyethylene.

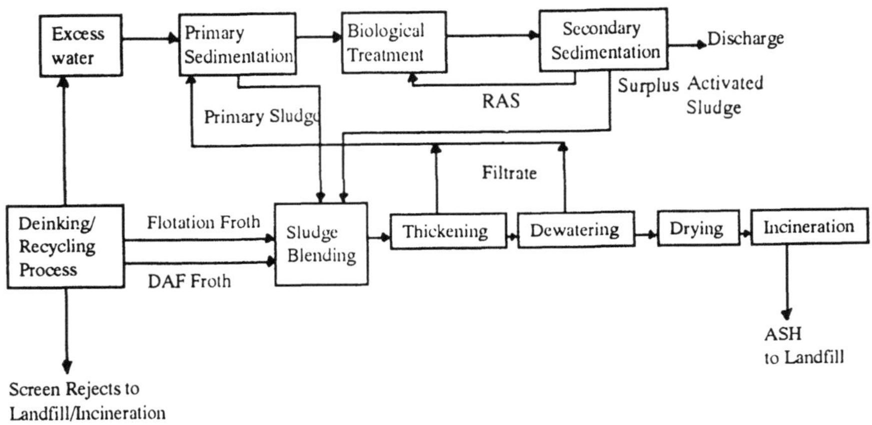

Figure 7.12 Schematic illustration of sludge treatment.

7.7.1 Sludge collection and blending

Unless primary sludges are reused within a board mill, sludge handling normally begins with blending, during which sludges from a variety of sources are mixed, to assist in subsequent processing. Volumes of primary sludge are usually considerably in excess of surplus biological sludges, but sludge disposal is made more difficult when biological sludges are present. A typical blending and treatment system is illustrated in Figure 7.12.

Primary sludges are relatively easy to dewater, flotation froths more difficult and surplus activated sludge more difficult again. Achieving a constant blend, by suitable storage and pumping rates is important, so that subsequent processing is on a blend without marked changes in composition. If this is not achieved processing is more difficult to control. However, blending is made difficult since excess storage time will result in hydrogen sulphide generation and odour, especially when surplus activated sludges are present. Sludges can be treated chemically to reduce biological activity.

Conventional gravity thickeners are very effective in thickening primary sludges and normal design considerations apply [11]. Where space is limited drum thickeners can be used as a preliminary dewatering stage, but blending is made more difficult. In some cases, processing can be improved if difficult to treat sludges are treated separately, for example, using centrifuges to dewater flotation froth, though this usually involves higher capital and operating costs.

Gravity thickening, which allows removal of 'free' water, produces a sludge with a high solids content (> 10%) when large quantities of mineral filler are present. If activated sludge is included, this can fall to 5% or lower. Extended retention in a gravity thickener inevitably results in anaerobic conditions and sludge can float to the surface on liberated gases.

Gas will also be liberated during subsequent pressing. Since hydrogen sulphide gas is explosive, very toxic and can result in severe corrosion problems, extended storage of sludge should be avoided, unless treated to reduce biological activity. 'Fresh' sludges are easier to dewater, whereas processing of stored sludges inevitably results in increased chemical costs and reduced sludge cake solids content. Although more costly, stainless steel presses will resist corrosion.

7.7.2 Sludge dewatering

The objective in using mechanical dewatering devices is to increase solids content, so reducing the weight and volume requiring ultimate disposal. Other reasons are that sludge becomes easier to handle and that dewatering is a form of stabilisation. Dewatering is essential prior to incineration.

Chemical conditioning is usually necessary, increasing the degree of flocculation of sludge solids, so releasing more water which can be removed easily. Factors which affect dewatering include sludge source, chemical type and addition point, temperature, age and feed solids content.

Since sludge type has the greatest effect on the necessary chemical dose it is important to use sludge blending to produce a consistent sludge, otherwise chemical conditioning needs vary with sludge composition. Good mixing of sludge and added chemical is important, but mixing should not shear flocs during or after floc formation. Retention times should be low, so that sludge reaches the dewatering stage rapidly after conditioning. Typical organic polyelectrolyte addition rates are in the range 2–6 kg polyelectrolyte per bone dry tonne of sludge. A number of devices are used to dewater sludge and the selection is based on space available, cost, etc. Devices available include:

- sludge drying beds,
- vacuum filtration (up to 20–30% solids)
- centrifuges (up to 20–40% solids)
- filter belt presses (up to 30–45% solids)
- flocculator, dewatering drum and screw presses (up to 60% solids).

The final moisture content target is frequently determined by the balance that exists between capital and operating costs, including transport and final disposal costs. Although total sludge weight is determined by moisture content, the effective density has a major influence on sludge volume, which is sometimes the basis of disposal costs, for example, if based on container size, rather than weight. Volume has a major impact on transport costs, since most containers reach their volume capacity before the vehicle reaches its weight limit. Effective density is affected by the type of process used to dewater sludge and the mechanism used to fill sludge containers. Screw presses, for example, tend to produce a sludge crumb which has a

Table 7.8 Volume and density at varying moisture contents of one tonne of sludge solids

Solids content (%)	Weight (t)	Typical density [a](g cm^{-3})	Volume (m^3)
5	20.0	1.0	20.0
30	3.3	0.95	3.5
40	2.5	0.9	2.8
60	1.7	0.85	2.0
80	1.3	0.8	1.6
100	1.0	0.8	1.3
Ash [b]	0.5 [c]	0.6	0.7

[a]Densities are typical, but can vary by ± 10%, or more at higher solids contents. [b]Typical tissue deinking sludge, at 50% filler content. [c]Weight of ash ignores thermal decomposition of $CaCO_3$

relatively low density compared to a belt press. Hence volume reduction is not always directly proportional to water removal. A relationship between density, weight and volume at various solids content is given in Table 7.8.

7.7.3 Sludge stabilisation

The objective of stabilisation is to produce relatively inert solids, to reduce organic carbon content, reduce volume and to minimise the potential for odour, leachate and other problems associated with sludge. Water removal is one form of stabilisation, but final disposal is made easier by another form of stabilisation. Landfill has been the most commonly used method which allows stabilisation to occur over an extended period of time. Land spreading on agricultural land reduces this period. Other techniques are stabilisation before landfill, for example by composting, digestion, incineration, wet air oxidation, gasification etc.

Water removal beyond that achieved by mechanical dewatering (about 60% solids) is usually accomplished by a drying process - frequently a prelude to incineration. Sludge origins generally determine the moisture content beyond which sludge burning can be independent of other energy inputs. Tissue deinking sludge normally has a higher ash content than that from newsprint or fine paper deinking, and thus self-sustaining combustion is more difficult to achieve without extended dewatering and drying.

7.7.3.1 Incineration. Incineration has been used for the stabilisation of paper mill sludges for a number of years and is being applied to deinking sludges, frequently with bark or another waste stream. Some of these installations have had less than satisfactory performance, largely due to improper selection of equipment, though there are other installations which have proved satisfactory, both in operational and economic terms [34, 35]. When sludge incineration is combined with combustion of other wastes, the

Table 7.9 Typical gross calorific values of some sludges and fuels

Fuel	Typical gross CV ($MJ\ kg^{-1}$)
Wood (dry, ash free)	19.8
Household refuse (dry)	11.6
Cellulose (dry)	17.5
Deinking sludge (55% ash, dry)	7.8
Deinking sludge [36] (20% ash, dry)	12.0

level of substitution achievable is determined not only by the moisture content of the sludge, but also by the relative ash and oxygen content. Combustion technology selected must be appropriate for the type of sludge and possibilities include [36] travelling grate, vibrating grate, fluidised bed and circulating fluidised bed.

Typical gross calorific values (CV) are given in Table 7.9 for a number of sludges. A more precise value should be obtained for any sludge considered as fuel. Gross CV is the total heat content on combustion, including the heat which would be obtained by condensing water vapour in the flue gases. Net CV is the gross value less the latent heat of vaporisation.

7.7.3.2 Land application. Due to the organic content of sludges the characteristics of some soils, especially heavy clay soils, can be improved by land application of sludge. Many trace elements are also present, so micronutrients required for plant growth are also supplied. Macronutrients, nitrogen and phosphorus are also present, but frequently supplements must be added to avoid immobilisation of nitrogen in soils. Odours from application can be avoided by sub-surface application, or by ploughing in, otherwise foul odours develop after a few days and persist, until the sludge is stabilised. This takes about 30 days in temperate climates.

A major drawback of this approach can be seasonal variations: autumn and spring may be seasons when land application is possible, but another stabilisation/disposal method is required when land application is not practical. Heavy metal buildup must be monitored, though in this respect deinking sludges typically have a lower heavy metal content than sludge from domestic sewage [4, 37].

Typical of land application programmes is that of Bridgewater Paper, a UK newsprint mill. Sludge analysis showed the sludge was suitable, since it revealed no red list materials and low levels of trace metals. Field trials demonstrated that sludge could be applied at levels up to $200\,t\,ha^{-1}$ (wet weight), equivalent to about $34\,t\,ha^{-1}$ of organic materials. Crop yield reductions through nitrogen immobilisation are avoided by the application of a nitrogen supplement, $40\,kg\,ha^{-1}$ of nitrogen per $100\,t\,ha^{-1}$ of sludge, or at a pro rata level. Before actual land application could start it was necessary to get permission from relevant authorities, and to meet their testing schedules. Benefits were:

- neutralising value, from the carbonate content;
- extra organic material;
- increase in available water capacity, of importance in dry seasons.

Bridgewater Paper Company produce about 3500–4000 t of sludge a week from deinking and paper production, and in 1993 about 95% of this was being applied to land, at about 25% of the landfill cost [37, 38].

7.7.3.3 Landfill. This has been the preferred disposal option, though stabilisation is slow, estimated at 30% over four years, in a temperate climate [39]. In developed countries, increases in landfill charges have had a marked impact on volumes disposed of by landfill. Even where mills have access to their own landfill sites, costs have increased substantially, due to increased regulatory needs governing landfill operation. Although there is not a shortage of potential landfill sites, strict control of licensing is leading to a shortage of landfill sites, especially close to urban areas where the need is greatest. Unless this licensing policy is reversed, costs will increase to the extent that other disposal options will be favoured by recycling mills, but some of these produce residuals which will still require landfill.

7.7.3.4 Other stabilisation processes. Due to the increased emphasis on sludge stabilisation and volume reduction many other types of processes are being evaluated. Ultimate disposal of residues is almost inevitably by landfill. Processes include:

- Composting – to produce a humus-like material, stable, free from nuisance odours. Sludge can be composted with other materials, yard waste, wood chips, straw, etc.
- Indirect steam gasification – gas generated is used as fuel [40].
- Supercritical wet oxidation – high pressures allow low-temperature oxidation to be used to stabilise sludge [41].
- Thermal conditioning – this produces a sludge which is easier to dewater and nitrogen and sulphur compounds in the sludge are removed.
- Anaerobic digestion – methane gas is produced, sludge solids are stabilised and an easily dewatered sludge is produced. Digestion could be with other organic wastes.

7.7.4 Beneficial uses of sludges

Some of the above methods of disposal are beneficial, for example, land application and incineration or other treatment for energy recovery. Other beneficial uses include:

- Use in low-grade board products, such as low-grade packaging papers;
- Use in building products, such as bricks, cement tiles, or as a component in cement production [42];
- Asbestos substitute [43];
- Cat litter production.

7.8 Chemical use in water clarification and effluent treatment

Many types of chemicals are used, but where processed water is recycled these can have a substantial effect on either wastepaper processing or machine performance and product quality. In tissue making, for example, polymers used to clarify deinking mill backwaters can have a negative effect on creping, whilst both inorganic and organic flocculants can have either a positive or negative effect on sizing, for example polyethyleneimine (PEI), polydiallyldimethyl ammonium chloride (PDADMAC), etc. promote alkyl ketone dimer (AKD) size cure [44]. Where defoamers are used to control foam, they can substantially reduce oxygen transfer efficiency in activated sludge systems, or reduce the efficiency of ink removal in flotation cells, as well as reducing the efficiency of DAF clarifiers.

7.8.1 Chemical use in freshwater clarification

In many cases this is not necessary, especially when underground water sources are used. When surface waters are used these tend to contain some coloured or suspended impurities which may have to be removed, dependent on the product being produced and the concentrations of impurities. In the case of board mills only suspended solids removal may be necessary, whereas if the recycling mill produces a woodfree market deinked pulp (DIP) then solids and colour removal will be necessary. There is likely to be considerable seasonal variation in surface water quality and hence treatment needs can vary with time, so it is necessary to have records for a full year when assessing treatment requirements and hence processes.

Natural colour is frequently derived from a group of substances known as humic acids, which give a characteristic brown colour to surface waters. Some of this material is in true solution, but high molecular weight fractions may have some colloidal characteristics. Suspended solids are present in a range of particle sizes, with the upper limit set by the turbulence of the surface water. Clay is frequently present and is usually present in the colloidal state ($< 1\,\mu m$ in diameter) and so cannot be removed efficiently by simple sedimentation or clarification. Where treatment is necessary, coagulants such as alum can be used as primary coagulants. In the case of humic acids, pH plays a major role in determining the coagulant dose necessary. As the concentration of humic acids increases the optimum pH falls, which is not the case with clay particles. A polyacrylamide polymer is frequently used to increase the size of the relatively small particles created by the primary coagulant, which increases the ease of removal of the suspended solids. Many other chemical types are used as coagulant aids.

7.8.2 Recycling mill backwater clarification

Recycling fibres is a water intensive process, but costs of high water use have become unacceptably high (in economic and environmental terms) and so most recycling mills rely heavily for a significant proportion of their process water on clarification of waste water which has already been used, for example, backwater from a washing stage can be clarified and used as dilution water. Reuse of water inevitably creates problems, since contaminants are recycled with water. The level of contaminants being recycled is dependent on the efficacy of the clarification process. As already noted, chemicals used in the clarification process can have a substantial effect on the recycling or paper making process. They are also a substantial component of operating costs. Hence the clarification process is of great importance, yet is frequently ignored or misunderstood, despite a considerable quantity of information being available on the basic chemistry of the process. The object of clarification is to produce a process water which can fulfil the requirements necessary for the process to be successful. It is not always necessary to produce a clear water, substantially free from suspended or colloidal solids. This standard is achieved only by increasing the quantity of chemicals used, which increases both operating costs and the risk that there will be a negative impact on the recycling or paper making processes. Hence the target for water clarity should be set by the requirements of the use of the clarified water.

Board recycling plants have the lowest quality requirements for process water. Backwater streams from board recycling plants tend not to have as high a load of mineral fillers as deinking mills, so that backwater clarification is relatively easy and can be accomplished by sedimentation or with a single chemical system of dissolved air flotation. Board mills also tend to have highly closed water loops. A high degree of closure can be achieved by collecting and clarifying effluent streams for reuse, but a closed system with zero discharge requires the removal of dissolved organics, which cannot be achieved by clarification. Biological treatment is necessary and after subsequent settling this water stream replaces the freshwater intake.

Deinking mill backwater streams tend to be difficult to clarify, though achieving a solids concentration of 200–400 $mg\,l^{-1}$ is relatively easy. Solids removal beyond this requires more complex chemistry, close control and increased chemical use. Residual suspended solids are due primarily to the presence of mineral fillers, especially clay, which has particles in the size range 0.1–1.0 μm, which form a stable colloidal suspension, which gives even clarified waters their typical milky white colour. Organic materials removed from wastepaper also forms colloidal suspensions. These two basic types of colloidal solids have very different properties, and so for a specific set of conditions, tend to be removed at different efficiencies, that is, optimal conditions for clay removal will not necessarily be optimal conditions for organic materials and hence COD removal.

7.8.2.1 Flocculation and coagulation. The efficient removal of suspended and colloidal solids is an important aspect of recycling, especially for deinking mills, so basic phenomena are outlined below. Aggregation of colloidal solids is an essential step in their removal, so their destabilisation is a necessary part in achieving this goal. In the case of backwater clarification this may be necessary when clarified water is to be used in an application which requires a low solids level, such as high-pressure showers, or when mill water systems are closed up, and freshwater consumption is low. Final effluent quality, after water treatment processes, is probably more important, since standards are usually legal requirements and high turbidity effluents are generally unacceptable.

Water molecules are oriented by all surfaces but especially by ionically active surfaces, such as those found on hydrophilic (water loving) particles, such as clays, and some organic molecules, such as polysaccharides and proteins. The surface charge is dependent on pH, but is usually negative, and the charge reduces (i.e. becomes more negative) as pH increases. Although there are many different types of particles present, they repel each other. Hydrophobic particles (primarily organic) have little attraction for water, since they do not have a strong ionic surface, and are stabilised by layers of water molecules which cover the surface of the particles. However, inert organic substances will adsorb anions, especially hydroxyl ions, and so acquire a negative charge. Other organic materials, ranging from microorganisms, proteins, cellulose etc. have a surface charge through the ionisation of, for example, carboxyl and amino groups.

Regardless of how the surface charge is developed, it contributes to the stability of colloidal particles. Similarly charged ions are repelled whilst oppositely charged ions (counter ions) become attached to the surface and are bound by Van der Waals and electrostatic forces forming an ionic double layer. These ions are ‘fixed’ to the surface and are bound strongly enough to resist thermal agitation. This arrangement is illustrated schematically in Figure 7.13. The predominance of one type of ion near the particle surface results in a potential energy gradient, illustrated in Figure 7.14 [45]. If the particle illustrated in Figure 7.14 (having a negative charge) was placed in an electric field, in an electrolyte solution it would be attracted to the cathode, pulling a cloud of ions with it. The effective charge of the particle is that which exists at the plane of shear, that is, at the surface of the cloud of ions. This value is the zeta potential. Use of the measured

Figure 7.13 Ionic double layer.

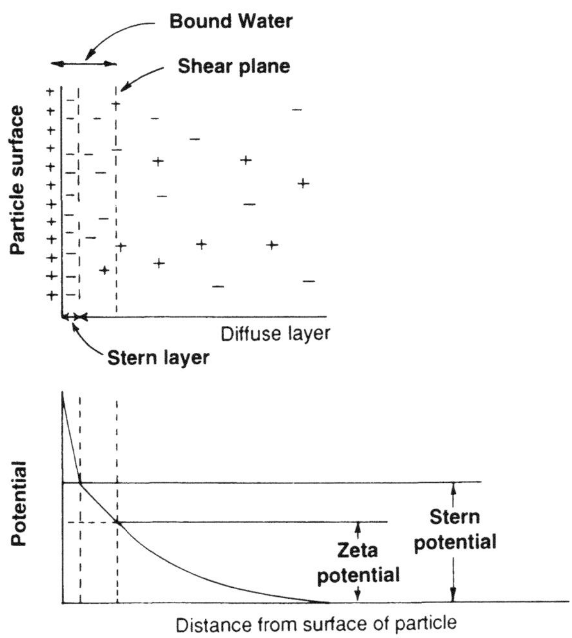

Figure 7.14 Stern model of electrical double layer [45].

zeta potential to predict colloidal destabilisation (the critical point) has not always been successful because the thickness of the layer of counterions varies with, for example, the nature of the solution components, temperature and pH, so its usefulness is limited.

Hydrophilic dispersions, which can be stabilised by similarly charged electric double layers are also stabilised by water molecules (bound water) around the particles. The innermost layers of bound water may not be penetrated by ions. Although the theory of hydrophilic dispersions is less well understood, it is clear that partial removal of the bound water around particles is necessary to destabilise these suspensions. In contrast, the theory of the stability of hydrophobic dispersions is well established, initially by the DLVO theory, which has been extensively reviewed [45, 46]. These dispersions are very sensitive to small amounts of additives which increase or decrease the range and/or the magnitude of the double layer.

There are two types of forces, repulsive and attractive, between colloidal solids. Repulsive forces are due to similarly charged electrical double layers surrounding the particles, or to particle–water interactions (bound water) or both. Attractive forces are Van de Waals forces between the particles. When a dispersion is stable, repulsive interactions between particles are greater than attractive forces, but to aggregate the particles the attractive forces must be greater than the repulsive forces. In a chemically complex situation such as exists in a backwater stream, changes in conditions will simultaneously reduce the stability of some particles, increase others and change inter-particle ionic interactions.

Repulsive forces, and hence the stability of a dispersion, are dependent on:

- particle size (radius), shape and thickness of the electrical double layer;
- the distance between particles;
- their surface potential;
- the ionic strength of the dispersing liquid.

These determine the total potential energy of interaction between repulsive and attractive forces, so that the curve for the total potential energy of interaction depends on the ratio of particle size to the thickness of the double layer, the electrolyte concentration and the surface potential of the particle.

Classical colloidal theory describes two forms of aggregation, flocculation and coagulation, illustrated in Figure 7.15 though these terms are normally used interchangeably. When the ratio of the particle size to the thickness of the electrical double layer is very large, particles may interact at a relatively large distance apart and reach a secondary potential energy minimum (S). This type of aggregation, which is relatively unstable and can be reversed by agitation is called flocculation, to distinguish it from aggregation at the primary minimum (P) which is called coagulation.

As electrolyte concentration increases or surface potential decreases the energy barrier to aggregation falls. When ions are adsorbed onto the surface and increase the surface potential, stability of the dispersion is increased, whilst adsorption of counter ions reduces surface potential and dispersion stability falls. The effect is dependent on the particular ion, since the effectiveness of a counter ion in colloid destabilisation increases substantially with charge. Thus the relative effectiveness of Na^+, Ca^{2+}, Al^{3+}, as measured by ionic concentrations required to cause coagulation, is usually about $1{:}10^{-2}{:}10^{-3}$. Overdosing counterions can result in charge reversal, leading to a stable particle of opposite charge, so careful matching of

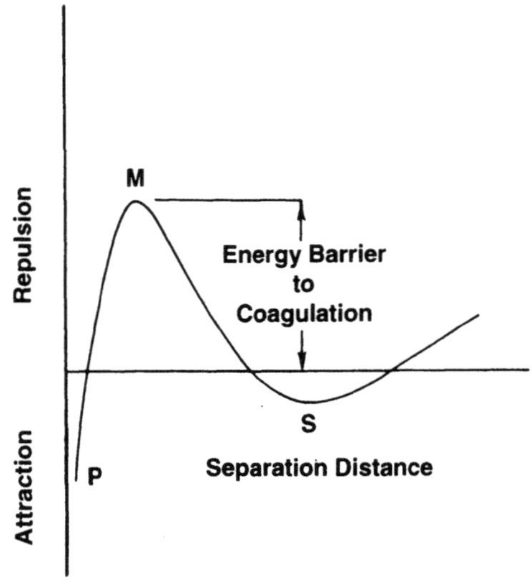

Figure 7.15 Net force between colloidal particles of like charge, as a function of separation distance.

coagulant dosage and particle charge is essential for colloidal destabilisation and coagulation. This is probably the most important mechanism in waste water treatment, but is the most difficult to control, since the nature of solids in the backwater feeding a typical clarification stage is constantly changing, with wastepaper changes. Fortunately, the feed to the final clarification unit, after biological treatment, tends to be more stable, so that good clarification is readily achievable.

Aggregation of dispersed particles (especially hydrophobic particles) can be achieved by:

- reducing particle size,
- reducing particle surface potential,
- increasing the ionic strength of the dispersing liquid,
- increasing temperature,
- increased colloidal concentration.

These explain why the specific COD (COD released per tonne of wastepaper input) falls as recycling water systems are closed up. The ionic strength of recycled water increases, due to increased electrolyte concentrations from dissolved salts as water systems are closed up, which also probably reduces surface potentials, and temperature increases. Under these conditions hydrophobic (organic) solids are more likely to aggregate and be removed as suspended solids, or be included within the fibre web on the paper machine.

Flocculation (coagulation) is thus a complex process and must be viewed as a sequence of events, including:

- Mixing of suspension with agglomeration chemicals,
- Adsorption of charge reducing compounds onto particle surfaces,
- Collisions between particles,
- Aggregation,
- Removal from suspension by sedimentation or filtration.

A number of possible mechanisms for agglomeration have been developed and reviewed. Mechanisms have been classified into groups [46], including:

- Charge neutralisation and charge reversal phenomena,
- The patch flocculation mechanism,
- Bridging flocculation (monopolymer systems),
  - adsorption flocculation
  - sensitisation flocculation
- Complex flocculation (multicomponent systems),
  - Dual polymer systems (cationic/anionic)
  - Microparticulate flocculant systems
  - Network flocculation phenomena.

In clarification of deinking effluents the most important is complex flocculation, using a dual polymer system. Cationic polymers are used to reduce anionic surface charges. Metal salts, such as cationic polyquaternary amines are common; for example, polydiallyldimethyl ammonium chlorides are used. Highly cationic materials can be overdosed easily, which can result in charge reversal, so that a stable dispersion is again formed. Since the total charge of the system is changing, with changes in materials from wastepaper, pH fluctuations, water recycling, etc. it is difficult constantly to achieve optimal polymer addition. On-line charge monitoring allows the total system to be monitored via electrophotometric monitoring equipment, though this does not measure the actual charge of various species, but measures the total charge. Using this gives a level of control to cationic polymer addition since the addition rate is controlled by the distance of the measured value from a set point.

It is frequently necessary to add a second polymer, for example, a high molecular weight polyacrylamide, to increase agglomerate size to assist removal by sedimentation or dissolved air flotation.

Colloidal particles not destabilised by the addition of coagulant aids may still be removed from suspension by their collection and enmeshment in the coagulated solids, as they form large, loose-matrix suspended solids. Since the effectiveness of this process, enmeshment, is dependent on colloidal concentration as well as coagulant dose, aids such as bentonite, activated silica, etc. are added to increase the probability of particle collisions.

7.8.3 Nutrients - nitrogen and phosphorus

Nutrients must be available for microbial growth: most, if not all recycled-fibre mills produce effluents which are nutrient deficient, primarily in

Table 7.10 Nutrient requirements for biological treatment

Treatment system	BOD	Nutrient ratio	
		N	P
Activated sludge [47]	100	5.0	1.0
Extended aeration [47]	100	0.8	0.2
Biological filtration [47]	100	5.0	1.0
Anaerobic treatment [47]	100	1.2	0.2
Activated sludge [7]	100	3.0	1.0
Anaerobic treatment (COD) [4]	100	1.0	0.3
Aerobic treatment (COD) [4]	100	2.5	1.0
Typical deinking mill effluent[a]	100	1	0.02

[a]Calculated from data in ref [8].

macronutrients nitrogen and phosphorus. Where biological treatment processes are used, macronutrient addition is necessary; both nitrogen and phosphorus must be added to avoid problems associated with nutrient deficiency, such as poor BOD removal, sludge bulking, etc. Various forms of nutrients can be added, which should be readily available for aerobic treatment systems; for example, nitrogen should be present as ammonium salts or organic nitrogen; the total of these is measured as Kjeldahl nitrogen. Phosphorus is also required, to a concentration about 20% of Kjeldahl nitrogen and when added as orthophosphates is readily available for biological metabolism. Other forms of phosphate are polyphosphates and organic phosphate.

Typical values of nitrogen and phosphorus requirements are given in Table 7.10, but actual addition rates will vary, dependent on the actual concentrations in the effluents and since the amount required for growth varies with the age of cells and environmental conditions, which are functions of the biological plant operation.

Nutrient requirements based on COD values should be used only after the relationship between a BOD and COD for a specific effluent has been established.

Micronutrients are also necessary for biological growth, and if absent can contribute to poor settling of sludges, sludge bulking problems, reduced BOD removal efficiency, etc. Detailed analyses are necessary to establish if micronutrient deficiency is a cause of problems, since the concentrations of micronutrients measurable in the effluent may not be the same as those available to the biomass [46]. Essential micronutrients are given in Table 7.11.

In order to avoid problems associated with eutrophication of receiving waters, e.g. algal blooms, or concerns arising from high concentrations in drinking water, nutrient addition needs to be carefully monitored. Some countries have introduced limits on discharges, e.g. in Germany total phos-

Table 7.11 Essential micronutrients [48]

Substantial quantities	Trace quantities
Sodium	Iron
Potassium	Copper
Calcium	Manganese
Magnesium	Zinc
Chloride	Boron
	Molybdenum
	Cobalt
	Vanadium
	Nickel

phorus limits are $3.0\,mg\,l^{-1}$, and for some effluents with concentrations $>0.1\ mg\,l^{-1}$ environmental taxes are payable [7].

When biological treatment of a recycling-plant effluent is in a combined domestic and industrial plant, the excess of nutrient available from domestic sewage means that additional nutrients may not be required, dependent on relative waste flows. Hence the total load on receiving waters can be reduced, since nutrients are consumed to produce microbial cells, which are removed as surplus sludge. When biological treatment of a recycling-mill effluent is not in a combined plant, the nutrient load on the receiving water is very low, provided care is exercised in nutrient addition. Added nutrient is removed as surplus cells and a high value in treated effluents indicates too high an addition rate.

References

1. *United States Patents*, Numbers 4,615,767 (1986) and 4,668,339 (1987).
2. Clarke, James d'A. (1981) *Pulp Technology and Treatment for Paper*, Miller Freeman Publications, San Francisco, p. 148.
3. Rasmussen, J. (1993) *Market Deinked Pulp*, Paper given at the 1993 ReC '93 International Recycling Congress, Geneva.
4. Bulow, C. and Kroi, B.H. (1990) The economic treatment of effluents in a papermill, in *Recycling Paper: From Fibre to Finished Product*, TAPPI Press Vol. I, pp. 254–271.
5. Hamm, I. and Gottsching, L. (1989) Schwermetallpfade bei der Altpapieraufbereitung, *EUCEPA Symposium* 1989, *Recycling of Fibres and Fillers in the Pulp and Paper Industry*, Ljubljana, pp. 83–103.
6. Miner, R., Berger, H., Fisher, R., Someschwar, A., Unwin, J. and Weigand, P. (1991) Environmental considerations and information needs associated with an increased reliance on recycled fibre, Focus '95 and Proceedings, *Landmark Paper Recycling Symposium*, TAPPI Press, pp. 343–374.
7. Diehn, K. and Zuercher, B. (1990) An integrated waste management program for high ash papermill sludge, *TAPPI Environmental Conference*, TAPPI Press, pp. 563–567.
8. Anon, *Best Available Technology for the Pulp and Paper Industry*, Report for Water Resources Branch, Ontario Ministry of the Environment, (February 1992), Appendix A, Queens Printer for Ontario.

9. Anon, *The Development Document for the Effluent Monitoring Regulation for the Pulp and Paper Sector*, Report for Ontario Ministry of the Environment (July 1989) Table 4. Queens Printer for Ontario.
10. Anon, US EPA, *Development Document for Effluent Limitation Guidelines and Standards for the Pulp, Paper and Paperboard Point Source Category*, EPA 440/1-82/025 (October 1982).
11. Metcalf and Eddy, revised by G. Tehobanoglous (1979) *Wastepaper Engineering: Treatment, Disposal, Reuse*, McGraw-Hill.
12. Clewley, J.A. (1983) Bridgewater - a new mill for the 80s and beyond, *Paper Technol. Ind.* **24**(1), 5.
13. Patterson, J., *Deinking at Bridgewater*, Pita Conference, PITA Paper Week 1987, Paper Industry Technical Association, UK.
14. Rosling, M. (1990) *Deinking and Newsprint*, North West Expo Conference, Vancouver.
15. *Best Available Technology for the Pulp and Paper Industry*. Report for the Water Resources Branch, Ontario Ministry of the Environment, (February 1992), Appendix C. Queens Printer for Ontario.
16. Fellkjaer, H.O. (1989) Fibre recovery and effluent treatment in a wastepaper mill, In Pira Conference Proceedings, *New Developments in Wastepaper Processing and Use*, Gatwick, Volume I.
17. Maat, D.Z. (1990) *Anaerobic Treatment of Pulp and Paper Effluents*, TAPPI Environmental Conference, Seattle, TAPPI Press, pp. 757–760.
18. Simon, O. and Ullman, P., Anaerobic treatment of effluents today, *Paper* (18th March 1985), pp. 42–44.
19. Habets, L.H. and Webb, L.J. (1987) *Anaerobic Effluent Treatment: Past Experiences and Future Opportunities*, Pita Paper Week Conference, York, Paper Industry Technical Association, UK.
20. Paasschens, C.W.M., *et al.* (Nov 1991) Anaerobic treatment of recycled paper mill effluent in the Netherlands, *Tappi J.*, **74**(11), 109–113.
21. Sarner, E., *et al.* (1988) Anaerobic Treatment Using New Technology for Controlling H_2S Toxocity, *Tappi J.*, **71**(2), 41–45.
22. MacDonald, M. (1990) Technology for Ecology, *Paper Technol.*, **21**(7), 38–40.
23. Webb, L.J. (1991) Wastewater characteristics, consent standards and treatment options, PITA Annual Conference Papex '91, *Environment 2000*, Manchester, Paper Industry Technical Association, UK.
24. Rowbottom, R.S. (1993) Risks of bacterial hydrogen generation in white water systems, *Tappi J.*, **76**(1), 97–98.
25. Robichaud, W.T. (1991) Controlling anaerobic bacteria to improve mill product quality and improve safety, *Tappi J.*, **74**(2), 149–153.
26. Mermillod, P., *et al.* (1992) Compact anaerobic/aerobic wastewater treatment at the Minguet and Thomas recycled paper mill in France, *Tappi J.*, **75**(9), 177–180.
27. Brown, S. (1991) Treatment of paper manufacturing effluent using the Colox™ process, PITA Annual Conference, Papex '91, *Environment 2000*, Paper 3.
28. McKinney, R.W.J. (1979) Deep shaft treatment of a deinking mill effluent, Pira conference, *Wastepaper Processing*.
29. Denton, R.S. (1992) Developments in secondary treatment, *Paper Technol.*, **33**(7), 21–24.
30. Thomas, R.G. and Wilson, C.M.W. (1990) *Use of Pure Oxygen for Biological Treatment of Strong Recycled Fibre Effluent*, PITA Annual Conference, Manchester.
31. Le Compte, A. (1974) A Pasaveer Ditch, American Jet Style. *29th Purdue Industrial Waste Conference*.
32. Bolton, R.L. and Klein, L. (1971) *Sewage Treatment and Trends,* Butterworth, 2nd Edn., London.
33. Guo, H. and Coller, B.A.W. (1988) Chemical characteristics of a paper mill - wastewater foam, *Tappi J.*, **71**(11), 167–172.
34. Louhimo, J.T. and Mullen, J.F. (1991) Sludge burning in fluidised bed boilers, *Tappi J.*, **74**(3), 119–172.
35. Nickull, O., *et al.* (1990) Burning mill sludge in a fluidised bed incinerator and waste-heat recovery system, *TAPPI Engineering Conference Proceedings*, TAPPI Press, 885–900.

36. Kraft, D.L., and Mitchell, G.W. (1991) Sludge amount, type are key factors in choice of incineration technology, *Pulp and Paper*, **65**(9), 167-169.
37. Stringer, D. (May 1993) *Recycling of Papermill Sludge to Agricultural Land - The Bridgewater Experience*, PITA Environmental Seminar.
38. Aitken, M. (1992) *Applying Deink to Farmland*, ADAS Paper, CJ116, Guide Notes for Farmers.
39. Board, N. (1986) Two stage treatment could solve the sludge problem, *Paper Technol. Ind.*, **27**(2), 91-96.
40. Kandaswamy, D.S., *et al.* (1991) Indirect steam gasification of paper mill sludge waste, *Tappi J.*, **74**(100), 137-143.
41. Modell, M., *et al.* (1992) Supercritical water oxidation of pulp mill sludges, *Tappi J.*, **75**(6), 195-201.
42. Soderhjelm, L. (1976) *Paperi ja Puu*, **9**.
43. Pira Report, (April 1973), TS116.
44. Thorn, I., *et al.* (1993) The use of cure promoters in alkaline sizing, *Paper Technol.*, **34**(1), 41-45.
45. Shaw, D.J. (1966) *Introduction to Colloid and Surface Chemistry*, Butterworth, London.
46. Lindstrom, T. (1989) *Some Fundamental Chemical Aspects on Paper Forming*, ed. C.F. Baker, Research Symposium, Cambridge, Vol. 1, pp. 309-412.
47. Gostich, N.A. (1990) The nutrient requirements in biological effluent treatment, *Paper Technol.*, **31**(8), 33-35.
48. Kimball, J.W. (1968) *Biology*, 2nd Edn, Addison-Wesley, Reading, Mass.

8 Manufacture of packaging grades from wastepaper

R.W.J. MCKINNEY

8.1 Introduction

Until the industrial revolution, packaging was a relatively unimportant product of the paper industry. In the UK, packaging board became important towards the end of the 19th century, especially in the industrial north west of England. Packaging board use was increased by the introduction of heavy weights of fibreboard for packaging in the early 1900s, which led to the installation of several new packaging board machines in the south of England, though these were initially based on imported virgin fibres. During World War I, wastepaper began to be widely used in the production of packaging grades and this continued after the end of the war. Through the 1930s about 800 000 tonnes per annum of wastepaper and board were collected, largely by local government authorities. Some of this was exported but most of the balance, about 600 000 tonnes per annum, was used in the production of packaging grades, with a utilisation rate of 30–40%. The use of fibrous packaging received another stimulus with changes in retailing and wholesale distribution, especially the trend towards department stores and multiple shops, which started between the two world wars. In 1938, packaging paper and board consumption in the UK was 1.4 Mtonnes per annum, out of a total consumption of 3.5 Mtonnes of paper and board products.

Wastepaper consumption during World War II increased to about 1 Mtonnes though its use in the production of grades other than packaging also increased. High rates of wastepaper use in packaging production continued after the second world war.

During the 1960s, the retail revolution gathered pace with a strong move towards larger department stores and supermarkets, which increased the demand for packaging board and corrugated packaging, used as transit packaging. As more women joined the labour force, packaged food grew in importance increasing the demand for packaging board and corrugated transit packaging. Hence increased demand for packaging was a consequence of changes in the organisation of society and the UK industry expanded to meet these demands. Expansion was based largely on wastepaper use, so that by the late 1970s many grades used in the production of corrugated containers and other fibrous packaging were produced entirely from wastepaper.

Table 8.1 Wastepaper use in the production of packaging grades in the UK, in 1992

Grade	Total production (Mt)	Total wastepaper use (Mt)	Utilisation rate (%)
Corrugated case materials	1.381	1.510	109
Packaging board	0.670	0.461	41
Packaging papers	0.092	0.038	69
Total	2.143	2.009	94

Table 8.2 Wastepaper use in containerboard production, in the USA, in 1992

Grade	Total production (Mt)	Total wastepaper use (Mt)	Utilisation rate (%)
Containerboard	27.7	15.1	35

Although the time frame of the development and use of fibrous packaging in other developed countries is different from that in the UK, the same general pattern was followed; for example, in the USA, the 'Pridham Decision' in 1914, broke the monopoly of the wooden box in packaging. In fibre-rich countries, such as the USA, wastepaper use was not widespread but in fibre-deficient countries, like the UK, wastepaper use in packaging became significant early in the life of the industry. Most of the wastepaper now used by the paper and board industry is recycled to produce packaging grades. Of the wastepaper recovered in 1992, for example, in the UK, about 65% was used to produce packaging: corrugated containers, solid packaging boards and packaging papers. Wastepaper use in these grades is illustrated in Table 8.1, and totalled 2.0 Mt, out of total wastepaper use by the UK paper and board industry of 3.1 Mt.

Of the total packaging consumption in the UK, about 56% was recovered and recycled in 1992. In areas with lower wastepaper utilisation rates, for example, the USA (Table 8.2) and Canada there has been considerable government pressure to increase wastepaper use, due to the significant contribution of paper and board packaging to domestic and solid waste streams, which is illustrated by the statistics given in Table 8.3.

In the USA and Canada this pressure resulted in a rapid increase in wastepaper use in board production, which has given rise to a number of new board recycling mills, located close to urban populations, rather than forests. These have been called 'mini' mills [7] since they have an appreciably lower capacity than mills based on wood. The increase in wastepaper utilisation rates, given in Table 8.4 for Canada, has been largely a result of political and social pressures, rather than for economic reasons.

Although the wastepaper utilisation rate in the USA and Canada is much lower than in many European countries, such as the UK and Germany, it

Table 8.3 Contribution of fibrous packaging grades to solid wastes (% of total)

	Commercial waste	Domestic waste	Municipal solid waste
UK	13.9[a]	6.5[b]	-
USA [1]	19.2	7.7	13.5

[a]Average of surveys in City of London [2], City of Westminster [2] and West Midlands [3]
[b]Average of Mole Valley [4], Ealing [5] and Richmond [6] surveys.

Table 8.4 Use of recycled fibre in packaging grades, in Canada [8]

	Wastepaper utilisation (%) in	
	1965	1992
Linerboard	18	31
Corrugating medium	9	44
Folding box-board	47	69

must be remembered that these latter countries import substantial quantities of kraft liner and other virgin packaging grades. Hence, direct comparisons are misleading, since fibre-rich countries produce their own virgin grades. Use of these, for example, kraft liner, is essential to provide some properties required for specific end uses.

8.1.1 Packaging products made from recycled fibres

There is a very wide variety of packaging products, ranging from multi-ply sacks through cigarette cartons to corrugated containers. Mechanical specifications preclude the use of recycled products in some cases, for example, due to high compression requirements. In other packaging grades, the contents of the packet may be in direct contact with the contents. Recycled fibres have a less certain history than virgin pulp, and so with food contact products, recycled grades are rarely used, for example, in sugar bags. In some countries there are restrictions on the use of recycled fibres in containers with food contact, especially moist and fatty foods; for example, in Sweden the National Food Administration regards recycled fibre-based products as unsuitable for use in direct contact with moist or fatty foods [9]. In the case of dry foods in general, production wastes (but not post-consumer wastes) can be used, though care should be exercised when dry foods have a fatty surface and migration from the packaging material to the fatty surface can occur. In these cases, the use of wastepaper is not recommended [9]. Even a virgin ply on top of recycled fibre is not considered adequate protection and a protective layer of plastic or aluminium is considered necessary [10]. However, even the adequacy of this protective layer is being questioned, since vapour-phase migration across the barrier layer can occur, because polyethylene film, for example, is permeable to many vapours [11].

Similar restrictions on recycled fibre use in food packaging materials are in use in Finland and Germany and the Council of Europe has published a draft Resolution on Paper and Board for Food Contact, which would set stringent purity requirements [11].

The most important packaging products produced from wastepaper are:

- components of corrugated containers - liner grades, sometimes called bogus, jute or test liner, and waste-based corrugating medium. White top liner grades have grown in importance as point of sale advertising develops, using the transit container. This white top, the white layer of the liner product, can be deinked fibre from woodfree wastepapers, or bleached virgin fibres.
- solid board - called folding box-board, containerboard, paper board. Solid board grades may also have a white layer and may be coated, for example, white lined chipboard.
- packaging papers - paper bags, wrapping papers, etc. Grades which do not use 'brown' wastepaper are relatively small and are not considered further in this chapter.

8.2 Wastepaper grades used in packaging production

Grades of wastepaper used have differing definitions in various countries, largely due to differences in terminology and collection systems. Using corrugated containers as an example, definitions for approximately equivalent grades, according to the various wastepaper grading systems, are:

- British definition from the British Paper & Board Industry Federation (BP&BIF): group 8: container waste (Old KLS):
  Used printed or unprinted corrugated boxes and solid fibreboard boxes. The proportion of solid board must not exceed 10% per bale. May contain a minimum of adhesive tape. Objectionable materials: 2% (but excluding wax, bitumen, plastic laminates, unshredded cores, egg boxes and kraft bags).
- Paper Stock Institute of America (PSI) grade 11: corrugated containers:
  Consists of baled corrugated containers having liners of either test liner, jute or kraft. Prohibitive materials may not exceed 1%. Total outthrows may not exceed 5%.
- European (CEPAC) A5: corrugated containers:
  Consists of used cases or sheets of corrugated board, with or without

kraft covers and a middle of straw or wastepaper, free of bitumen, waxed and plasticised paper.

- German (BVP) group 1: grade B19:
  Wastepaper from retail outlets, used packaging and wrapping, but containing at least 70% corrugated material, solid board and packaging papers; a maximum of 1% contains foreign bodies or grades which may be detrimental to production.
- Japan Paper Recycling Promotion Center definition: grade 22: old corrugated containers:
  Old corrugated containers. Objectionable materials maximum is 3%.

Although the wastepaper defined is used to produce equivalent grades in different countries, the differences in definitions mean that quality and hence processing needs can be very different. In the UK (BP&BIF) and Germany (BVP) definitions, containerboard (solid board) is permitted, though the fibre quality of this grade is usually lower than corrugated case material, due to a lower proportion of softwood kraft fibres. In the BP&BIF and CEPAC definitions, waxed board is not permitted, but this is present in old corrugated container (OCC) waste in the USA, though individual mill definitions normally exclude waxed OCC. Prohibited materials allowed ranges from 1% (USA, Germany), to 2% (UK) and 3% (Japan), with no limit defined in the CEPAC grade definition.

Similar differences in definitions exist for other grades; for example, in mixed papers the proportion of short fibres permitted varies; in the USA < 10% of groundwood is acceptable in grade 3, super mixed paper; whereas in the CEPAC definition of mixed papers and boards No. 1 (A2) it must contain < 15% short fibre papers; in the UK (group 9) mixed papers must not contain more than 10% of news and PAMS and the German grade B12, sorted mixed wastepaper must contain < 40% old news and PAMS.

Since brown wastepapers are the most important grades traded internationally, differences in definition are very important. In 1992, 2.6 Mtonnes of OCC were exported from the USA, as well as 0.8 Mtonnes of mixed papers. Given these differences in grade definitions it is not surprising that there are occasional problems with respect to different definitions of acceptable qualities.

Other grades used include double-lined kraft, multi-ply kraft sacks, kraft cuttings, box-board cuttings and trim, double sorted corrugated, etc. Trims from converting processes are generally very clean.

Prices for these grades vary according to their fibre content and market conditions, that is, the balance between supply and demand, which results in considerable price variations, as illustrated in Figure 1.2, in chapter 1. Differences between grades are due to varying levels of contaminants and relative fibre quality. The most costly grades are those which contain high proportions of softwood kraft and no, or very few, contaminants, such as 'new double lined kraft'. Old corrugated containers quality varies according to their source. In 1992, for example, in the USA 65% of corrugated materials use virgin kraft liner, whilst in Germany only 26% of corrugated materials use kraft liner. Hence old corrugated containers from the USA contain a greater proportion of first cycle softwood kraft fibres than those from Germany, and so OCC from the USA has a higher price than its equivalent German grade. Softwood kraft fibres are desirable in liner grades, due to their strength properties.

Fibres used in corrugating medium are relatively short, and in virgin medium are produced by a number of processes, including alkali carbonate, green liquor and neutral sulphite semi-chemical pulping, usually from hardwoods, though straw is also used, for example in Spain. Ideally, recycled products would be made from similar fibres, liner grades from long fibre (kraft softwood) and corrugating from short fibres from hardwood. However, it is not possible to be specific in the choice of fibres when using wastepapers, since these inevitably contain fibre mixtures. Differences in fibre furnish are thus one of the main reasons for differences in properties between recycled and virgin grades, discussed in section 8.7.

8.2.1 Wastepaper use in liner grades

In virgin kraft liner production, filler fibres can be used, up to about 15–20%, without adversely affecting strength properties. Frequently old news or similar grades are used, without deinking, though centrifugal cleaning and screening may be necessary, to remove contaminants such as grit and stickies. Up to 20% wastepaper use still permits the liner board to be classed as virgin liner board in the USA. If a higher proportion of wastepaper is used, new double lined kraft can be used, which again requires minimal or no processing. However, if lower quality grades of wastepaper are used, more sophisticated processing is necessary. Quality levels of the liner grade also falls, though an acceptable liner for many applications can still be produced. Test, jute or bogus liner is usually made from 100% waste, but from grades which contain some softwood kraft fibres, which are essential to provide required strength properties. A top ply may be added to provide a cleaner appearance, by using high quality wastepaper, or used with white fibre to give a mottle or solid 'white top'.

White top liner can be produced from 100% wastepaper. In this case, the white ply is produced from woodfree grades, using conventional deinking processes, as described in chapter 3. Since the requirement is for a good quality finish, to accept printing for point of sale advertising, it is essential to produce a clean, bright layer, with good printing properties. Thus the two fibre preparation systems (white and brown liner) need to be completely separate, including water systems, though excess backwater from the white layer preparation system can feed the brown stock system, but not vice versa.

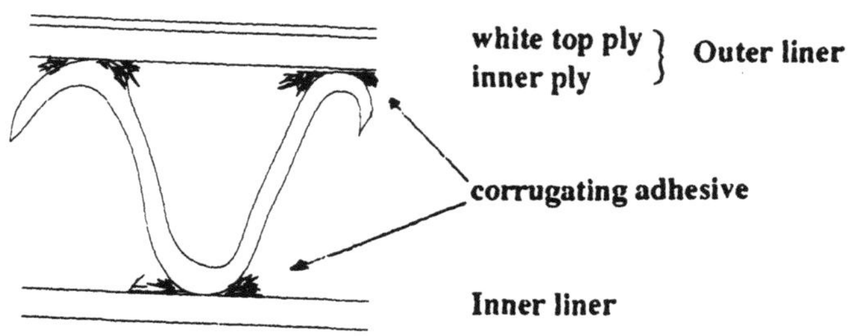

Figure 8.1 Structure of combined board.

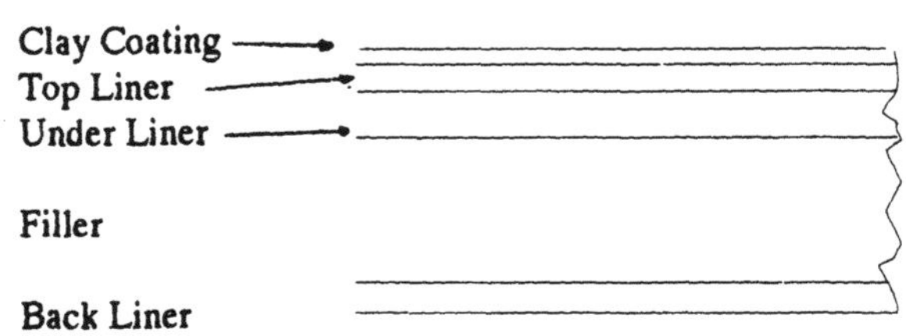

Figure 8.2 Coated packaging board (box-board).

8.2.2 Wastepaper use in corrugating medium

Stiffness is the most important criterion for corrugating, which is why hardwood fibres are used. In Europe, most medium is produced from 100% wastepaper, which in the USA is sometimes called 'bogus' medium. Initially, wastepaper grades used were clippings but this was replaced by old corrugated containers to reduce costs. In Europe, high proportions of mixed wastepapers are frequently used to produce medium, blended with OCC or equivalent wastepaper grades.

The structure of combined board, medium and liner combined, is illustrated in Figure 8.1.

8.2.3 Wastepaper use in box-board grades

This grade has to be stiff, able to fold at scores and usually has to have good printing properties. Short fibres enhance stiffness properties, but long fibres are required to give strength and runnability properties. Short fibres in the top ply provide good printing properties, which are improved by coating the top ply. Heavyweight grades of solid board are normally produced on multi-ply machines and wastepaper grades used normally reflect the requirements of the ply. The structure is illustrated in Figure 8.2.

Inner or filler plies tend to be produced from low-quality grades, such as mixed, whilst outer plies (top liner and back liner) are made from higher quality grades, though the back liner can be made from a 'clean' filler furnish. If a white top or coated board is produced, a news ply can be added between the white top and the brown grade, called the under liner ply, to reduce the weight of the white ply necessary to cover the brown and so minimise costs. Very high quality is required in the outer ply if it is coated.

8.2.4 Wastepaper use in packaging papers

Strength specifications of grocery bags are high and unless a lower strength is acceptable, wastepaper use cannot be pushed beyond the use of filler fibres, similar to kraft liner. Old news can be used, up to a maximum of about 20%. If higher proportions of wastepaper are used, then normally old news would not be included in the furnish, but grades of wastepaper with a high proportion of kraft fibres, such as new double lined kraft (NDLK) would be used. Old corrugated containers can also be used, but these require extensive cleaning and screening and some strength loss is unavoidable, at equivalent basis weights, if high proportions are used. Provided lower strengths are accepted, grocery bags can be made from 100% wastepaper, using OCC and higher quality grades.

8.3 Wastepaper processing of brown grades

Brown grade wastepaper systems are very different from those used in processing news and woodfree wastepaper grades and so are discussed separately in this section. There are differences in processing between geographic regions, some of which are explained by differences in wastepaper quality, as previously described, though some differences are due to the use of distinctly different technology; for example, soaking towers are common in Asia, but less common in Europe and North America.

Different applications also use different processing lines, though there are generally unit operations common to all systems. If recycled box-board is being made from 100% wastepaper there can be four separate quality requirements, each with a separate processing line, in addition to a virgin pulp stock preparation line, though this would usually be used instead of the top ply wastepaper system. The top ply in box-board and liner board has the highest quality requirements, filler plies and corrugating medium have the lowest, with the quality requirements of bottom ply in box-board somewhere in between. However, some grades of medium require higher quality, so either higher quality waste grades or more extensive processing are used.

Most of the wastepaper used by the paper and board industry is not deinked, but is used in packaging production, illustrated by figures given in Table 8.5. About two thirds of total wastepaper consumption is used in

Table 8.5 Wastepaper use in packaging grades, in 1992

	Total wastepaper consumption (Mt)	Wastepaper use in packaging grades (Mt)	%
USA	23.6	16.6	70.3
UK	3.1	2.0	64.5

packaging production. Hence there is a considerable amount of experience in processing these grades. As utilisation rates increase and quality demands remain the same, a level may be reached at which the quantity of new kraft fibres entering the system will not be adequate to maintain strength values, unless processing methods change.

Some of the wastepaper grades used, especially mixed, can contain a considerable amount of gross contaminants, which is why wastepaper processing systems are different from those used in recycling other wastepaper grades. In addition, there has not been a need to reach the high cleanliness levels required for other grades. This may change as demands for higher quality increase, so that packages can carry a good quality printed image of the package contents.

8.3.1 Pulping systems

A wide variety of pulping options exist, including:

- vertical low consistency, continuous pulpers;
- horizontal low consistency, continuous pulpers;
- soaking drum before pulping;
- soaking tower after pulping;
- high-consistency batch pulpers.

Low-consistency pulpers are the most common, with consistencies usually in the range 3–6%. Lower consistencies allow the use of a ragger rope in the pulper. Bale wires are cut (when baled waste is used) when wastepaper is on the pulper feed conveyor, but are not removed. These wrap around the ragger rope and entangle other contaminants such as plastic strips. However, the action of bale wires and other heavy debris can result in severe wear of the pulper tub when these are not removed by trash removal systems, such as the ragger, junk trap etc. Heavy rejects can be too large to be accepted through the perforated extraction plate in the pulper, so that a mass of rejects accumulates in the pulper bowl. To prevent this from occurring, a trash removal loop has been introduced to pulping systems to increase the efficiency of debris removal and to prevent the build up of a large mass of rejects. If a large mass of rejects does build up, removal is possible by use of an overhead crane.

In low-consistency pulping, mechanical forces are very strong, which results in the breakdown of many of the contaminants present. The pulper rotor rotates at high speed and imparts its high momentum to paper in contact with the rotor. Baffles are used to turn the wastepaper in towards the rotor. In the space between the extraction plate cutting bars on the pulper floor and the rotor, mechanical forces are high; these are known as attrition forces. Stock is withdrawn continuously from the pulper via perforations in the extraction plate and so is subject to these forces, which are effective in defibering, but also in breaking down contaminants. Nevertheless, the flake content is high, since some of the material passing through the extraction plate has been in the pulper for a relatively short period of time, enough to be able to pass through the extraction plate, but not enough for complete defibering. Normally, flake contents are in the range 8–15% by weight.

The usual choice is between vertical or horizontal low-consistency pulpers, with the added choice of a soaking drum before pulping. In one UK mill, the vertical pulper was preferred over the horizontal, due to a belief that ragger operation is better in the vertical pulper [12]. A soaking drum is similar to the initial stages of drum pulping, but there are no perforations in the drum and hence no separation into accepts and rejects. Wastepaper is fed into the drum by a conveyor and dilution water added at the wastepaper feed end to give a consistency of about 30%. Very gentle wetting and defibering energy is supplied by the impact of the wet wastepaper falling from the roof to the floor of the drum as it rotates - similar to the action of a cement mixer. The soaked wastepaper is then discharged into a low-consistency pulper. Advantages of using a soaking drum before pulping include increased pulper capacity (reduction in pulping time) and better pulping (no floating bales).

Retention time in the drum is important and is adjusted by altering the horizontal angle of the drum. Although increased retention time improves soaking, too long a retention time allows ropes to form, which block the discharge end. If the holding time in the drum is not compatible with the feed rate, blockages can occur at the drum inlet. Baling wires are not normally removed, since these are essential to form a ragger rope, if a ragger is used in the low-consistency pulper. When a soaking drum and ragger are both used, baling wires must be cut only once, since short lengths of baling wire can get trapped in internal buffles and perforations within the soaking drum. In addition, if the baling wire is cut too short it will not form a strong ragger rope, but will form solid lumps in the pulper. In automatic systems, a sensor detects the ragger and cuts it automatically to a preset length, otherwise this is manual. Ragger ropes are disposed of as solid waste. A soaking drum in a low-consistency pulping system is illustrated in Figure 8.3.

As well as the operation of the ragger there are other systems for removing contaminants in low-consistency pulping systems. These include:

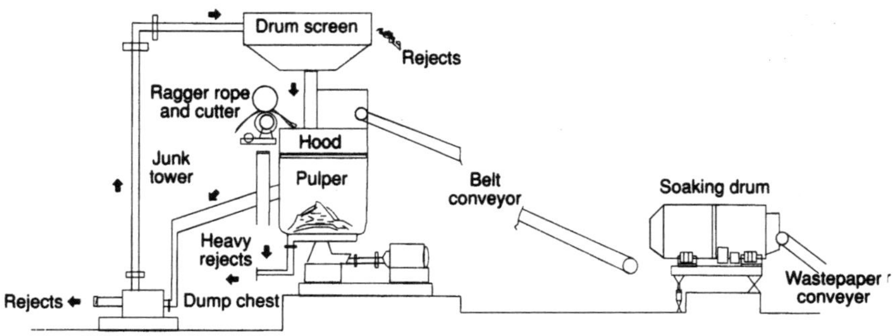

Figure 8.3 Low-consistency pulping system with soaking drum and purge loop.

- junk traps - gross heavy contaminants;
- extraction tubes - removal of heavy contaminants at the bottom and light contaminants by an overflow over a weir at the top of the extraction tube;
- a purge loop, which incorporates screening to prevent the build up of trash in the pulper and possibly a defibering system. Accepts are returned to the pulper. Normally this treats about 15–30% of the pulper capacity, though this can be varied, dependent on the level of contamination of the wastepaper.

Of these, the purge loop is the most recent and was first introduced in the early 1980s to reduce wear of the pulper tub, and the rotor extraction plate, as well as to allow increased pulper operating time, due to less down time for cleaning. Contaminants removal is more efficient, since these are removed before being broken down to a size which allows them to pass through the pulper extraction plate, which is usually located under the pulper rotor. Extraction plate hole size is determined by downstream screening equipment and the layout of the screening system of the purge loop. Typical hole sizes are in the range 10–18 mm. Wastepaper addition is (semi) continuous, with the rate of addition either under computer or operator control. Machine white water, or backwater from a thickening stage, is recycled for use as pulper dilution and is continuously added to the pulper.

Defibered stock is continuously removed from the pulper, but inevitably it contains a substantial proportion of fibre bundles, up to 15%, or even more if wastepaper grades used include some with high wet strength. Soaking towers were developed to follow low-consistency pulping. Stock may be cleaned in a high density centrifugal cleaner, then thickened, before being fed into a soaking tower. Chemicals such as caustic soda can be added. Retention times can be as high as 24 hours. Advantages claimed for soaking towers include:

- improved fibre strength properties;

- reduced fibre losses;
- improved contaminant removal;
- lower power requirement for secondary defibering.

This approach has been adopted extensively in Asia and its use has extended from brown grades into woodfree wastepaper processing. In brown grades, benefits have not always been gained. In some cases in Asia, soaking is in water, in others under caustic conditions. A prolonged soak in water would allow fibre flakes to be defibered more readily, without severe attrition of contaminant particles, which occurs in conventionally operated deflakers. Better removal of contaminants would improve some strength properties. A caustic soak has a more fundamental effect on fibres. Since caustic treatment helps to reverse some of the effects of recycling, it is this action of caustic which is responsible for improvement in strength properties.

One study found that 2% caustic soda was the optimal concentration and that many strength properties are improved. Over a 16 hour soak period, the short span compression index (STFI) increased by about 17%, with 10% achieved in the first hour, showing the response is fast and that normally, extended soaking would not be required. It was also shown that this effect could be reversed in the presence of multivalent cations and salts. Magnesium chloride and barium were shown not only to reverse improvements in short span compression but also to reduce the strength of untreated OCC [13]. These interferences may explain why another study was unsuccessful in demonstrating increased short span compression strength with caustic treatment. Over a series of tests rather erratic results were reported [14].

However, caustic treatment substantially increases the organic strength of mill effluents and would probably increase the dispersion of pressure sensitive adhesives, perhaps leading to increased problems with stickies. Despite these difficulties, caustic treatment may prove useful in extending the replacement of virgin kraft liner with recycled test liner.

Chemicals are not normally used in pulping, though if wet strength grades or waxes are present, chemicals may be added. Wax is a very common additive to corrugated grades; it is used to give water resistance, so providing wet strength or a barrier to water vapour. In 1990, about 175 000 tonnes of wax were used by the converting industry in the USA, but it is less common outside the USA. If wax is present in OCC and is not removed, it will dramatically reduce the coefficient of friction (COF) of liner and adversely affects strength, glueability and printing properties, as well as causing spots in the sheet. Other than specifying in wastepaper definitions that waxed board is unacceptable, there are two approaches to controlling wax: high temperature (>60°C) pulping, with the addition of surfactant, to disperse wax; and low temperature (<50°C) pulping, with no chemical addition, to prevent dispersal of wax for more efficient removal during later cleaning operations.

It is evident that these approaches are contradictory. The most common is to keep pulping temperatures low and use low-density cleaners to remove wax particles, though this inevitably results in the loss of fibres still associated with wax.

In the absence of wax, pulper temperatures are usually high, to increase pulper capacity, typically > 60°C.

High-consistency pulping has not been adopted by the producers of packaging grades, primarily because of the higher energy consumption and capital costs of these pulpers. Normally, stock is discharged from a high consistency pulper through a detrashing screen and rejects are retained in this until the pulp dump sequence is over, when rejects are washed and discharged from the detrash screen. In the case of heavily contaminated brown grades the body of this type of screen would have to be very large, to accommodate all the rejects. However, conventional high-consistency pulping would result in better defibering, leading to improved downstream contaminant removal. This is because the defibering forces in high-consistency pulping are less severe than in low-consistency pulping. Rotor speed is relatively slow and defibering forces are provided by fibre to fibre shear, rather than by mechanical forces or attrition. As recycling rates increase, improved contaminant removal will probably become more important and this pulping method more widely adopted.

Irrespective of the pulping method used, as stock is dumped from the pulper it is pumped through high density centrifugal cleaners, for the removal of large pieces of high-density materials, such as staples, grit, fragments of bale wire, nuts, bolts, etc. These are low-pressure units, typically with a diameter of 30–45 cm, operating at about 140 kPa. Accepts are usually sent to a stock chest and rejects to a sand separator, to dewater rejects, before being transferred to a rejects conveyor.

8.3.2 Cleaning and screening systems

These are similar in concept to those used in deinking systems, though there are variations. A typical system is schematically illustrated in Figure 8.4.

The number of cleaning stages used is dependent on both the quality of wastepaper and the product; for example, lower quality medium and middles may not use either the fine screening or reverse flow centrifugal cleaning stages. In systems where quality demands are high, for example a top ply, all of the stages illustrated would be used, (other than with very clean wastepapers) and accepts may go to further treatment stages, such as dispersion or kneading, perhaps via a fractionation stage.

8.3.2.1 Coarse screening. Coarse screening is by use of perforated pressure screens and a typical system is illustrated in Figure 8.5.

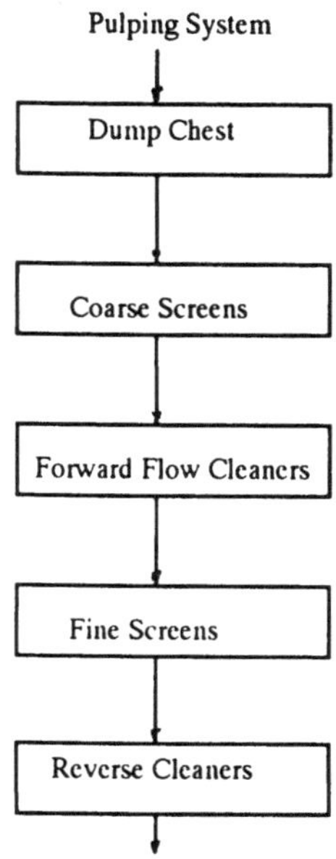

Figure 8.4 Schematic illustration of screening and cleaning systems.

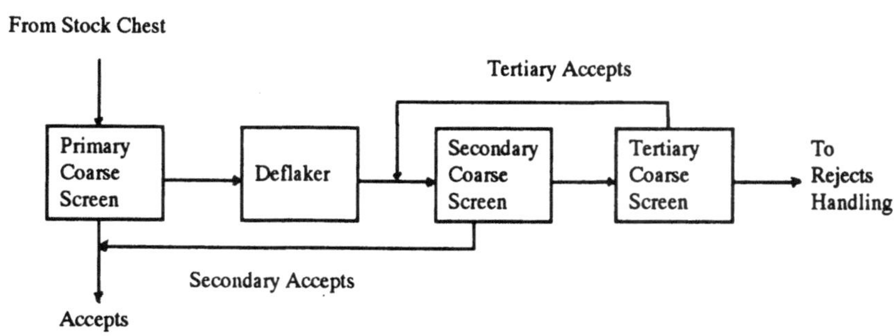

Figure 8.5 Schematic illustration of coarse screen system.

Since the stock usually contains a proportion of undefibered flakes due to the use of continuous pulping, rejects from the coarse screen contain a high proportion of fibre flakes, so deflaking is necessary, unlike most deinking systems, which use high-consistency pulping. Typical screen perforations are 1–2 mm, and sometimes a slightly lower hole size is used in the final screening stage.

There is considerable debate regarding the best way to handle secondary screen accepts, since they inevitably contain a higher proportion of contaminants than primary screen accepts. In some systems these are returned to the primary screen feed chest, though this can lead to a reduction of contaminant size, by the repeated action of agitators, pump impellers, etc. Feeding accepts forward avoids this and also provides a significant increase in throughput. One means of achieving cleaner secondary screen accepts, to feed forward, is to have a slightly smaller hole size in the secondary screen, with a further reduction in hole size in the final stage. A purging pressure screen is best in this position, to minimise fibre losses. In older systems, vibrating screens were used in this position, but these are very inefficient screens and led to the recirculation of contaminants, rather than their removal. Use of vibrating screens in this position should be avoided. A purging pressure screen discharges rejects on a controlled timed interval, after a washing cycle, which helps minimise fibre losses.

8.3.2.2 Forward flow centrifugal cleaners. These are similar to those used in deinking and are multiple stage systems, with the number of cleaners and stages dependent on cleaner capacity reject rates. Low capacity, low-diameter cleaners, with high-pressure drops, usually give best cleaning results. Cleaners are designed to remove sand, grit, glass, clay particles etc. Their efficient operation is important in all grades, but especially in those which are to be coated or medium. If these cleaners are ineffective, medium with a high abrasivity can be produced, resulting in, for example, accelerated wear of corrugator rolls. In a study of the abrasiveness of European fluting papers it was found that the most abrasive were those mediums containing sand particles $< 200\,\mu m$. Mills equipped with only high-consistency centrifugal cleaning produced the most abrasive fluting whereas the least abrasive fluting was produced in mills with multistage low-consistency cleaners [15].

Having low-consistency centrifugal cleaners is no guarantee of low abrasiveness. An earlier examination of the efficiency of centrifugal cleaning in mills revealed that in many cases poor maintenance led to poor performance of centrifugal cleaners, with some cleaners blocked, worn tips, low-pressure drops, incorrectly balanced flows, treatment of rejects, etc. [16]. Maintenance of cleaners is thus very important, to be checked at regular intervals for wear and replaced as necessary. Recycling plant operators need to check cleaners frequently, to ensure there are no blockages. Although narrow diameter cleaners tend to achieve better cleaning efficiencies, they also tend to be more prone to blockage, which leads to accelerated wear.

8.3.2.3 Fine slotted screens. To avoid abrasive wear resulting from the presence of grit and other materials, fine slotted screens usually follow forward centrifugal cleaners. However, this means that they must operate at a low consistency and so more screens are required, hence capital costs of low-consistency screening are high, although screening efficiencies are also higher. An alternative is to have medium consistency (2.5–4.5%)

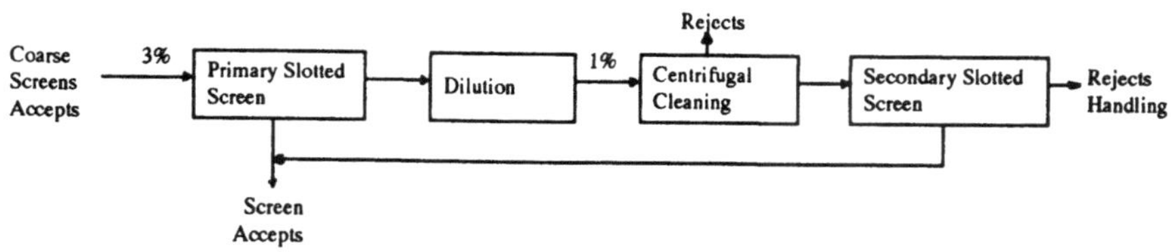

Figure 8.6 Schematic illustration of high-consistency fine slotted screening system.

slotted screens, before forward flow centrifugal cleaners. To minimise abrasive wear, rejects are diluted and cleaned using forward flow centrifugal cleaners, before the second stage of fine slotted screens. This is illustrated in Figure 8.6. Rejects from the secondary screen should be treated in a purging pressure screen and not a vibrating screen.

Secondary screen rejects are illustrated as being fed forward; again these can be fed back to the primary screen feed, though consistency differences limit the volume which can be recycled. Although capital costs are reduced in the type of system illustrated, unless full flow centrifugal cleaning is included as a later process stage, the removal of heavy particles is inefficient, which could result in high abrasivity. If another stage of high-density centrifugal cleaning is included capital savings from medium consistency screening are reduced.

Slot size is a key in determining contaminant removal efficiency, though other factors to do with screen design, for example rotor speed, screen basket design, pressure drop, temperature of operation, velocity through the slot, power input via the rotor, etc. are also important. Generally, narrower slots will produce a cleaner accepts stream, but possibly at the expense of increased long-fibre losses, through fractionation by the screen. Typical slot sizes are 25–40 mm, but slots as narrow as 20 mm are available [17]. New mill installations have used 0.25 mm slots [18]. Slotted screens remove stickies, small particles of dirt, etc. Problems due to screen blockage can be minimised by operating screens under pressure drop control. In this system the pressure drop across the screen is monitored by a pressure sensor, linked to a computer control system. If the pressure drop increases above a preset level (an indication that the screen is blocking) the rejects valve opens and the increased reject rate allows the screen plate to clear. After a preset time, the rejects valve returns to its previous setting.

Slot size selected is influenced by the quality required, since smaller slot sizes inevitably mean that more screens are required. Due to the smaller open area of the screen basket with narrower slots capacity falls. Medium consistency operation also requires more open slots than low-consistency slots. The very best screening efficiency is achieved by having medium consistency slots followed by low-consistency slots later in the system, but of course this has a high capital cost. High overall screening efficiencies are achieved, since no screen has a 100% cleaning efficiency, hence two stages achieve better results than a single stage.

In a study in the UK it was found that in normal mill situations slotted screens were the single most effective fine cleaning stage, especially in the removal of stickies. However, efficiencies were in the range 50–70% of stickie removal with 0.25 mm slots, operating at a low slot velocity, at a low consistency [16]. The condition of screen baskets also plays a major role in determining stickies removal efficiency. Baskets should be checked at regular intervals to monitor wear and damage, checking slot orifice width with feeler gauges. Baskets should be repaired or replaced as required, to maintain high cleaning efficiency and product cleanliness.

8.3.2.4 Low-density (lightweight) centrifugal cleaners. These are identical to systems used in deinking, though there is normally a higher quantity of debris to be removed, including polyethylene film, polystyrene (from packaging inserts) and waxes.

The efficiency of wax removal is dependent on system temperature, especially pulping temperature. In a study examining the effect of wax contamination on the angle slip of linerboard, it was established that a pulping temperature of 60°C effectively dispersed wax throughout the system, since many paraffin waxes have melting points in the range 50–60°C. Dispersed wax was a stable suspension of minute particles, which could not be removed effectively. The study showed that pulping at 50°C formed ‘free wax’ – large discrete wax particles which float, which could be removed by low-density cleaners, such as reverse cleaners or a rotary cleaner [19]. Another study found tacky wax flakes were formed when the pulping temperature was close to the melting point of the wax, whereas at lower temperatures non-tacky wax flakes were formed. Tackiness of the wax ‘stickies’ was governed by pulping pH, consistency, wax viscosity and fibre-release additive (chemicals to release wax from fibres during pulping) content. This research objective was to derive a formulation for a recyclable wax, which would produce large flakes for easier removal by screening and cleaning [20]. Given the low density of wax, it is probable that this will be predominantly in lightweight centrifugal cleaners.

There are three types of lightweight cleaners, using centrifugal forces as the separating force, which are:

- Rotary cleaners, such as the Gyroclean, which exerts high centrifugal forces due to its rotation at high speed;
- Core cleaners or combination cleaners – normal centrifugal cleaners, but having a central outlet in the core of the accept stream, where lightweight particles are concentrated;
- Reverse flow cleaners – these have reversed flow patterns to those of a normal cleaner – the heavy fraction is the accepts and the light

fraction, rejects. Through-flow lightweight cleaners have also been developed, with the light rejects discharged at the bottom of the cleaner.

Normally, these centrifugal cleaners operate at a low consistency, in the range 0.7–1.2%, to achieve optimal efficiency. Operation at higher consistencies would reduce capital and operating costs. Cleaning at up to 2% consistency, with maintained efficiency, has been reported [15].

Removal of contaminants is again affected by cleaner condition and so cleaners must be routinely inspected to check for wear, etc. Continual operator attention is also required, to ensure cleaners are not blocked, and are operating under design conditions.

8.3.3 Rejects handling

Yield is dependent on wastepaper grades used, the design of the processing line and product grades, but is typically in the range 85–90%. As recycling increases, this may fall, since increased fines loss is inevitable and may be necessary to maintain strength properties. A substantial proportion of losses are dissolved solids, such as starch, creating high effluent loads, discussed in chapter 7. Solids are rejected at various process stages and their treatment is dependent on the disposal method used. In general terms, it is more convenient to dewater reject streams, where necessary, and divert all resultant solid reject streams to a central area, for further processing, or disposal. Conveyors are used to transport reject streams, including sludge, to a central area.

Dilute reject streams can be dewatered by the use of inclined wire screens and further thickened by screw dewatering. Heavy rejects can be dewatered by the use of a sand separator or degritter.

Rejects and sludge may be separated into separate streams if incineration is a possibility, for example, in a wood waste boiler. Heavy rejects are combined, from, for example, ragger rope lengths, junk trap rejects, rejects from high-density, high-consistency centrifugal cleaners, and possibly rejects from high-density, low-consistency centrifugal cleaners. These contain high proportions of iron scrap, sand and grit, etc. Iron can be removed by magnetic separators and sold as scrap, leaving a residue for disposal by landfill. Other rejects, from screens and low-consistency, low-density centrifugal cleaners, etc. can be combined with effluent sludges, dewatered, thickened and burned, for example, with wood waste, at a controlled addition rate.

It is essential to ensure that reject dewatering does not allow the recirculation of rejects with recovered backwater, which would allow a contaminant loop to be established. Final polishing of backwater can be achieved by filtration through inclined fine mesh screens, though these may need to be washed at regular intervals to remove scale deposits, to maintain capacity.

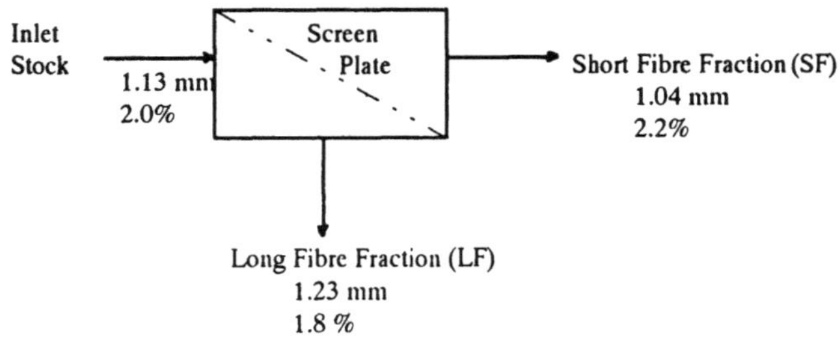

Figure 8.7 Schematic illustration of a fractionating device.

8.3.4 Fractionation and dispersion/kneading

When quality levels achieved by cleaning and screening are not adequate for a particular product, additional techniques are used, which include fractionation and dispersion. These are not always used together; for example, a top liner ply can be treated by a disperger without prior fractionation, or cleaned stock can be fractionated without the subsequent use of a dispersion stage. However, they are frequently linked and so are discussed together. There are no rejects from either dispersion or fractionation.

8.3.4.1 Fractionation. Fractionation or classification is a process which divides a mixture of fibres into two or more classes, usually according to their length. However, there is a strong relationship between fibre length and coarseness, so that differences in properties of the fractions, or classes, are due to both fibre length and coarseness differences. Both are fundamental fibre properties and have a significant effect on properties. Since no papers published describing fractionation in use in secondary-fibre mills have included coarseness measurements, incorrect conclusions may sometimes have been drawn regarding the cause of changed properties, resulting in some controversy regarding the success, or otherwise, of fractionation when practised on a commercial scale. It should also be noted that fractionation is an enrichment process, which does not provide a sharp division in fibre length or coarseness. In wastepaper systems, there is a great diversity in fibre types present, so that differences in flexibility, curl and external fibrillation prevent a sharp cut from being achieved. Fractionation is illustrated schematically in Figure 8.7. Weighted average fibre length and fines percentage given are from a fractionation unit in Australia, using wastepaper from supermarkets and households. The fractionation unit had 1.6 mm holes [21].

The degree to which the fines fraction is separated from the long-fibre fraction can be controlled by adjusting the volume of the flows from the fractionator. If the shorts flow is increased, then for a given stock inlet, the longs become freer and on average, longer. Similarily, if the longs flow is increased, the longs become less free, with a higher concentration of short fibres and fines. Fractionation efficiency has been defined differently by a number of researchers, and one commonly used is:

Fraction efficiency, E_F is given by [22];

$$E_F = \% C_S + \% C_L$$

where, $\% C_S$ = concentration of short fibres

$$= \frac{\% \text{ short fibre in short fraction} - \% \text{ short fibre in feed}}{\% \text{ short fibre in shorts}} \times 100$$

and $\% C_L$ = % concentration of long fibre

$$= \frac{\% \text{ long fibre in long fraction} - \% \text{ long fibre in feed}}{\% \text{ long fibre in longs}} \times 100$$

Usually, the percentage of longs and shorts is measured by the use of a laboratory fractionator, or by a fibre-length measuring instrument, such as a Kajaani.

Many devices used in recycling mills are fractionating devices; for example, washers are effective fractionators, since they separate fibre fines, fillers and small ink particles from longer fibres. Early fractionators included centrifugal systems [23], fast rotating discs [24] and modified pressure screens [25,26]. The process (using pressurised screens) was first applied in waste systems on a commercial scale in the mid 1970s [27] and was also used in virgin-fibre systems [28]. In the latter case, fractionation of a high-yield kraft pulp gave both top and base stocks of liner board from a single cook. Despite these early applications in the USA, the process has been adopted more widely in central Europe, especially in Germany, primarily in waste processing for packaging board and liner production.

In a 1989 survey of fractionation in Germany, of 13 mills which had installed fractionation units, four had shut them down. Details were reported for six mills and of these, two were using the fractions on separate machines, to produce test liner (long-fibre fraction) and medium (short-fibre fraction); one produced test liner and the remaining three produced medium, after recombination, following separate treatment of the long-fibre fraction. All used 100% recycled wastepaper to produce these grades and modified pressure screens for fractionation [29].

Equipment offered by a variety of manufacturers is primarily modified pressure screens, with either holes or slots in the screen basket. Hole diameters range from 1.2–1.8 mm and slot widths from 0.20–0.55 mm. Differences in performance between holes and slots have been reported. In an early comparison, which traced the development of a fractionator, it was found that holes gave a better fractionation performance than slots [22]. Another evaluation reached a similar conclusion, and found that slots gave

Table 8.6 Two stage fractionation pilot plant trial results [32]

	First stage fractionation	Second stage fractionation
Ratio SF/LF[a]	50	60
°SR		
Inlet	26	20
Long-fibre fraction	20	17
Short-fibre fraction	43	44

[a]SF = short fibre; LF = long fibre.

a better cleanliness to the short fraction, by concentrating debris in the long fraction [30]. When various hole sizes were compared (1.4, 1.6 and 1.8 mm) better separation of longs and shorts were found with 1.4 and 1.6 mm holes, but operational considerations resulted in 1.6 mm holes being preferred [26]. Another study confirmed better fractionation efficiency with lower hole size; 1.4 mm was found to be best, and 1.2 mm holes and 0.6 mm slots approximately similar, but the plate with 1.2 mm had operating problems, hence a preference was expressed for either 1.4 mm holes or 0.6 mm slots when fractionating mixed waste. When two slot widths were compared, (0.25 and 0.35 mm) fractionation was better with the larger slots, but cleaning efficiency best with the narrow slots [31]. In the German study of installed fractionators, 0.55 mm slots were found to be best and 1.4 mm holes better than 1.6 mm holes, in retaining long fibres in the long-fibre fraction. However, overflow rates were not equal and it is possible that if the overflow rate (55%) of the 1.4 mm holes had been the same as 0.55 mm slots (45%) then the 1.4 mm hole fractionator would have performed at least as well as the 0.55 mm slotted fractionator [29].

Normally, one stage of fractionation has been installed, but in some cases two stages in series have been used, to give a better separation of long fibres. Published results indicate an improvement, in for example, freeness, suggesting a gain in long-fibre concentration. Some pilot plant results are given in Table 8.6 [32].

This system was planned to include dispersion for the top ply of liner only, with no other stock being so treated, and the proposed system is illustrated in Figure 8.8.

In this Dutch mill, a total of three fractionation stages were installed as illustrated in Figure 8.8 and a fourth planned in another line.

Benefits from fractionation have been reported by a number of wastepaper using mills [27,32,35] and also projected by equipment suppliers [25,26,30,31]. Benefits include:

- Energy savings – from refining and perhaps dispersion of only the long fraction;
- More efficient fibre use – either through using two fractions in two

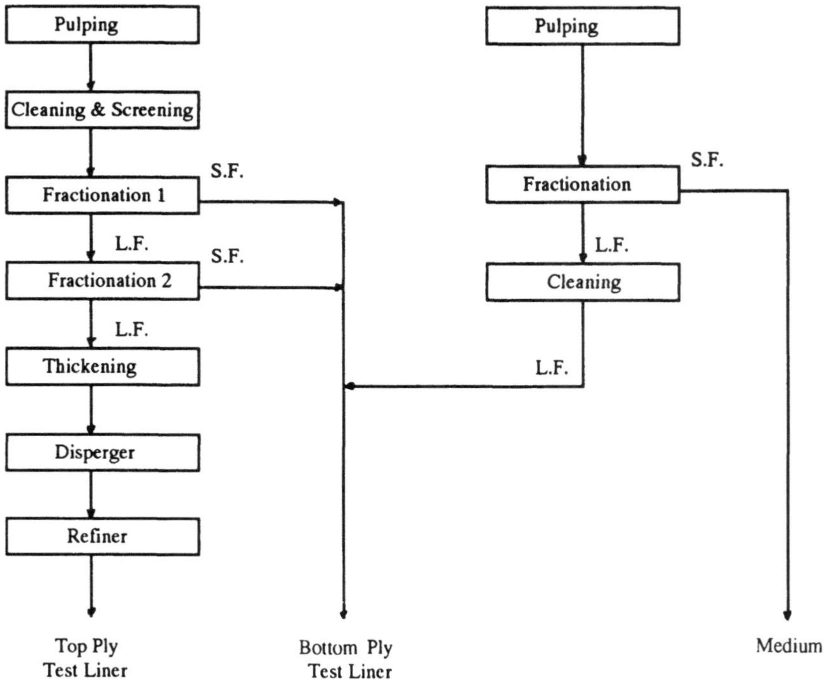

Figure 8.8 Double fractionation installation [32].

different products, or two different plies of the same product; fibre engineering;

- Use of lower quality and cost grades of wastepaper - achieved by fibre engineering, when combined with dispersion;
- Improvement in optical properties - visual appearance improved, but only if fractionation is combined with dispersion;
- Control of short run fluctuations in waste quality - dependent on design of wastepaper treatment process.

There have also been several papers which report no benefit from fractionation [36, 37] even though the performance of the fractionation unit was not faulty. Mills involved reported some benefits, but in one case these were not adequate to justify the capital and energy costs involved in fractionation [36]. In this instance, fractionation was followed only by refining, before recombination, so energy savings were small. In the other case, unexpected characteristics of the wastepaper processing system prevented benefits from being realised [37]. This was largely due to further fractionation by pressure screens, which prevented an enriched long-fibre stock from being fed to the top ply, in a multicylinder box-board mill. In the German survey, four mills had shut their fractionation units down [29], which

Table 8.7 Energy consumption (total waste line) before and after fractionation

	Specific energy consumption (kWh t^{-1})	
	Before fractionation	After fractionation
kWh t^{-1} test liner [32]	350	187
kWh t^{-1} test liner [35]	402	240

suggests they were not achieving expected benefits. Other mills reporting benefits illustrated energy savings of up to 50%, summarised in Table 8.7.

In both these cases substantial power savings were derived from dispersion and refining of long fibres only. However, although some mills have reported product strength benefits [33, 35] others have reported that strength benefits are small, less than expected [36]. Improved machine runnability, due to better strength from the reduced production of fines during refining and cleaner stock (when combined with dispersion) [32, 35] was also reported.

Control of fractionation is important in gaining benefits associated with fractionation, especially of short run variations in fibre mix. It is achieved by a variety of methods, including flow control, power input (load control) and consistency of the overflow.

Of these, consistency of the overflow is the most common. Flow control uses flow meters, controlled to a set ratio. Load control provides an opportunity to provide a constant ratio of long fibres to short fibres, by a system which allows compensation for short run fluctuations in relative fibre proportions. This is achieved by linking a signal from the fractionator impeller motor to the flow control valve. When wastepaper with a high long-fibre content is in use, their concentration in the fractionator and on the screen plate will increase and so the motor load will increase. If the load exceeds a set point, then the control valve (short-fibre fraction) will close and a greater flow will be 'rejected', that is the long-fibre fraction flow will increase. Conversely, when there is not a lot of long fibre, the impeller motor load will fall and in this case, a greater flow will pass the screen plate, as short-fibre fraction, when the control valve opens. This is analogous to the system to prevent screen plugging in fine screening. Provided the two fibre fractions were held in separate chests and then blended together in a set ratio (if there is recombination after separate processing) short run fluctuations in fibre content can be controlled. This type of system is illustrated in Figure 8.9 and is illustrated in a schematic process flow in Figure 8.10.

Benefits from fractionation are largely dependent on separate fibre treatment, as indicated in Figure 8.10, though it is unusual for the short-fibre fraction to be treated. However, as recycling rates increase, it may become necessary to remove some fines, to maintain mechanical strength properties of packages; this is illustrated in Figure 8.10. One mill reported strength properties improvement of 8–14% by deashing the short-fibre fraction [29].

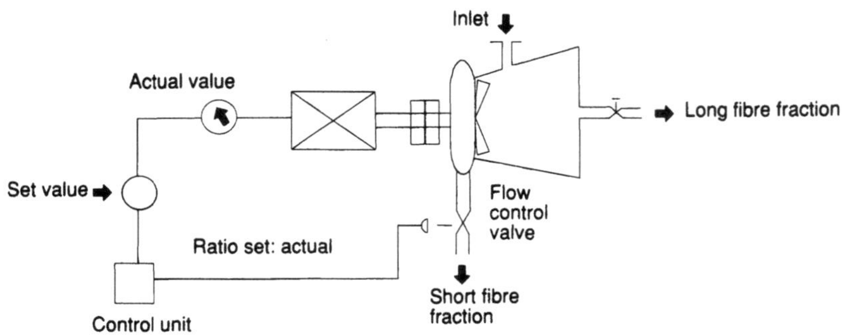

Figure 8.9 Fractionation load control system.

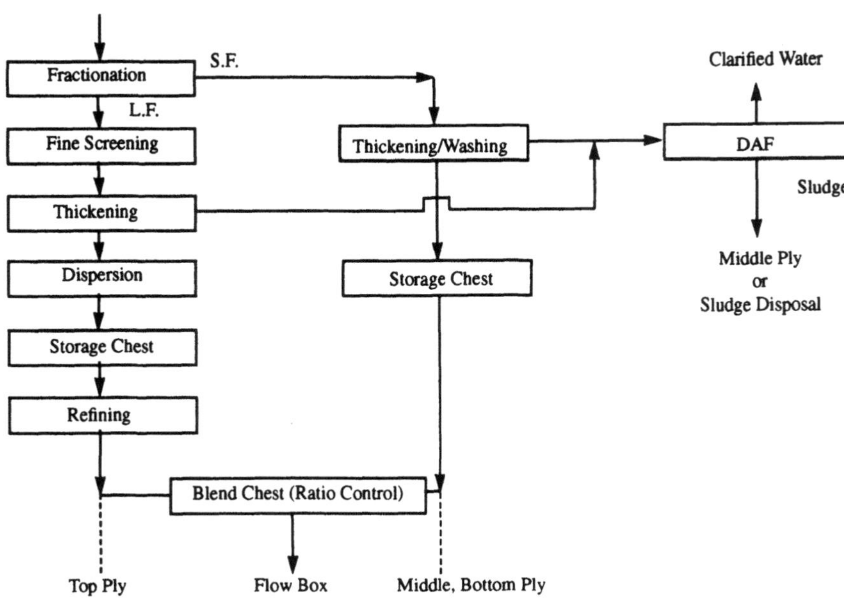

Figure 8.10 Schematic illustration of separate treatment of fibre fractions.

Another mill with a short-fibre fraction deashing step reported an increase in freeness, to a level greater than that of the long-fibre fraction. Their fibre length analyses are illustrated in Figure 8.11 [21].

Washing or deashing itself acts as a fractionation process by the separation of fibre from mineral filler present, crill, fines, etc. It may be possible to reuse the 'short fraction', that is, the material in the wash water, in a filler ply. However, there is also an increasing trend towards printing on the top ply, either in liner or packaging board. Controlled addition of this

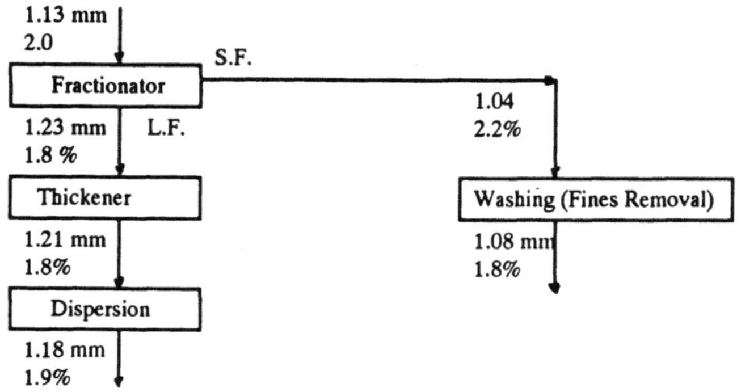

Figure 8.11 Weighted-average fibre lengths and percentage fines [21].

mineral filler and fines mixture would improve the printing properties of the top ply. If the flocculation of the filler and fines is maintained (this is necessary for their removal in dissolved air flotation (DAF)) filler and fines retention on the wire will be high and freeness of the top ply will be affected to a lesser extent than without flocculation.

If the bottom ply in Figure 8.10 were a backliner for packaging board, it may be necessary to include additional screening and cleaning, to ensure the back ply is free of stickies. This is to ensure there is no transfer of stickies from the back ply to the top ply, when the board is reeled up at the machine. This is frequently a problem with this grade and results in increased contamination of the printing blanket during the printing process.

Strength benefits from separate treatment of the long fibre have been documented by several mills [32, 34–36], as well as a number of researchers [26, 29]. In one case, a white top multi-ply packaging board, the fractionator was used to provide a long-fibre stream for refining and subsequent use in the middle ply. The short fibre was used as bottom liner, since it was relatively clean. Strength comparisons before and after fractionation are given in Table 8.8, for equivalent furnishes [34].

Another multi-ply box-board mill used the reverse of this; short fibre in the filler ply and long fibre in the back ply, and increased stiffness of the board by 13%. They also found that drying was more efficient and that caliper increased, which gave about 5% of the stiffness increase, so that 8% was due to the change in sheet structure. An additional benefit was better creasing properties with long fibre in the back ply [21]. Their results are given in Table 8.9.

Since both mills claimed benefits from different uses of the two fractions, it is clear that the objectives from fractionation need to be clearly defined, so as to install the system which will yield desired benefits. The results presented by the two mills are not contradictory; one compared properties before and after fractionation, the other compared two possible sheet structures.

Table 8.8 Quality of multi-ply board, before and after fractionation [34]

	Before fractionation	After fractionation	Change (%)
Basis weight (g m^{-2})	309	307	−0.7
Thickness (mm)	0.42	0.41	−2.7
Burst strength (psi)	89.0	100.4	+12.8
Puncture (Beach units)	45.0	50.4	+12.0
Tensile strength (kg/15 mm)			
MD[a]	16.8	19.8	+17.8
CD[b]	5.1	6.1	+19.6
RCT (lb)[c]			
MD	72.0	90.4	+25.6
CD	109.0	132.2	+21.3

[a]MD = machine direction; [b]CD = cross direction; [c]RCT = ring crush test.

Table 8.9 Changes in sheet properties with variations in sheet structure [21]

	MD stiffness (mN)	CD stiffness (mN)	Thickness (μm)
Long fibre in backliner[a,b]	452	171	450
Short fibre in backliner[a,b]	397	160	440

[a]short fibres in middles; [b]weighted-average fibre length, long fibre, ex machine chest, 1.15 mm; [c]long fibres in middles; [d]weighted-average fibre length, short fibre, ex machine chest, 1.00 mm.

With liner, even with a reduction in quality of wastepaper used, Mullen and tensile increased, by 8% and 13%, respectively [35] and others reported that reductions were made possible in wastepaper quality [32], or starch use [33].

With medium, fractionation followed by refining of long fibre prior to recombination with the short fraction resulted in an increase of 2.8% in concora (concora medium test, CMT) and 2.0% in crush (concora crush test, CCT). Air permeability was improved by 60% [36]. These smaller increases in strength following recombination have been confirmed in other reports. In the survey of fractionation in German mills, comparisons of properties following various treatments were made with the long-fibre fraction from one mill – these were untreated, refined, dispersed and dispersion and refining combined. This study showed that when compared with refining, dispersion had a smaller detrimental effect on freeness, with lower fines production, but with better flake content reduction. When subject to both treatments, strength properties of the long-fibre fraction increased by 3–27%, including breaking length, burst, CMT, RCT (ring crush test) and STFI, compared to the original long fraction. When recombined with short fibres, and compared to the original whole stock (without treatment), breaking length, burst, CMT and STFI had increased by about 10% and RCT by about 3%. Bending stiffness was essentially unchanged [29].

However, in a study in the UK, it was shown that when whole stock is refined and compared with a recombined short and long fraction, following refining of only the long fraction, at equivalent total refining energy inputs, there was no consistent pattern to show strength benefits were gained from fractionation followed by refining only the long fraction and then recombination [37]. This demonstrates very clearly that the advantages of fractionation, with respect to strength benefits, are best realised by separate use. If dispersion is included in the long-fibre treatment, strength benefits can be realised, with other advantages, energy saving, visual improvement, etc.

Other treatments for the long-fibre fraction are also possible; in Figure 8.10, an additional screening stage is included, before thickening for dispersion. Since most contaminants are collected in the long-fibre fraction and dispersion does not remove these, the removal of some of these prior to dispersion will help improve sheet properties, since contaminants prevent inter-fibre bonds from being made and are responsible for weak points in the sheet. The importance of good contaminant removal, to gain strength properties in recycled paper board, was shown, by a study which included light and heavy contamination; RCT of the heavily contaminated was about 18% lower in liner and burst about 25% lower [38].

Separate treatment could include chemical treatment. Many added chemicals, including starch, are adsorbed onto the surface of the fines fraction, which has a much greater surface area, but this contributes little to strength increases. When cationic starch (0.5%) was added to the long-fibre fraction, the effect on breaking length was greater than if added to the whole stock; increases were 13–17% and 10% respectively [39].

8.3.4.2 Dispersion/kneading. Dispersion is used primarily as a means of improving optical characteristics by disguising the presence of contaminants. Even intensive screening and cleaning is not successful in the removal of all contaminants, so that visible contaminants remain in the sheet. If there is a need for the product to be visually equivalent to a virgin product, then dispersion helps to achieve this, as specks are reduced to a size not readily visible to the naked eye.

In high-speed disc dispersion units, all contaminants are reduced in size; contaminants such as stickies usually cause fewer deposition problems on machines and in some cases dispersion units were installed to protect machines from stickies. One other benefit of dispersion is that under typical dispersion conditions, burst and tensile properties are enhanced, similar to low-consistency refining, but without the freeness loss caused by cutting in a low-consistency refiner [29]. Under high-temperature conditions fibre cutting in a disc dispersion unit is further limited.

In brown grade processing, the high-speed disc dispersion unit is the most

Table 8.10 Effect of beating on mechanical properties of OCC [40]

Time (min)	CSF (ml)	Caliper (μm)	Ring crush 16/6	CMT 16	STFI ($kN\ m^{-1}$)	Tear (mN)	Tensile ($kN\ m^{-1}$)
0	580	188	30	25	2.4	700	4.4
10	430	196	48	52	3.2	600	7.4
20	270	171	51	58	3.2	700	7.8
0	600	238	41	25	2.3	1400	4.2
10	420	197	55	54	3.2	1400	6.8
20	200	177	63	63	3.5	1400	7.8

popular. The intensity of its action is determined by the design of the disc, energy input (40–80 kWh t^{-1} is a typical range) temperature, residence time and the gap between the discs. Narrow disc gaps (0.1 mm or less), low temperature and high residence time all increase the intensity of dispersion, but also increase freeness loss, but allow good dispersion to be achieved.

High-speed dispersion and low-speed kneading can introduce curl into treated fibres. Slow-speed kneaders have a higher retention time and it appears these introduce more curl, which affects, for example, tearing strength, dimensional stability, bulk, etc. Specific circumstances will determine if curl creates a problem or enhances desirable properties.

8.3.5 Strength development by refining

Refining is the most commonly adopted method of improving the strength properties of recycled fibres; drainage falls, but strength properties increase, since fibre bonding increases. Over-refining is easier to reach with secondary fibres, resulting in a dramatic reduction in drainage rates.

Results from two laboratories, from beating OCC clippings in a Valley Beater are given in Table 8.10 and show that mechanical properties develop rapidly, but freeness also falls rapidly [40], reaching a value after 20 min which would be unacceptable in a mill.

Refining in a conical refiner showed that specific edge loads has a marked effect on property development at energy inputs greater than about 40 kWh t^{-1}, with a lower edge load being preferred. Tear strength is most sensitive and falls more rapidly at higher specific edge loads. Refining of old containers (predominantly OCC) at an edge load of 1.0 Ws m^{-1} in a conical refiner gave the results given in Table 8.11. These show that at low edge loads and low-energy inputs some development of tear strength is possible, without sacrificing drainage rates. Breaking length is slower to develop.

In a German mill survey, the refining energy of the long-fibre fraction, after fractionation, was 12–30 kWh t^{-1}, using disc refiners. In one mill, a comparison was made between refining and dispersion, though conditions

Table 8.11 Effect of refining on mechanical properties of old containers

Energy input ($kWh\ t^{-1}$)	SR °SR	Burst index ($kPa\ m^{-2}\ g^{-1}$)	Tear index ($mN\ m^2\ g^{-1}$)	Breaking length (km)
0	32	2.2	8.2	3.4
25	39	2.5	8.7	3.0
50	45	2.9	8.2	4.3
75	51	3.1	8.5	4.8
100	59	3.3	7.4	5.3

Table 8.12 Refining and dispersion of a long-fibre fraction [29]

	(Initial values = 100)		
	After refining	After dispersion	After refining and dispersion
Breaking length	111	110	117
Bending stiffness	94	95	98
Burst strength	123	127	124
CMT	113	113	113
RCT	103	103	109
STFI	108	104	108
°SR	126	117	122
Flake content	80	70	80
Fines content	103	109	96

were not given. Since freeness increased by 26% after refining, energy input must have been low. Results from this single evaluation are given in Table 8.12 and are indexed, so that the initial value of the long-fibre fraction equals 100 [29].

8.3.6 *Water treatment*

As discussed in chapter 7, many European board mills operate with low or no liquid effluent discharge, normally using a combination of anaerobic and aerobic effluent treatments, before water is recycled. In other geographic areas, there is also pressure to reduce specific water consumption, to reduce effluent volumes requiring treatment and to limit freshwater consumption. In many cases effluent charges are based on flow, suspended solids and organic strength and so in these cases it is desirable to have at least primary treatment, to remove most of the suspended solids. Recirculation of some of the clarified water, to substitute for freshwater in some mill duties serves to reduce the volume discharged from the mill.

Primary clarification is usually by dissolved air flotation, due to its low retention time. If sedimentation is used, it is essential to design a settling system with a low retention time and rapid sludge removal, to prevent the

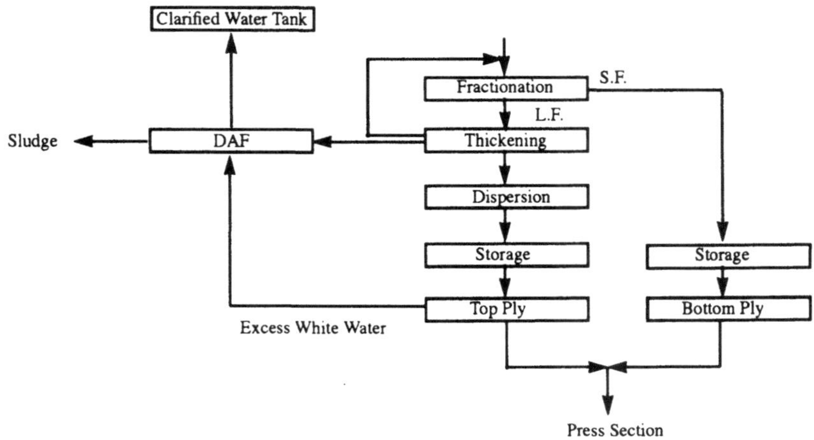

Figure 8.12 Water clarification in board processing.

system from becoming anaerobic, which results in foul odours and floating sludge. Clarifiers should be designed with a high overflow rate (typically $1.1\ m^3\ m^{-2}\ h^{-1}$) with a shallow side wall depth, to give minimal retention. Dissolved air flotation avoids these problems, but has higher operating costs, due to the requirement for flocculant chemicals.

Recirculation of clarified water may provide opportunities for energy savings, though if it is necessary to run pulpers cold to reduce wax or stickies problems, this may create problems in keeping system temperatures low. With recirculation the organic strength of the system increases and probably additional chemicals, to control biological activity and prevent deposition in stock approach systems, are required. Corrosion may become a problem, due to much higher dissolved solids concentrations. If anaerobic conditions are allowed to develop, board can be produced which has a sharp odour, due to the production of simple acids.

All or some sludge removed during primary clarification can be recycled. If cleaner rejects have been discharged to effluent via floor drains this will create problems, since the water loop would become a reservoir of contaminants. If sludge recycle is planned, it is essential to ensure no rejects can find their way to the effluent system. Sludge may also adsorb oil, if oil leaks to drains, and this can have a significant impact on board properties, including the coefficient of friction. In multi-ply board systems recovered fibre may be used in filler ply, or lower performance grades of medium or liner. In higher grade products, little or no sludge can be reused and sludge is further dewatered, prior to its disposal.

A typical installation is illustrated schematically in Figure 8.12, processing excess white water and backwater from a thickening stage. Other examples are given in chapter 7.

8.4 Board machine runnability with recycled grades

In common with many other products made from wastepaper, one of the major problems is stickies, causing contamination of board machine clothing, holes and tears in the product and contamination of printing blankets. Good design and operation of screening and cleaning systems, combined with dispersion, have been very effective in minimising stickies problems. Another runnability problem is the slower drainage of recycled fibres. As recycled levels increase, the concentration of fines also increases, so that drainage rates fall, unless a substantial reduction in yield is taken to remove fines. Since board grades can have very high basis weights, drainage limitations can considerably reduce machine productivity.

8.4.1 Box-board

Machine configurations are similar for virgin or recycled grades, but vary appreciably according to the type of product. White lined chip and box-board machines usually use vat formers. In Western Europe in 1989 there were 100 white lined chip and box-board machines, of which 92 used vat formers, although newer machines were multi-fourdriniers [41]. A major disadvantage of these with respect to recycled-fibre grades is a problem with poor ply bonding. Starch can be applied as a spray between two plies to improve ply bonding, but this can create a dirty wet end. Fines are important in forming strong ply bonds, and although recycled sheets carry more fines, some of these have been recycled and are inert, and do not form inter-fibre bonds, thus acting as a hindrance to good ply bonding. Inert fines removal reduces yield, but helps improve ply bonding. At couching, fines must be able to migrate between plies, but the web must not be wet enough to allow crushing.

In a rebuild of a waste-based box-board mill, vat formers were replaced by three fourdriniers, four head boxes and two top formers, which enabled improved quality standards and increased production to be achieved [41]. This enabled high ply bonding strengths to be reached, as well as increased machine speeds, through better water removal.

8.4.2 Liner board and medium

Liner board machines are fourdriniers. Newer machines tend to produce two (or more) ply liner, which is achieved by the use of, for example, a secondary headbox, two wires or a multilayer headbox. Twin wire gap formers are beginning to be used due to their potential for high speed operation, good formation, etc. Where different plies are formed separately, thermal stability problems can occur if one layer is treated by hot dispersion and this steam is significantly warmer than other layers. Corrugating medium and grocery bag paper is usually made on a fourdrinier, and frequently fluting and liner are made on the same machine, but when corrugating is being produced, lower grades of wastepaper are used.

Over the last ten years there have been major improvements in press section efficiency, due to the use of presses with longer nips or heated press nips. Both are used extensively in liner and corrugating machines. For recycled fibre they have the advantage of increasing sheet density and hence they promote better inter-fibre bonding, so increasing strength properties, as well as increasing the efficiency of water removal. Normal strength/density relationships are maintained. In the case of press drying the strength/density relationship changes; higher strengths are achieved at equivalent densities. Although this has not reached commercial application, it has considerable potential for recycled-fibre grades, since it results in new strength/density relationships. In press drying, the press nip is hot, so that high sheet temperatures are reached in the press nip. This enables dense sheets to be produced, but also allows softening of hemi-celluloses and residual lignin, with probably some redistribution of lignin, forming more inter-fibre covalent bonds than would otherwise be the case. It is this latter mechanism which changes the strength/density relationship. Another advantage of press drying is that the performance of press dried cases under high humidity conditions is very much improved [42].

As with other recycle grades, special attention needs to be given to machine cleanliness, with cleaning programmes for forming fabrics and press felts. Continuous or intermittent chemical cleaning of these is normal. High-pressure needle showers, either single or multiple are used to assist in the prevention of deposition in the wires. Cleaning doctors on felt and wire rolls help to prevent these from transferring deposits or from driving deposits into the open wire or felt. Different roll covering also help to prevent deposition on rolls; for example Teflon composite roll coverings prevent the transfer of stickies from felt or wire surfaces, to the roll and back on to the wire or felt. Dryer felt permeability can be reduced by deposits resulting in increased drying costs. Higher solids from press sections (a drier sheet) can help reduce dryer felt contaminition.

When chemical conditioning to prevent deposition on wires and felts is used, problems with deposits may be moved down the machine. Deposits may occur on drying cylinders and doctoring of these is necessary to prevent buildup of deposits. Double doctoring of the first cylinder may be required. Periodic caustic boil outs of the stock approach system reduces problems from deposits building up and breaking away when a critical mass is reached. Monitoring machine and reeler breaks can provide early indications of problems, from poor quality wastepaper or processing failures.

When doctors or sprays are used to remove stickies, it is essential to ensure that they are not allowed to fall within the machine or re-enter the white water system. Doctor blades geometry should allow the installation of trays to collect stickies and other deposits. Double doctor blade cleaning is more effective than single doctors. Water flushing of the tray can be used to keep it clean, with the contaminated water being directed either to the effluent system or to a rejects handling system, for dewatering and removal of contaminants. Where solid deposits are removed, for example from drying cylinders, they must not be returned via a broke bin. Following a machine clean and system boil out, care needs to be exercised to ensure water systems have been flushed out and contaminants are not recycled in water systems.

8.5 Effect of the use of recycled fibres on product properties

When recycled fibres are used, it is inevitable that strengths are lower, since inter-fibre bonding is reduced by a combination of mechanisms, including:

- With successive recycles, fines build-up. Recycled fines do not have good bonding properties and can act as an inert coarse filler, so interfering with inter-fibre bond formation.
- During the first cycle, fibres collapse and without treatment during subsequent cycles do not swell to their maximum potential, which reduces internal and external bonding.
- Hornification at the surface of the fibre, which increases stiffness, so that fewer inter-fibre bonds will be formed, due to reduced flexibility.
- Non-fibrous contaminants accumulate - clay, ink particles, stickies, chemicals, etc. Their presence may prevent inter-fibre bond formation and may result in localised weak points in the sheet, leading to erratic test properties.
- Some multivalent cations and salts may bond to carboxyl groups of fibres, preventing inter-fibre bond formation. Inks and other additives may be a source of these interfering cations and salts.
- Due to freeness considerations, recycled fibres may not be refined to their full strength potential.

There is a constant compromise between cleanliness, yield of wastepaper and strength development. Other means to improve strength properties are:

- increased basis weight to give equivalent performance to virgin grades;
- use of strength aids, such as starch, which is very widely used in Europe, but less so in the USA;
- use of higher grades of fibre, for example, softwood in corrugating medium production, from wastepapers used.

Tear strength is a property which is difficult to achieve with recycled grades, especially packaging papers such as grocery bags. One study showed that tear strength is a complex property, related to fibre strength, length and coarseness. In weakly bonded sheets, such as recycled sheets, there is an increased link with fibre length, which is probably due to longer fibres being better anchored in the sheet, so that in the tear zone short fibres pull out, but better anchored fibres break. Fibre length is more important than fibre strength in weakly bonded sheets [43]. This explains why tear strength is enhanced by fractionation.

8.5.1 *Starch use in recycled grades*

Starch is the most important dry strength additive used in brown grades, and can be >5% of the final sheet weight, excluding starch used as a corrugating adhesive. Starch improves inter-fibre bonding through adsorption and the creation of new bonding sites on fibre surfaces, with stronger than original fibre-to-fibre bonds. Various forms of starch are used, usually classed as native or modified starch, and can be applied in a variety of ways. However, its use has several disadvantages, which include:

- addition increases drying requirements, especially when added via a size press, which considerably reduces machine speed, given fixed drying capacity;
- organic strength of effluents is much higher, so that treatment costs increase substantially;
- machines tend to run 'dirtier' with starch use, especially with starch sprays;
- if starch is not made up and stored correctly then retention can be much reduced, so that starch use is inefficient, and performance is variable;
- dependent on the type of starch used, control of wet end chemistry can become more difficult;
- internal sizing can be impaired, together with formation.

Despite these problems, starch is widely used in recycled grades, especially in Europe, since it provides a low-cost method by which recycled grades can reach necessary strength properties, and its use allows lower cost wastepaper grades to be included in the furnish mix. It also provides increased surface strength; increased pick resistance, which is important in grades to be printed and in multi-ply grades, increases inter-ply bonding. As a wet end additive, cationic starches increase retention. Starch types are classified according to charge, and include anionic, cationic and amphoteric.

Cationic starches are adsorbed almost irreversibly with cellulose fibres, whereas the adsorption of most amphoteric and anionic versions is fairly reversible. The adsorption is roughly proportional to the available surface area, which leads to greater adsorption on fines than fibres. Anionic trash increases cationic starch demand, so that strength values can be difficult to achieve, until the demand from fines and anionic trash is satisfied. Also, when using cationic starches, a level is reached (when the available surfaces have reacted) beyond which further starch has no beneficial effect. However, if more starch is added, for example, to reach a strength target, charge reversal can occur, leading to serious retention and deposition problems. Nevertheless, cationic starch is widely used as a wet end additive, due to its good retention.

Amphoteric starches have both negative and positive charges, so that higher uptake rates can be reached, giving improved strength, and providing less likelihood of charge reversal. It also reduces the dependency on the high molecular weight cationic used to improve retention.

Three basic methods are available to incorporate starch into the sheet: wet end addition, spray system and surface application. Any combination of these can also be used.

In wet end addition with recycled grades, and closed water circuits, a very high level of anionic trash is normally present, so that a set level of cationic starch addition may not give desired strength values. Wet end addition is common in all packaging grades, including multi-ply box-board, where it is used to increase the strength of liner plies, as well as to improve inter-ply bonding. In test liner and corrugating medium it increases dry strength properties. Amphoteric starches can replace cationic starches and give better performance without sacrificing retention [44].

Spraying is usually used to improve inter-ply bonding but sheet strength improvement is low, since add-on rates of starch reached are limited. Air turbulence, especially at high speeds, can disturb the spray pattern, leading to uneven sheet properties. Nozzle blockage can also be a problem, especially with poor cooking of starch, which will also lead to poor uniformity of sheet properties.

Size press application is very common and is a method of achieving high add-on levels. However, machine speeds have to be reduced, compared with machines of similar drying capacity, with no size press. The only properties which can be changed are the starch concentration and molecular weight, both of which affect viscosity. These also affect machine speed, so that the additional drying requirements are determined by the volume of water added as a carrier for starch, the water retention of the starch - the higher the molecular weight, the more difficult to dry, and the degree of penetration of the sizing solution into the base sheet - the deeper the penetration, the more drying required.

Speed is also restricted by splashing in the size press pond, especially when narrow diameter size press rolls are in use. Roll diameters of 120–150 cm help to counteract this. Using a deeper pond improves runnability at higher starch viscosities. Metering of starch flow to the size press rolls, by blade or rod elements, reduces some of the problems of conventional size presses, and can reduce the drying energy requirements, so partially overcoming speed penalties [45]. Application units are illustrated in Figure 8.13.

In a comparison of a conventional size press and a rod metered size press

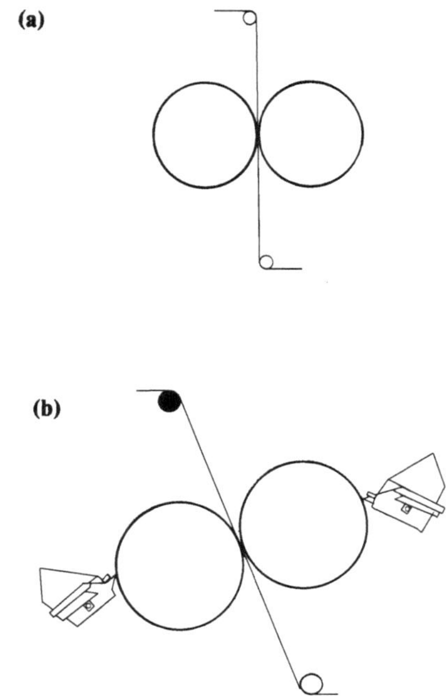

Figure 8.13 Size press units: (a) horizontal size press; (b) blade metering size press.

(also called a film press) it was found that the method of application and the type of starch had a significant effect on the development of various sheet properties and drying requirements. The concora medium test (CMT) increased by between 60 and 80%, with a 3–5 g m^{-2} starch coating. With this property, better results were achieved with a high-viscosity starch. In three cases there was an improvement of about 80% in CMT, achieved at different starch add-on levels. There was 30% penetration from the wire and 35% penetration from the felt side of the medium, using a high-viscosity starch, at a coat weight of 4.9 g m^{-2}, on a conventional size press. Using a low-viscosity starch, a 5.9 g m^{-2} coat was necessary to give the same improvement (80%) and complete penetration of the medium was achieved, on a conventional size press. On a pre-metered size press (rod metered) using a low-viscosity starch, 80% improvement in CMT required 6.4 g m^{-2}, which also reached almost complete penetration in the medium. The basis weight of the medium was 111–121 g m^{-2}. These results showed that complete penetration into the medium is not necessary to achieve good CMT improvements - penetration of about 30% on both sides is adequate. Incomplete penetration should help improve stiffness, by the I-beam effect. Since deep penetration may reduce tear [45], a pre-metering size press which reduces penetration could help to avoid the loss of tear strength.

Starch addition, using the pre-metered size press, gave increases in burst strength, up to about 60%. Using a high-viscosity starch, the improvement was 45–60%, for a coat weight of 4.5–6.4 g m^{-2}. Using a conventional size press the high-viscosity starch gave better results, at equivalent add-on rates; for example, to get a 40% increase in burst, about 3.5 g m^{-2} high-viscosity starch coat weight was needed, whereas with low-viscosity starch, 5.4 g m^{-2} coat weight was necessary. At high starch concentrations (>24%) there was a sharp fall in burst values.

Ring crush test (RCT) improvements were also evaluated. Using a conventional size press, high-viscosity starches gave better improvements, at equal add-on rates. A 50% improvement in RCT was reached with about 4.2 g m^{-2} high-viscosity starch compared to about 5.3 g m^{-2} of low-viscosity starch. Using a pre-metered size press, higher add-on rates of low-viscosity starch gave higher improvements, up to about 85%, with a 9.2 g m^{-2} coating.

In this study it was found that at identical starch concentration or viscosity:

- add-on rates were always higher with low-viscosity starches, at identical starch concentrations;
- a conventional size press gave higher add-on rates than a pre-metered sizes press, at identical starch concentrations;
- dry add-on rate increases with starch solution viscosity.

It was concluded that the selection of starch or size press type would help to reduce the reduction in machine speed made necessary by the use of a size press. High-viscosity starches provide the most efficient use of starch, at the cost of reduced machine speed. A low-viscosity starch allows higher machine speeds though higher starch add-on rates are necessary to gain equivalent property benefits, compared to a low-viscosity starch. The use of a pre-metered size press enables further speed increases to be gained, again at the cost of increased starch add-on, and in this case it is necessary to use a low-viscosity starch, with adequate penetration properties [46]. Thus there is a balance between machine efficiency, starch costs and waste-paper costs. Higher machine speeds can be achieved, at the cost of increased starch use. However, with high concentrations of starch (>20%) sharp reductions in machine speed can occur, since starch develops very strong water binding capacity [46].

8.6 Properties of recycled fibre grades

As in other paper making grades, it is more difficult to meet product specifications when using recycled fibres rather than virgin pulps. Essential properties vary according to grade.

8.6.1 Solid fibreboard, box-board

Finished boxes must have stacking strength, to avoid collapse or excessive bulge, and are frequently printed, so that appearance and printability may be as important as protection. Stiffness is the most important mechanical property, which is a function of the cube of caliper. Since the caliper of recycled board, at equivalent basis weights, is usually lower than for virgin pulp board, stiffness tends to be lower. However, stiffness is not just a function of caliper, but is also related to fibre bonding and fibre stiffness. Fibre bonding improves with refining and although caliper falls, stiffness increases [40]. With recycled grades, inter-fibre bonding can be improved by the use of starch or other chemical additives, by sheet densification through increased wet pressing, or from refining of some of the layers making up box-board.

Another required mechanical property is the ability to fold when scored, without cracking the top surface. This is related to the tensile of the top ply, which can be improved by wastepaper selection or fractionation, to get longer fibres or by refining or dispersion.

When four-colour printing is used, a strong surface is needed, with the ability to accept high printing and converting speeds and also to give readable bar codes. Smoothness can be achieved by calendering, but this can substantially reduce caliper, which results in reductions in stiffness and compressive strength. High-temperature calendering, either one or two sided, using one or two rolls at a surface temperature of about 200°C has been shown to produce surface properties equivalent to conventional calendering, with a much reduced caliper loss [47]. This enables recycled board grade properties to get closer to virgin grades.

Visual appearance is also important; a specky top ply is unattractive and can lead to inaccurate reading of bar codes. Appearance is determined by wastepaper selection for the top ply, wastepaper treatment, separation of stock and water systems for different plies and by good machine cleanliness. If the box-board is uncoated, the use of a relatively clean underliner ply helps prevent dirt in the filler ply from being visible as a 'shadow' beneath the top ply and also avoids the cost of increased top ply basis weights. If the top ply is produced from white wastepaper, especially any which include thermoplastic printed papers, then it is essential to have a good deinking system for speck reduction and ink removal. Dispersion or kneading, followed by flotation and washing is the best mechanical solution to electrostatic inks [48].

8.6.2 Corrugating medium or fluting

Waste-based fluting competes with semi-chemical fluting made, for example, by the neutral sulphite semi-chemical process - defined by the EC as "containing not less than 65% of unbleached semi-chemical hardwood pulp, calculated on a total fibre content and having a concora medium test crush resistance exceeding 20 kPa (200 N)" [49]. Normally waste-based fluting is made from 100% recycled paper. In the UK most of the waste-based fluting produced is in the lower basis weight grades; 80–90% is 105 and 112 g m^{-2}. Although basis weights range between 105 and 240 g m^{-2} the predominant weights are 105, 112, 127 and 150 g m^{-2} [50].

Since medium is hidden by liner plies, its appearance is rarely an issue, but mechanical properties and runnability on the corrugator are major performance criteria. Corrugated boxes are frequently subjected to high compressive loads and so compressive strength is the single most important requirement for corrugated boxes. This has been shown to be related to edgewise compressive strength (ECT) and flexural stiffness, with the former being about three times as important as the latter [51]. ECT is largely dependent on the compressive properties of the components of combined board, whilst in combined board (liners and corrugated fluting) flexural stiffness depends on the elastic moduli of the liners and to a limited extent of the medium. Box performance is also dependent on the quality achieved on the corrugator. Various tests are used to measure properties related to strength and stiffness, which include:

- Concora medium test (CMT)
- Ring crush test (RCT)
- STFI short span compression test, or index STFI
- Edgewise compression test (ECT)

After corrugating, the combined board must be able to maintain the board caliper during converting operations, such as printing and during transport and storage, since stiffness is related to caliper. CMT is a measure of the resistance to crushing of the flute after it is formed by the corrugator. CMT is also related to impact resistance. RCT is a measure of the contribution of the fluting to the stacking strength of the finished corrugator box, though the use of the STFI strength index has been found to be a more accurate indicator of combined board performance than RCT, under some circumstances [52]. ECT of combined board is frequently predicted from the compressive strengths of the component liners and medium.

When recycled mediums are compared with virgin, they are usually weaker, with lower caliper and hence lower stiffness – virgin flutings gain stiffness by the use of hardwood. Stiffness of recycled can be as low as 50% of those of a virgin medium and properties are variable due to an inability to control the fibre blend exactly, as well as residual contaminants giving rise to localised weak spots in the medium. In one comparison of virgin NSSC (mixed hardwood, refined to 400 ml CSF) with recycled, two types of recycled sheets were made, each using two sample sources. One set was from clean corrugated wastes and the other from commercial sources, using post-consumer wastes. Starch was not added when 127 g m^{-2} medium was made on a pilot paper machine. Compared to virgin NSSC, the clean corrugated was approximately equal in RCT, but the post-consumer average was about 18% lower; in CMT, clean corrugated was about 25% lower, but the post-consumer was > 40% lower [38]. These results are a clear indication of the effects of contaminants in strength reduction.

Starch is the normal additive used to increase strength and stiffness, which is normally applied with a size press, at about 5% by weight. In a comparison of properties of different flutings it was found that the waste-based (starch treated) medium was equivalent or better than virgin flutings at 105 g m^{-2} and 112 g m^{-2} basis weights, for both CMT and RCT. At higher basis weights, virgin mediums had higher CMT and RCT values, thought to be due to a reduced contribution from the chemical reinforcement by starch [50].

Another method of improving strength properties is increased wet pressing, leading to a more dense sheet. In a study on virgin mediums, it was shown that increasing density by increased wet pressing substantially increased the ECT of the combined board and gave higher box compression, as well as higher flat crush, so reducing crushing during conversion and end use. Recycled mediums also increase in density with higher press pressures and nip residence time, so that extended pressing is another mechanism to improve strength and stiffness properties of recycled mediums, though these presses are used mainly to increase solids content after pressing.

Where starch is used to improve strength, porosity is an indicator of starch uptake. If the medium has a low porosity, it may not be able to absorb sufficient starch; for good starch penetration good porosity is required. Starch viscosity can also be changed to improve starch absorption, but a reduction in viscosity may be at the expense of reduced machine speed, if this is achieved by reducing starch concentration.

Runnability of fluting through the corrugator is also important, and a wide range of tests are used to assess runnability. These include:

- caliper;
- abrasiveness;
- MD tensile strength and stretch;
- moisture profile, and content;
- absorbency;
- coefficient of friction, against a heated metal plate.

None of the treatments to improve recycled medium strength properties improve caliper, and although the corrugating operation can accommodate some caliper differences, low basis weight fluting, 105 g m^{-2}, can give rise to poor flute formation by increasing high/lows, if caliper falls below 200 μm [50]. This observation is at variance with other findings, which suggest that higher MD tensile strength, higher stretch, lower hot coefficient of friction and lower caliper promote higher corrugating speeds without excessive high/lows [53]. However, in reaching this conclusion, tests were carried out on virgin 127 g m^{-2} and 161 g m^{-2} fluting, so that the critical minimum caliper of 200 μm would not have been reached. This suggests that a 112 g m^{-2} waste-based fluting would have some advantages over a virgin fluting, provided other parameters were equivalent, due to its lower caliper.

With a comprehensive wastepaper cleaning system the abrasiveness of fluting should not be a problem. High abrasiveness results in accelerated wear on corrugator rolls and facers and is normally due to poor removal of grit and coarse filler particles. Problems can arise with lower quality medium, or when centrifugal systems are not adequately operated or maintained, or when rejects are recycled within mill water systems. Effective centrifugal cleaning will reduce abrasiveness, as discussed in section 8.4.2.

MD tensile strength is necessary to cope with tensile stresses during corrugating, due to high MD forces from the unwinding operation, wraps on pre-heater drums, brakes, bending, friction and speed. If tensile stresses exceed tensile strength of the medium, visible fractures in the flute occur. With recycled grades, tensile strengths are achieved by selecting a wastepaper blend which includes 10–15% long-fibre content. Higher proportions increase tensile, but at the expense of stiffness.

With moisture profile and content there is nothing to distinguish recycled medium from virgin, since both are controlled by machine conditions and operation. However, it has been shown that virgin NSSC-based flutings retain higher tensile stiffness and compression strength under high humidity conditions than waste-based paper. This was attributed to more moisture resistant bonds formed by sulphonated lignin in NSSC during high temperature and moisture content conditions in the corrugator nip, when sulphonated lignin softens [54]. Under high humidity conditions, or cyclic changes in humidity, starch may begin to hydrolyse, resulting in reduced compression strength. High temperature and moisture contents in the corrugator also produced a more stable flute profile from the NSSC medium, but at low temperatures and moisture contents, waste-based fluting was better [54].

Absorbency is important, since the water absorbency rate is related to the receptivity of the medium to the corrugating adhesive. The quality of the bond formed between the liner board adhesive and the fluting is a critical factor in corrugated case performance. In a study using a variety of mediums, it was shown that bonding was enhanced by mediums that were more porous and wettable, under normal conditioning conditions [55]. In the case of waste-based mediums, absorbency tends to be lower and the presence of wax as a contaminant substantially reduces absorbency. Both porosity and absorbency can be achieved by control of the waste paper furnish, refining and chemical addition.

Kraft liner and semi-chemical fluting are less likely to give problems with adhesive bonds. Fibre composition and surface properties also affect glueability. The higher proportion of fines in recycled fluting reduces glueability and fibre composition is more variable, which can also create problems. Very dense sheets produced by extended or hot pressing on the paper machine can also reduce glueability, especially of recycled sheets. Penetration into the sheet, both liner and medium, is also important in achieving strong adhesive bonds, but it is penetration of water, not starch which is important [56]. Water penetrates as liquid and steam into fibres and pores, so dense sheets inhibit penetration. Recycled fibre sheets have higher densities at similar basis weights and so penetration is lower. One estimate is that the Bendtsen porosity should be 400–500 ml min^{-1} [57].

Wettability is the time needed before penetration occurs, which is proportional to the roughness of the surface, and surface energy. Recycled sheets tend to have lower wettability.

Since some of the factors which affect glueability are more difficult to control and optimise, it is more important with recycled grades to have ideal corrugating conditions, and adhesive composition. The temperature of the medium, for example, has a significant effect on adhesive bond strength and is optimal when the medium is at 100°C as adhesive is applied. Starch concentration and viscosity can be adjusted to get adequate penetration, without an undue requirement for drying. Adhesive temperature should be in the range 35–38°C [57].

The coefficient of friction varies substantially in different mediums. It decreases with surface temperature and increases with moisture content. Most mediums have lower coefficients of friction under hot conditions. Cold and hot measurements are not well related [53]. A lower hot coefficient of friction permits higher corrugating speeds without flute fractures and excessive high/lows. Recycled mediums usually have low coefficients of friction, especially if wax was present in the waste supply, though some virgin grades, for example, birch NSSC, have been shown to have equally low friction coefficients, at temperatures > 100°C. This was believed to be due to residual waxes in the NSSC medium; birch is high in waxes and fatty acids [54].

Waste-based flutings are normally appreciably lower in price than virgin, which greatly influences the choice of which to use. The end use of the box is an important consideration; for example, virgin medium would be more appropriate in corrugated containers destined to be stored in a damp atmosphere. Types of fluting could be considered complementary, rather than competing, since there are many circumstances under which waste-based performance, especially of lower weight grades, is acceptable. A relatively poor performance of recycled at equivalent basis weights can be compensated for by providing a higher weight recycled grade, but there are applications for which waste-based mediums are less suitable than virgin, especially when high performance is required.

8.6.3 Liner board, test liner

Waste-based liner, usually called test liner, though sometimes called 'jute' liner, competes with kraft liner, produced primarily from softwood. A definition of kraft liner is "kraft liner means machine finished or machine glazed paper or paperboard, in rolls, containing not less than 80% of chemical sulphate softwood pulp calculated as total fibre content, weight more than 100 g m^{-2} and having Mullen burst ratio of not less than 35" [58]. Since that definition, there has been a recognition that the burst requirement does not adequately reflect properties essential for corrugated container performance, illustrated by the revision of Rule 41 of the Uniform Freight Classification, in the USA, which set minimum burst strengths for various liner weights, for example, 689 kPa for 205 g m^{-2} liner. This was changed in 1992, by introducing the Alternate Rule 41, which allows the use of the edge crush test, ECT. However, the use of the short-span compression test (STFI) has been reported as being more accurate in predicting box compression strength than the use of ECT on combined board [59].

Recycled liner board (test liner) made from 100% wastepaper has been shown not to meet original burst strength standards. When liner was made from clean OCC and from commercial sources of post-consumer OCC stock and compared with virgin kraft liner (kappa number, 50; southern pine, refined to 570 ml CSF) burst strength of the clean liner was 22% lower and contaminated averaged about 50% lower. Ring crush of the clean OCC was only a few percent lower than the kraft liner, but the post-consumer was almost 20% lower in RCT [38].

In the same study properties of combined board were compared, and again there were substantial differences in the properties of the post-consumer grades. Flexural stiffness, perpendicular and parallel, of the clean corrugated were both about 16% lower than the combined kraft liner and NSSC medium, but the contaminated stock was almost 45% and 35% lower, respectively (perpendicular and parallel). Compressive strength was down by about 15% when clean combined board was compared with kraft and NSSC, at both 50% and 90% relative humidity (RH) but the contaminated stock was down by > 30%, at both relative humidities [38]. In another study, corrugated boxes made from recycled liners and medium were shown to have 11% lower compression strength at 50% RH and 28% lower at 85% RH, compared with boxes made from kraft liners and semi-chemical medium [60]. Kraft liner at equivalent basis weights to test liner was said to have a better bending stiffness and higher resistance to creasing and cracking, as well as giving higher case compressive strengths at high humidities [61].

The internal strength of recycled liner also tends to be lower, which can cause weaknesses at glue joints in the corrugating case, leading to case failure. It is improved by refining, by chemical use, high-intensity wet pressing, wastepaper grade selection, etc.

Caliper tends to be lower with test liner, at equivalent basis weights, but adjustments in corrugating can accommodate this, provided the caliper is uniform. Calendering, to improve smoothness, reduces caliper, as does high intensity pressing, refining, etc.

Compressive strength (RCT or STFI), which contributes to stacking strength of the finished box is lower with test liner. One study which examined the effect of various chemical treatments, including sodium hydroxide; hydrogen peroxide plus sodium hydroxide; and oxygen plus sodium hydroxide, reported no increase in compressive strength. However, increases in STFI of about 11% over the value of never dried kraft pulp were achieved, using alkaline treatment, alkaline refining, increased wet pressing and drying under restraint [14]. Other studies using caustic soda treatment have been more successful [13] (see section 8.4.1).

Due to its lower strength properties, but lower cost, test liners were first used as the inner liner, to give cost savings in box production. Since then, developments have allowed test liner to substitute for kraft over a much wider range of applications. Techniques which have been adopted to improve mechanical properties of test liner include:

- starch or other chemical additives or treatment;
- increased basis weight;
- increased mechanical pressing;
- wastepaper processing techniques, such as fractionation;
- wastepaper grade selection.

Starch use will give comparable results to those discussed for medium. Glueability is as important with liner as medium and the use of starch must not negatively impact on glueability. One report suggested starch-treated liner can show brittle adhesive bonding during corrugating [56]. However, this is not considered to be a problem in Europe, where starch-treated test liner is used extensively.

Sodium silicate has been shown considerably to increase the compressive strength of linerboard; for example a 21% addition was shown to almost double the STFI compression of a 205 $g\ m^{-2}$ kraft liner. Sodium silicate also improves the fire resistant properties of containers [62]. The use of caustic soda has already been discussed, which can be used to improve mechanical properties of liner [13, 14].

When the basis weight of test liner is increased, strength advantages of kraft liner are diminished; for example, 180 $g\ m^{-2}$ test replacing 150 $g\ m^{-2}$ kraft was shown to be equivalent in compressive strength and puncture resistance of cases. The greater moisture sensitivity of test was eliminated by using a 140 $g\ m^{-2}$ test to replace a 125 $g\ m^{-2}$ kraft, which gave equivalent RCT and compressive strength over a humidity range of 65% RH to 90% RH [61].

Increased wet pressing, for example using a long nip or extended nip press, or similar, has been shown to increase density and strength properties, excepting tear strength. The reason for installing these presses is primarily to increase sheet dryness after the press. Recycled fibre, which has a lower water retention value than unbleached kraft, responds well to mechanical pressing and will normally give higher press exit solids than a virgin kraft sheet. In one comparison, for example, recycled-fibres press exit solids averaged 47% compared to 43% for virgin kraft. Exit solids in excess of 51% were observed with recycled grades. Press conditions were the same for all tests [63]. This allows refining to be reduced which will help to open up the sheet, since fines production is reduced. This is important in determining glueability of the liner.

In one survey of the effects of increased pressing on recycled and virgin liners, it was shown that typical increases in burst and RCT were 5–10%. Increased pressing was also said to give smoother surfaces [64] though press felt marking can counteract this effect.

Wastepaper processing techniques, for example, fractionation, to provide longer fibres to a test liner machine are very successful in improving strength properties. Dispersion can replace refining, since dispersion improves strength properties without generating as many fines as refining, which gives a more open sheet (see section 8.3.5). However, dispersion or kneading can induce curl and so it is important that this is removed, otherwise fibre orientation can vary, leading to warp in combined board.

Mechanical properties of test liner are evaluated by similar tests to those for medium, with the exception of the CMT, which is not relevant to liner.

Other properties of liner which are important include visual, printability and runnability properties.

Visual properties are only important for outside liners. In unprinted or printed liners the appearance should be even, with no obvious blemishes, similar of that of a kraft liner. Dirt specks are generally removed by dispersion, possibly of just the top ply in a multi-ply liner.

Printability is a very complex property, but is becoming more important. As retail patterns change, goods are left in their transit (corrugated) containers, which reduces handling and so is more efficient for retailers. Since consumers are selecting goods at the display point, packages need to give details of their contents by providing a description and high-quality illustrations. In applications with high-quality print images, corrugated cases compete with plastic containers, which can carry very high print quality. Most liner printing is on flexographic presses and developments in flexographic printing presses, anilox rolls and inks have enabled good quality print to be achieved on liners. For the best print quality, clay coated liners are required, with white top and mottled liners giving diminished, but acceptable performance for most applications. If a waste-based white top liner is coated, it is essential that no contaminants are present in the top ply, since these can cause coating streaks, hence a comprehensive white wastepaper processing system is required.

Bar code printing is also increasing in importance. There are several different types of bar code readers, but for all there must be good contrast between the print and background, with no contaminants between the bars, or near the bars. In one study it was found that a contrast of 70% is needed between the background and print, with a minimum 58% ISO brightness, using a simple bar code reader [65]. A comparison was made between the printability of test to kraft liner and it was concluded that:

- colour of test liner can be less predictable than kraft, and can often have a grey shade;
- the surface of test liner can be more uneven and it absorbs more ink, so that more ink is required for the same print density;
- test liner can have variable absorbency which results in colour unevenness.

No details of the type of test liner evaluated were given [65] but it is clear that there is a wide variety in test liner quality and that these deficiencies can be avoided. The problem of colour is partially due to increased printing on corrugated board, using flexographic inks. On repulping, these break down to very small ink particles and so inevitably they have an impact on appearance. Flexographic inks are very difficult to remove. However, a top white ply on liner will disguise any difficulties associated with off colours, or shade variation, arising from flexographic inks, or any other source.

Surface structure is the most important determinant of print quality, which can be measured by, for example, Bendsten roughness. In some respects, test liner has better potential for achieving higher quality print than kraft. The experience of the printing industry is that softwood fibres, which predominate in kraft liner, do not give a good surface for printing, or coating. Since test liner has a higher proportion of fines, a smoother surface should be achievable. Most recycled grades also contain hardwood fibres, which also contribute to a smoother surface. Finally, recycled grades also contain some mineral filler, which helps fill voids between fibres, again giving a smoother surface. The use of bleached chemical pulp, either virgin or recycled, to give mottle or white top, also improves surface smoothness. Calendering will improve smoothness, though at the cost of the loss of stiffness and some strength. Outer liner needs to have a smooth outer surface, for printing, but an open inner surface, for glueability, so one-sided high-temperature (thermal) calendering seems especially appropriate, though a smoother wire side (inner side) has been shown to be better for adhesive bond development [55].

Drying of liner under high temperature and simultaneous high z-direction pressure (press drying) using a condebelt process also gives a smooth finish to one side, and can also increase performance (in terms of CD, RCT) to high performance containerboard levels, even though this process results in substantial loss in caliper, using a mixed waste furnish [66]. Printability or glueability assessments of the dense liner board produced were not made.

Another important surface property is surface strength; linting and picking at the surface must be avoided, to avoid fouling the printing blanket, which leads to print defects. Recycled liner has a lower surface strength than kraft, but chemical treatment, with starch or CMC (carboxymethyl cellulose) increases surface strength and helps to avoid linting and picking problems.

Wettability affects ink transfer. Lateral spread which causes dot gain must be avoided and it is necessary to have even wettability. Since wettability is also important for glueability in the corrugator, this is sometimes measured and one comparison was that a typical wetting time for kraft was 0.1 s, whereas test could be up to 0.4 s. Wetting time has been reported as being proportional to the roughness and correlates with Bendsten roughness [57]. However, surface chemistry and surface free energy play an important role in wettability. High wetting times may indicate some contamination with wax in recycled grades.

Good formation helps to avoid mottled print. Due to the lower percentage of long fibres in recycled liner, formation can be better, though the use of high concentrations of cationic starch can lead to strong flocculation of fibres and reduce formation quality. The use of gap formers improves formation, and in the case of multi-ply liner, the use of a top wire former or mini-fourdrinier improves formation. A gap former also helps to negate the lower freeness of recycled fibres.

Higher RCT and lower spread of test results have been claimed with gap formers on recycled grades [67].

Runnability of test liner in corrugating and converting has several important aspects, which include glueability, friction properties and creasability.

Factors which affect glueability of liner are similar to those affecting glueability of medium. Outer liner is however, a more demanding product, in that if printed the outer surface has to have quite different properties from the inner surface, so that outer liner is a highly engineered product. The inner surface (wire side) water penetration rate, measured for example by the Cobb test, has to be able to absorb water - the adhesive carrier. However, if this is too rapid, insufficient water is available to swell the starch granules in the adhesive, resulting in weak bonds. This is rarely a problem with recycled liners. However, if water is absorbed slowly, penetration may inhibit both water and adhesive penetration, resulting in weak bonds or slower corrugator speeds. This is more likely to be a problem with dense recycled liner boards, following extended wet pressing. It can be controlled by reducing press pressure, reduced refining, use of dispersion rather than refining, reduced size addition, etc.

Moisture content and profile evenness influence glueability - if low the receptivity to adhesive can be reduced, resulting in weak bonds [68]. Although this is a function of machine operation and should be as controllable for recycled as virgin, in practice machine clothing is more likely to be contaminated, leading to moisture profile variations, with recycled fibres. Another potential problem with uneven moisture profile is warp, which occurs when the two liners in combined board have differing moisture contents. Recycled grades are also more affected by relative humidity changes and so storage conditions or wrapping of recycled roll stock needs more attention than with kraft liners.

Porosity affects glueability, since it influences the rate at which water vapour can be removed from the liners as well as the receptivity to the adhesive. Recycled liners tend to have relatively low porosities, since they have a more dense sheet, with a higher proportion of fines and short fibres, especially if subject to a high degree of wet pressing. Reduced refining, or fines removal from the recycled-fibre stock, for example by fractionation, or washing, will help to increase porosity. However, in some converting or box filling operations, where these are fed by vacuum suction pads, a lower porosity is an advantage, since a high porosity prevents the development of adequate suction.

Friction properties have a significant effect on the ease of use of liner board in converting and subsequent operations. Normally this property is discussed as coefficient of friction or slide angle. The coefficient of friction is the resistance to movement of one sheet of liner when placed against another. The two terms arise from the test method used to test this property (TAPPI Test Method T 815) in which one sample is attached to a plane and a second attached to a sled. The plane is tilted until the sled begins to slide - the angle at which this occurs is the slide angle and the coefficient of friction is equal to the tangent of the slide angle. Both terms describe the same property.

Recycled liner boards, especially in North America, tend to have lower coefficients of friction than kraft, due largely to the much greater use of wax in the USA to control moisture resistance of containers. European definitions of acceptable wastepaper quality do not allow waxed boards and this is much easier to achieve in Europe than in the USA, so this is less of a problem in Europe. When slip angles are low problems occur, such as handling of roll stock with clamp trucks, instability of combined board stacks on pallets and of filled containers during transport, and handling. In a study in Canada the average angle of slip for seven kraft liner boards was 28.6°, whereas the average for three recycled linerboards was 11.7°. A strong relationship with surface free energy was shown, with one exception, rosin size; this reduced surface free energy, but not the angle of slip. Even a monolayer at the surface of the liner affects the surface free energy. Increases in surface free energy can be got with anti-skid treatment, for example, colloidal silica. Surface roughness was shown not to be a factor in increasing friction. Friction increased with increased moisture content [69].

Another study in the USA showed that the contamination of OCC with 1% of waxed OCC reduced the angle of slip by 6°. The minimum acceptable angle of slip was given as 18° [19].

Wax is not the only material to reduce the angle of slip. In newsprint, residual fatty acids, or other pulp extractives reduce the angle of slip, as could contamination on a machine calendar. Angle of slip is increased if an absorptive (porous) filler is used, which either adsorbs or immobilises angle of slip reducing materials in the body of the sheet, not by increasing friction at the surface of the liner [70].

Creasability is important in converting, in that seams are formed by scoring, so that boxes can be erected but without cracking. This is affected by long-fibre content, so that recycle liners can have lower tolerance to creasing. However, fractionation allows long-fibre content in a top ply to be increased.

It is clear from the preceding discussion that liner needs to be able to perform many different functions, especially outer liner. Some of the functions require competing properties, for example, printability versus glueability. To produce an acceptable test liner, its end use will determine how wastepaper preparation systems, machine conditions, etc. should be operated. Test liner is complementary to kraft liner and in many applications is a lower cost substitute, especially at weights $< 200\,\mathrm{g\,m^{-2}}$, provided the case will not be used in an environment where its susceptibility to moisture reduces case performance. When deciding which liner, or medium to use, the basic criterion is the end use of the corrugated case.

8.7 Use of OCC to produce bleached board

Restrictions imposed on forest harvesting rates in some areas, for example, the northwest USA and Canada have resulted in a shortage of wood chips in some regions. Consequently, there has been a resurgence of interest in upgrading OCC to produce a bleached chemical pulp. Several systems of this type are operating and one of these, in the USA, provides a bleached pulp for solid bleached board, from a post-consumer wastepaper, OCC. This pulp is reported to meet the US Food & Drugs-Administration standards for direct food contact, even with moist and oily foods [71]. Details of the process have not been released, but there have been several studies on delignification of OCC [72–75].

An early study examined oxygen–alkali delignification of OCC and other wastepapers. In the case of OCC it was concluded that this would allow a substantial increase in the amount of recycled fibres used in liner board, without sacrificing quality standards [72]. Another study using oxygen–

alkali delignification showed that removal of some contaminants by centrifugal cleaning before delignification could produce a good quality pulp, when delignification was followed by bleaching [73].

The use of kraft mill liquor for delignification has also been evaluated and this approach is being developed in Canada to produce a fully bleached pulp, for use in printing and writing papers [74].

Comparisons of OCC fibre content and kappa number illustrate differences between European and North American OCC. A sample of European OCC had 55% softwood and 45% hardwood (25% NSSC birch) with a kappa number of 70 [72]; whereas in the USA fibre content was 60–80% softwood and 20–40% hardwood, with kappa numbers in the range 80–100 [75].

Both delignification processes reduced kappa numbers to about 30, dependent on alkali charge. Yield was shown to be approximately proportional to active alkali, down to a kappa number of 30 [75].

A comparison of pulp properties of oxygen-alkali delignified OCC (without bleaching) showed that at the same drainage rate mechanical properties were significantly improved, tear increased by 15%, strain by 35%, tensile by 10% and burst by 20% [72].

An economic analysis showed that those processes, under some circumstances, could be very competitive with kraft pulping and would be a low-cost method for producing a post-consumer waste-based chemical pulp [75]. If the technology proves to be widely adopted, it would have an impact on OCC supply and price for traditional users of OCC, producing waste-based packaging grades.

References

1. Hickman, H.L. (1992) Looking at solid waste composition and mandatory removal rates in north America, *ISWA Times*, **2**, 2.
2. Papworth, R. (November 1991) *Analysis of Samples of Waste Arisings from the Cities of London and Westminster*, Warren Spring Laboratory.
3. Anon, *Commercial Waste Arisings in the West Midlands County 1989*, MEL. Waste Research Report 90/01.
4. Anon (1992) Mole Valley District Council, Draft Waste Recycling Plan.
5. Anon (1992) London Borough of Ealing, Draft Waste Recycling Plan.
6. Anon (1992) London Borough of Richmond on Thames Waste Recycling Plan.
7. Sackellares, R.W. (1993) Environmental issues play key role in planning, siting urban mini-mills, *Pulp and Paper*, **67**(9), 83.
8. Anon (1993) CPPA Fibre Furnish Survey. Canadian Pulp and Paper Association.
9. Kardell, A., Sandberg, E. and Svensson, K. (July 1992) SLV Statens Livsmedelsverk in *Reply to Query Regarding Recycled Fibre Use in Food Packaging*.
10. Svensson, K. (20 November 1991) National Food Administration of Sweden. Information Sheet.
11. Chamberlain, J. (October 1993) Iggesund Paperboard, Personal Communication.
12. Muir, J. (1986) A Modern Approach to Waste Paper System Design, Pira Conference, Recent Developments in the Use of Wastepaper in the Manufacture of Paper and Board.

13. Freeland, S. and Hrutfiord, B. (1993) Caustic treatment of old corrugated containers (OCC) for strength improvement during recycling, *TAPPI Pulping Conference Proceedings*, TAPPI Press, Vol. 1, pp. 127-134.
14. Springer, E.L., Klungress, J.L., Spangenberg, R.J., Minor J.L. and Tan, F. (1993) *TAPPI Recycling Symposium*, pp. 163-171.
15. St Amand, F.J. and Bernard, E. (1992) Centrifugal cleaning of brown grades, *Paper Technol.*, **33**(3), 34.
16. McKinney, R.W.J., Cathie, K. and Voss, G. (1987) *Efficiency of Stickie Removal Systems*, Pira Project Reports PB/MC/85/3.
17. Chivrall, G.B. (1992) Stock preparation for brown grades - new developments, *Paper Technol.*, **33**(4), 8.
18. Fleming, S. (1993) The new OCC line at Weyerhauser, Valliant, *TAPPI Recycling Symposium*, TAPPI Press.
19. McDonnell, W.T. (1993) Wax is source of low surface friction in kraft/recycled linerboard. *TAPPI Recycling Symposium*, TAPPI Press.
20. McEwan, J.G.E. and Wang, B. (1993) OCC recycling: improving the repulpability of waxed coated corrugated paperboard, *TAPPI Pulping Conference*, TAPPI Press, pp. 493-501.
21. Lock, R. (1993) Fibre fractionation in practice. *World Pulp & Paper Technol.*, pp. 71-74.
22. Leblanc, P. (1975) Tracing the development of a new product, *TAPPI Secondary Fibre Conference*, TAPPI Press.
23. Anon (21st June 1982) Noss New Products, *Paper*, p. 53.
24. Duffy, G.C. (1980) High consistency pulp fractionation with an atomiser (NIRO), *TAPPI Engineering Conference*, TAPPI Press.
25. Seifert, P. and Long, N.L. (1974) Fibre fractionation - methods and applications, *Tappi J.*, **57**(10), 69.
26. Musselman, W. (November 1980) *In-line Fractionation of Wastepaper*, UNIDO International Experts Group Meeting, Manila.
27. Mearsman, T.A. (1976) Fibre classification in a multi cylinder mill, *TAPPI Secondary Fibre & Pulping Conference Proceedings.*
28. Dobbins, R.J. (1977) High kappa grocery bag production by means of stock fractionation, *TAPPI Alkaline Pulping & Secondary Fibre Conference.*
29. Putz, J.P., Torok, I. and Göttsching, L. (1989) Industrial fractionation - application and fibre quality improvements, Pira Conference *New Developments in Wastepaper Processing and Use* Vol. 1.
30. Selder, H. (1992) Fractionation - the technology of the future, *Paper Technol.*, **33**(3), 13.
31. Kohrs, M. (1992) The application of fractionation technology, *Paper Technol.*, **33**(3), 10.
32. Lunsing, H. (1987) Fibre fractionation of waste, *PITA Conference Proceedings.*
33. Manges, W. (1984) Wastepaper fractionation is the key at PWA's Redenfelden mill, *Pulp and Paper*, **58**(3), 118.
34. Ibrahim, H. (1989) Fibre fractionation - an appropriate technology for upgrading recycled fibres and saving energy, *EUCEPA Symposium*, Ljubljana.
35. Culicchi, P. (1989) The upgrading of stock furnishes by the fibre fractionation system in corrugating papers production, *EUCEPA Symposium*, Ljubljana.
36. Larsen, G. and Nielson, S. (1987) Production scale trials with fractionation of recycled fibres, In *Pulp, Paper and Board*, Eds I.F. Hendry and W.J.H. Hanssens, Elsevier Applied Science.
37. Claydon, M.W. (1986) UK demonstration project on fractionation of freely available grades of wastepaper, BP&BIF, Final Report.
38. Fahey, D.J. and Bormett, D.W. (1982) Recycled fibres in corrugated containers, *Tappi J.*, **65**(10), 107.
39. Howland, P. (1987) Chemical treatment of fractionated stock from freely available grades of wastepaper, in *Pulp, Paper and Board*, Eds I.F. Hendry and W.J.H. Hanssens, Elsevier Applied Science.
40. Estes, T.K. (1986) Important properties of recycled fibres relating to paperboard, *TAPPI Pulping Conference*, TAPPI Press.
41. Marley, M.E. (1989) Fiskeby invests in the latest board-making technology, *Paper Technol.*, **30**(3), 21.

42. Horn, R.A. and Bormett, D. (1985) Press drying recycled fibre for use in paperboard, *Tappi J.*, **68**(12), 78.
43. Seth, R.S. and Page, D.H. (1988) Fibre properties and tearing resistance, *Tappi J.*, **71**(2), 103.
44. Anon (1993) Colthrop uses amphoteric starch to improve internal sheet strength, *Paper Technol.*, **34**(2), 22.
45. Klass, C.P. (1990) Trends and development and size press technology, *Tappi J.*, **73**(12), 69.
46. Glittenberg, D. (1992) Using the film press to increase speed on a fluting machine, *Paper Technol.*, **33**(11), 34.
47. Gratton, M.F. (1989) Temperature gradient calendering of recycled boxboard, *Tappi J.*, **72**(3), 87.
48. McKinney, R.W.J., Cathie, K. and Staves, J. (1989) *Deinking Efficiency Improvement*, Pira Project 32/PB/88/8.
49. Anon *Official J. European Communities*, 31/12/1976.
50. Lyddon, M. (1990) *PITA Annual Conference*, PITA.
51. McKee, R.C., Gonder, J.W. and Wachuta, J.W. (1963) *Paperboard Packaging*, **48**(8), 149.
52. Whitsitt, W.J. (1988) Papermaking factors affecting box properties, *Tappi J.*, **71**(12), 163.
53. Whitsitt, W.J. (1987) Runnability and corrugating medium properties, *Tappi J.*, **70**(10), 99.
54. Back, E.L. and Salmen, L. (1989) The properties of NSSC-based and waste based corrugating, *Paper Technol.*, **30**(10), 16.
55. Batelka, J.L. (1992) Development of green bond strength in the singlefacer, *Tappi J.*, **75**(10), 94.
56. Kroeschell, W.O. (1990) Bonding on the corrugator, *Tappi J.*, **73**(2), 69.
57. Bradatsch, E. (1990) Bonding on the corrugator; the energy aspects, *Tappi J.*, **73**(1), 81.
58. Jonson, G. (1987) *PRIMA Conference.*
59. Rennie, G.S. (1992) quoted by Kroeschell, W.O., *Tappi J.*, **75**(10), 78.
60. Young, R.A. (1992) quoted by Kroeschell, W.O., *Tappi J.*, **75**(10), 78.
61. Martin, J.H. (1989) Upgrading test liner to compete with kraft liner, *Pira Conference; New Developments in Wastepaper Processing and Use.*
62. Walthy, G.J. (1987) Paperboard chemical enhancement for strength and other benefits, *Tappi J.*, **70**(10), 35.
63. Sorma, O.P. (1989) The Tampella long nip press - five years of practical experience, *World Pulp and Paper Technol.* p. 161.
64. Lange, D. (1988) An update on extended nip pressing, *World Pulp and Paper Technol.*, **29**(5), 265.
65. Haglund, N. (1988) Improving the printability of kraft liner, *Paper Technol.*, **29**(5), 265.
66. Lehtinen, J. (1993) The basic condebelt process can be modified to satisfy a variety of special quality demands, *World Pulp and Paper Technol.*, p. 105.
67. Salrai, N. (1989) *New Beloit Forming Technology for Corrugating Medium and Testliner Grades from Recycled Fibres*, ATP Conference, Bordeaux.
68. McGrattan, W. (1990) Key characteristics of linerboard, corrugating medium and roll stock mechanical condition and their influence on the manufacture of corrugated products, part I, *Tappi J.*, **73**(11), 99.
69. Inoue, M., Gurnagul, N. and Aroca, P. (1990) Static friction properties of linerboard, *Tappi J.*, **73**(12), 81.
70. Gunderson, D. (1993) Friction properties of paper and paperboard, *TAPPI Recycling Symposium*, TAPPI Press.
71. Anon (1993) *Tappi J.*, **76**(3), 24.
72. de Ruvo, A., Farnstrand, P.-A., Hagen, N. and Haglund, N. (1986) Upgrading of pulp from corrugated containers by oxygen delignification, *Tappi J.*, **69**(7), 100.
73. Markham, L.D. and Courchene, C.E. (1988) Oxygen bleaching of secondary fibre grades, *Tappi J.*, **71**(12), 168.
74. Nguyen, X.T., Shariff, A., Earl, R.F. and Eamer, R.J. (1993) Bleached pulps for printing and writing papers from old corrugated containers, *Prog. Paper Recycling*, **2**(3), 25.
75. Bisner, H.M., Campbell, R. and McKean, W.T. (1993) Bleached kraft pulp from OCC, *Prog. Paper Recycling*, **3**(1), 27.

9 Manufacture of newsprint using recycled fibres

J.K. HUSTON

9.1 Introduction

In 1991, just over 32 million tons of newsprint was produced world-wide, with an estimated 8.5 million tons of wastepaper going into the production of this newsprint [1]. Historically, the majority of newsprint producers who used recycled fibres were in fibre limited regions. Fibre limitations, in regions such as the Far East and western Europe, will continue to contribute to the growth of recycled fibres in newsprint production. However, new factors, including legislative requirements, strict energy usage requirements, and technological advances in both the processing of wastepaper as well as paper making, will result in newsprint becoming one of the fastest growth markets for the use of recycled fibres. It is estimated that by the year 2005 the use of recycled fibres in the production of newsprint will have increased by more than threefold to about 30 million tons [2]. This dramatic increase will be largely driven by increased consumption in North America and western Europe.

This chapter will discuss the raw material being recycled, the technology currently available for recycling, and the economics associated with the use of recycled fibres in newsprint production. An understanding of these fundamental areas is essential if we are to successfully increase the consumption of recycled fibres in the production of newsprint.

9.2 Wastepaper grades used

9.2.1 Definition of wastepaper grades

Standard definitions for wastepaper grades have been established in every major region of the world. These standard definitions include criteria such as acceptable types of paper in the furnish, maximum contaminants allowable, and trade practices such as purchase agreements, transportation charges, etc. An example of the standards which are currently used to govern the wastepaper market are summarized in Table 9.1.

The specific grades which have been developed in each region are influenced by the economies and method of wastepaper collection, as well as the technology available to process the wastepaper. The grades, while

Table 9.1 Wastepaper quality specifications

Region	Specification title	Agency responsible for specification
United States (and Canada)	Guidelines for Paper Stock (PS)	Institute of Scrap Recycling Industries
Europe	European Standards Qualities of Wastepaper	CEPAC
Germany	List of German Standards for Wastepaper Quality	Verband Deutscher Papier-fabriken (VDP)
United Kingdom	Descriptions of Standardized Grades of Wastepaper	British Paper and Board Industry Federation (BPBIF)
Japan	List of Japanese Standard Qualities of Wastepaper	Paper Recycling Promotion Centre (PRPC)

similar, do differ from one world region to the next. The majority of the grades used in the manufacture of newsprint consist of old newspapers, commonly referred to as ONP, and old magazines and coated flyers, commonly referred to as OMG or PAMS.

9.2.2 *Old newspapers*

As the name implies, old newspapers (ONP) consists of newspapers which have been published, distributed and then collected for recycling. A summary of the ONP grades used for the manufacture of newsprint is found in Table 9.2.

The characteristics of the ONP fibre are based on the newsprint manufacturing process, as well as the printing process that is used. Generally speaking, ONP has a high percentage of mechanical fibre, groundwood or thermomechanical pulp. Chemical pulp, such as kraft and sulphite pulp, can be as much as 30% by weight (30 wt%) of the furnish. As the recovery rate for ONP continues to increase, fibre which has been recycled at least once is starting to find its way back into the recycle stream.

In addition to the fibre component of the furnish, additives required for the production of newsprint are also present in the ONP stream. These additives include starch, inorganic fillers and dyes for colour control. ONP is relatively low in these additives, with ash ranging from 3 to 12 wt%.

Ink is introduced in the press room. The main processes used in the printing of newsprint is letterpress, offset and flexographic. Ink makes up 1 to 2 wt% of the ONP furnish.

Contaminants associated with ONP are generally introduced through the collection process and may include unbleached fibres, such as boxes or grocery bags, plastic and miscellaneous trash material.

Table 9.2 Definitions of old newspapers grades

Grade	Reference	Definition
Special news	PS 7	Baled newspapers, less than 5% other papers. Contaminants <2 wt%
Special news deink quality	PS 8	Baled, sorted newspapers, free from magazine and over-issue. Contaminants <0.25 wt%
Over-issue news	PS 9	Unused, overrun regular newspapers, baled or in bundles. No contaminants allowed
Crushed news	BPBIF (Group 5)	Newspapers which have been read. No magazine or PAMS allowed. Contaminants <2 wt%
Over-issue news	BPBIF (Group 5)	Unsold newspapers free from pins, coated or sized papers. Contaminants <1 wt%
Once-read news	CEPAC B1	Old newspapers with <5 wt% coloured inserts. Contaminants <1 wt%
Over-issue news	CEPAC B2	Unsold daily newspapers, free from coloured inserts or illustrated material. Strings allowed
Old news	PRPC 16	Old newspapers with <2 wt% contaminants

9.2.3 Magazine

Magazine is a generic term which generally refers to coated paper that is bound with staples or glue. The specific definitions are summarized in Table 9.3.

Magazine is a highly variable raw material. The fibre component of the magazine can range from 100% kraft pulp to 100% groundwood. An individual magazine may have several grades of paper included in its production, thus resulting in a highly variable furnish component. A single magazine may include coated free sheet, coated groundwood, uncoated mechanical fibre, and card inserts made of hardwood groundwood.

In addition to variability in the fibre component, the additives are also highly variable. Fillers such as clay, alum, and precipitated calcium carbonate (PCC) are added in the paper making process to improve the sheet characteristics. In magazine stock this inorganic portion of the furnish can range from 10 wt% in the uncoated sheets to as high as 50 wt% in a sheet that is coated on both sides.

Table 9.3 Definitions of magazine grades

Grade	Reference	Definition
Mixed PAMS and magazines	CEPAC A6	Mixed, once-read pamphlets, magazines, catalogues, directories and newspapers
Over-issue PAMS and magazines	CEPAC A7	Unsold pamphlets and magazines, with or without adhesive binding
Mixed news and PAMS	CEPAC A9	A mixture of newspapers and pamphlets (at least 50 wt% news) with or without adhesive binding, strings allowed
Over-issue PAMS and magazines free from adhesive bindings	CEPAC A8	A mixture of unsold newspapers and pamphlets, free from adhesive binding, strings allowed
Old Magazines	PS 27	Dry, baled, coated magazines, catalogues and similarly printed materials that may contain a small percentage of uncoated news-type papers. Contaminants <3 wt%
Old Magazines	PRPC 17	Old magazines with <2 wt% contaminants
News and PAMS	BPBIF (Group 5)	Newspapers and magazines free from latex backed or bound books, periodicals, telephone directories, or plastic laminated covers. Magazine content <30 wt% of each bale
Over-issue PAMS	BPBIF (Group 5)	Over-issue periodicals printed on mechanical coated or uncoated paper, stapled but not bound in a latex glue

The addition of dye is also common in the production of magazine-grade papers. As many as 30 000 different dyes are used. Since some of these are sensitive to pH, and colour stripping of recycled fibre can be expensive, this dye addition can interfere with the recycling process.

Contaminants associated with magazine grades are introduced in the converting process. Adhesives associated with the bindings, thermal plastics and hot melts can all contribute to stickies. Ultraviolet cured inks, common on magazine covers, are difficult to deink. Ink which is printed on coated paper can also present a removal challenge in the deinking stages. Ink can range from 1 to 7 wt% in magazine grades of wastepaper.

9.3 Recycled fibre processing

9.3.1 Major processing steps

The processing of ONP and OMG for the production of newsprint actually begins with the wastepaper collection process. This has been discussed in chapter 1 and will not be addressed here.

The purpose of the recycle facility is to produce a pulp with optical and physical properties which are acceptable for their end use. When recycled fibres are processed to produce newsprint, the recycle facility is referred to as a deink plant. While the exact configuration of the process equipment varies, most deink plants will have the following unit operations:

- pulping
- screening
- cleaning
- deinking
- dispersion
- bleaching
- solid-waste handling
- water clarification

9.3.2 Process flow sheets

The deinking technology that is utilized in the production of newsprint from recycled fibres is well established. However, the technology which developed in North America differs from the technology developed in Europe or Asia. Major factors that influence this technology development and process design include:

- product quality requirements
- resource availability (power, water, etc.)
- wastepaper characteristics
- per cent of recycled fibre in furnish
- environmental regulations

In Asia, product quality requirements for newsprint include cleanliness and runnability. The language in this part of the word is very complex and an ink speck can significantly change the meaning of a character. The Asian pressrooms are highly efficient, measuring breaks per 1000 rolls. Any stickie-type contaminant can result in pressroom breaks and is therefore not acceptable. Water conservation is required in most regions of Asia, and as a result, closed water loops are common in Asian deink facilities.

A typical example of an Asian deink process is summarized in Figure 9.1. Continuous drum pulpers are utilized to maximize contaminant removal

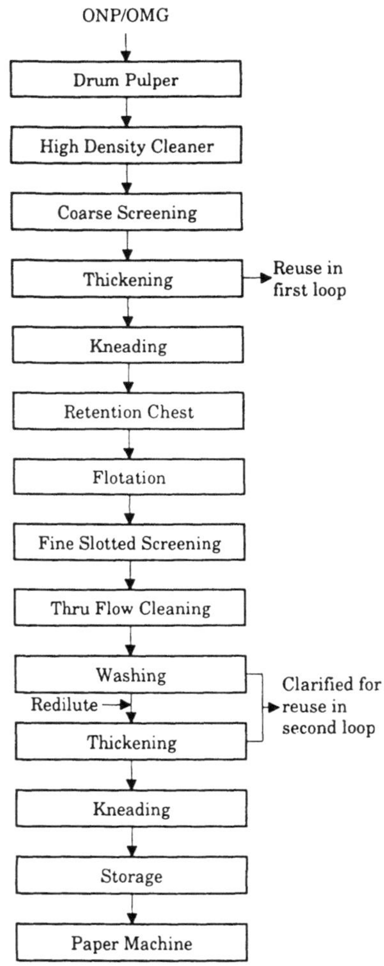

Figure 9.1 Asian deink process flow sheet.

and minimize fibre loss. A kneading (low intensity dispersion) stage takes place prior to flotation. The kneader has two counter-rotating shafts, where each of the latter has flights and rotates at low speeds. The pulp at ambient temperature feeds the kneader at high consistency. Chemicals such as caustic, silicate and peroxide are often added at this stage. The interfibre friction results in a separation of ink and contaminants from the fibre, thus reducing the particle size with minimal freeness drop [3].

Preflotation kneading is often followed by a retention chest or soak tower. This retention time (one to four hours) at high consistency aids in the removal of ink particles in the flotation process [4]. A washing step follows the low-consistency loop. The washer filtrate is clarified using

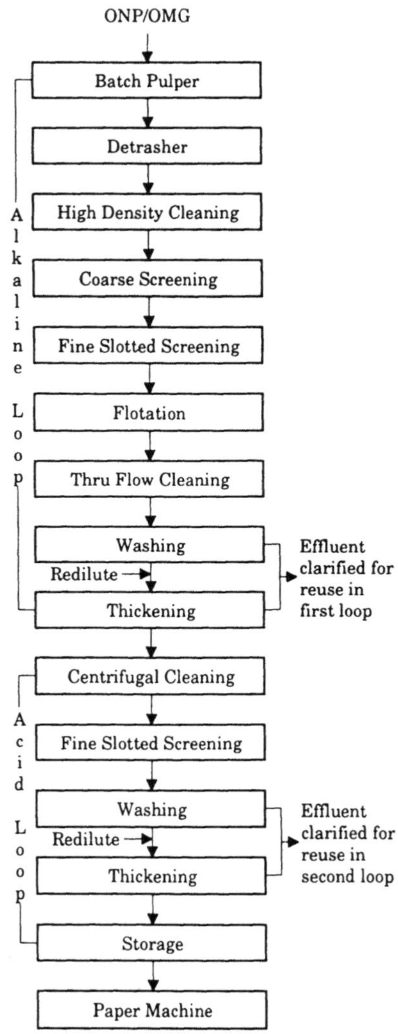

Figure 9.2 European deink process flow sheet.

microflotation and reused to minimize water usage. A second stage of kneading, which includes bleach addition, constitutes the last process step. This additional kneading reduces the size of any remaining contaminants and serves as an excellent bleach chemical mixer.

In Europe, deinking technology developed which allowed the newsprint manufacturer to economically produce his product from recycled fibres. This included minimizing fibre loss in the system, and the conservation of both energy and water.

A western European deink process is summarized in Figure 9.2.

In this process a high-consistency batch pulper is used to defibre the wastepaper. High-consistency pulping optimizes any chemical addition to the pulper and minimizes contaminant attrition. The initial processing occurs in an alkaline environment, with the pH in the range 9–11. Coarse screening may consist of both a perforated and slotted screen system.

The deinking stage, flotation, follows the coarse screening. A two-stage thickening process allows fine ink particles, clays and fines to be removed along with the alkaline water stream. The water is clarified and reused in the alkaline process loop.

The pH is then adjusted to ~5–6 in order to create an acid shock. Colloidal contaminants, including stickies, agglomerate as a result of this pH shock, thus allowing for their removal by using slotted screens and cleaners [5, 6]. The stock is diluted to less than 1% and abrasive contaminant removed with centrifugal cleaners, while fine slotted screens are used to remove agglomerated stickies. In the next step, the stock is thickened once again and the filtrate treated. The clarified effluent stream is reused in the acid loop, so minimizing the need for any fresh water make-up in the system. Finally, the thickened stock is bleached and stored for usage on the paper machine.

A variation on the two-loop system described above was first implemented in Scandinavia [7]. Summarized in Figure 9.3, this system utilizes the two-loop system to minimize the carryover of chemicals and dissolved solids to the paper machine. The initial pulping is high-consistency in order to optimize chemical addition. The gentle action of the drum-type pulper which is used here maximizes contaminant removal early on in the system. The first loop contains all of the screening and cleaning equipment and flotation deinking. Pulp is then thickened in a two-stage thickening process using dilution washing principles. The filtrate is processed in a microflotation clarifier and reused in the first loop.

High-consistency dispersion follows the first process loop. A disc or tooth-type unit is used, generally operating at temperatures above 80°C. The rotational speed in this type of dispersion can be as high as 1800 rpm, compared to 100 rpm for the kneader. The disperser reduces the larger contaminants in size and also releases attached ink from the fibre. The pulp is diluted to ~1% and additional deinking then occurs in the post-dispersion flotation step. A two-stage thickening process removes small ink particles and residual chemicals from the pulp, after which the filtrate is clarified and reused in the second loop.

In North America, deinking for newsprint has been based historically on wash deinking. This technique can be effective if the raw material is 100% ONP and if limitations on water usage do not exist [8]. However, the majority of the new deink facilities coming on line in North America combine washing and flotation deinking. They are influenced strongly by new technology developed in both Europe and Asia, and so may include some

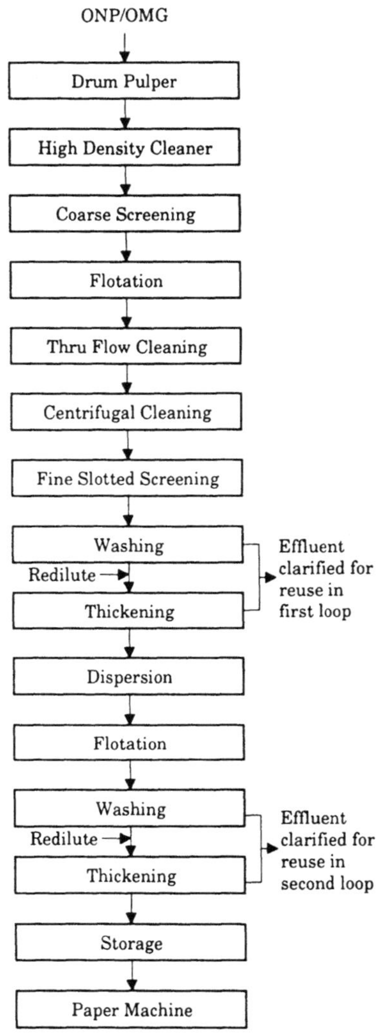

Figure 9.3 Scandinavian deink process flow sheet.

type of dispersion or kneading, post dispersion flotation or two-loop processing step.

9.4 Production of newsprint from recycled fibres

The technology to produce newsprint from recycled fibres is well established. However, the use of recycled fibre does contribute to a slightly reduced paper machine efficiency. Twenty-three modern newsprint machines were compared worldwide and the comparison indicates that at a minimum recycled fibre level of 30 wt%, machine efficiency is reduced by an average of about one-half of one percent [9].

In order to minimize the impact of contaminants in the recycled fibre on machine efficiency, improvements to the machine clothing, cleaning and conditioning systems are often required. In the press section, felt conditioning is important. High-pressure showers, running continuously, assure cleaning of the press fabric. Reducing the cross-direction spacing of the showers to 4–8 cm assures an adequate cleaning time. Location of the showers on the sheet side, and as close to the fabric as possible, minimize potential fabric damage [10]. To handle the excess water associated with felt cleaning an adequate suction is required. Suction dewatering should be capable of 2.9–3.7 $m^3\,min^{-1}\,cm^{-2}$ (16–20 $cfm\,in^{-2}$) of Uhle box open area. A thicker, more open felt makes handling this volume of water easier and also contributes positively to the sheet properties.

The wet web strength of recycled fibres is often less than virgin pulp. As a result, problems with sheet threading can occur. This can be minimized by using sheet transfer equipment to improve sheet stability. The areas with open draws, such as the press section, and between the press and dryer section, are the most sensitive.

The dryer is an area which is sensitive to the buildup of ink, ash, stickies and fines. Ink builds up in the fibre cross-overs and also accumulates fines and inorganic materials, while stickies build up in the initial sections of the dryer and also on the dryer felts. This buildup can result in nonuniform drying, picking and reduced runnability. To minimize buildup on the dryer fabrics, a 100% monofilament fabric is recommended. While monofilament fabrics are subject to buildup, they are easier to clean by mechanical methods, steam showers or chemical addition [11].

As contaminants build up on the dryer surface, picking occurs. This picking can be minimized if small particles, such as fines, are set in the sheet prior to contact with the hot dryer surface. This can be accomplished by gradually increasing the dryer temperature, i.e. gradually ramping up the dryer pressure.

Dryer surfaces are generally kept clean by using doctor blades. These or other surface cleaning devices are recommended on the dryer surfaces that first make contact with each side of the sheet. Up to the first third of the dryer section may have doctor blades installed [12].

Calendering is used to reduce the sheet bulk and also improve smoothness. Because recycled fibre has both a lower bulk and a higher density than virgin fibre nip pressures often have to be reduced to maintain caliper. This can have a negative effect on the printing properties. Soft nip calendering makes it possible to improve smoothness without significantly reducing caliper.

Uniform roll quality can be a challenge when using recycled fibres. Newsprint manufactured from recycled fibres has a lower bulk (and caliper) and also a lower coefficient of friction. These sheet properties can contribute to winder breaks, crepe wrinkles and the telescoping of roll ends. Producing tightly wound, large diameter rolls with recycled fibre-based newsprint can be difficult. Roll density can be effectively controlled by using a single drum winder, where nip pressure can be reduced by changing the angle between the winder drum and the paper roll.

White-water management is also a key technique for maximizing quality characteristics and machine runnability. The small particles that are removed on the paper machine wet end include stickies, ink, fines and inorganics. Clarification, or some type of contaminant purge, of this white-water stream is often required.

9.5 Quality and performance of recycled newsprint

In general terms, newsprint produced from recycled fibres has a higher density, a lower caliper, is more absorbent, and has a lower coefficient of friction than virgin newsprint. Recycled fibre-based newsprint also has a potentially better smoothness and porosity, while its strength properties are generally equivalent to the virgin product [13]. Optical properties, such as scattering coefficient and opacity, are generally improved when recycled fibre is incorporated into the newsprint sheet. In addition, brightness can be slightly lower and shade variability increases. Each of these quality characteristics influences the performance of recycled newsprint in the press room. The reduction in the coefficient of friction associated with recycling, and the increased inorganic content of the recycle pulp, can result in improper press feeding. Snaking and misregister problems in the press room can result from this.

Linting is a term used to describe the tendency of fibres and fines to be removed from the newsprint sheet surface and accumulate on the blanket, and is a major runnability concern in press rooms. Fibre quality, such as bonding characteristics and fibre stiffness, are major factors contributing to the printing process [14].

Data on the impact of multiple recycling on fibre quality indicates that the IGT pick strength can be reduced by as much as 25% after four recycles. Press room data substantiate this. Two recent surveys, involving one of the press rooms in New York, and one of all Gannett pressrooms, both indicate that problems with linting of recycled newsprint had a negative impact on both runnability and printability [15].

Excess water absorbency, particularly in multicolour, offset printing presses, can result in curl and folder wrinkles. Newsprint produced from recycled fibres has a higher water absorption than virgin newsprint sheets. This increased absorption appears to be due to the removal of self-sizing agents, such as natural resins and fatty acids, in the deinking process.

Table 9.4 Comparison of units of consumption for virgin and recycled fibres

	Unit/ADMT[a] finished pulp		
Variable	Unit	TMP	Deink
Raw material	ADMT	1.042	1.176
Energy			
Electricity	kWh	2200	600
Steam	MJ	1.00	1.00

[a] ADMT, air dry metric tonnes

Historically, recycled newsprint has been considered inferior to virgin newsprint when considered on the basis of pressroom runnability, printability and brightness [16]. Technological advance in the production of recycled fibres, as well as in the newsprint manufacturing process, has eliminated this concern. The technology available today makes it possible to produce recycled newsprint which is comparable to a virgin sheet. As a result, newsprint produced from recycled fibres has gained press room acceptance throughout the world.

9.5.1 *Economics of wastepaper usage - capital costs*

The capital cost associated with a 100% recycle newsprint mill is about equivalent to a 100% thermomechanical pulp (TMP) newsprint mill. The capital associated with the newsprint machine, shipping and environmental treatment are the same, with the primary differences occurring in the areas of solid-waste handling and pulp processing [17].

9.5.2 *Variable manufacturing costs*

When evaluating the variable manufacturing cost of recycled fibres and TMP, the two most significant cost components are raw material and energy. Both the price of raw materials (wastepaper and wood chips) and the price of electricity are very site specific. Table 9.4 summarizes the units of consumption associated with these variables for both the TMP and deink processes. The estimated cost associated with the production of TMP and recycled fibre, excluding raw materials and energy, is summarized in Table 9.5. The impact of raw material and energy cost on the variable manufacturing cost is shown in Figure 9.4 for TMP and in Figure 9.5 for deink pulp. Keeping in mind that the non energy and raw material variable cost to manufacture TMP pulp is ~$50/ADMT, Figure 9.4 demonstrates that energy can range from ~20 to 45% of the variable cost of TMP production. Similarly, the price of the raw material can account for between ~30 and 60% of the variable manufacturing cost. The deink pulp non energy and raw material variable cost is estimated at ~$65/ADMT. Energy costs are

Table 9.5 Comparison variable manufacturing costs for virgin and recycled fibres

		$/ADMT[a] finished pulp	
Variable	Unit	TMP	Deink
Labour (fully loaded)	$25/effort hour		
Pulp process		9.00	10.25
Maintenance		2.25	1.75
Chemicals	$/ADMT	10.00	25.00
Supplies	$/ADMT		
Pulp process		3.00	3.00
Maintenance		5.50	2.50
Miscellaneous operating expense	$/ADMT	15.00	10.00
Overhead	$/ADMT	5.00	5.00
Solid-waste disposal	$/ADMT	0	8.75
Total (excluding raw material and energy)		49.75	66.25

[a] ADMT, air dry metric tonnes

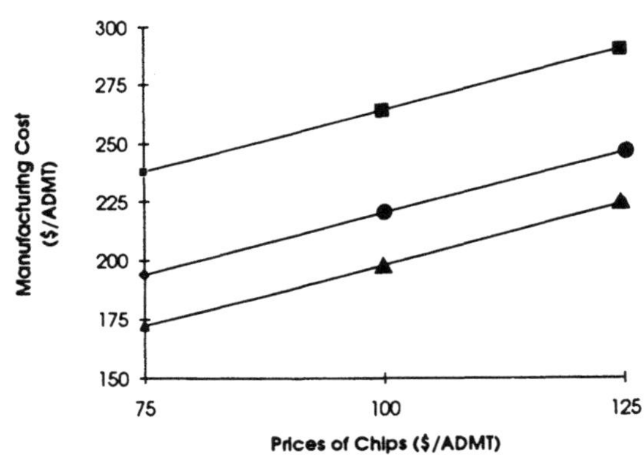

Figure 9.4 Impact of raw material and energy cost on the total variable manufacturing cost in TMP production: (▲) $0.02/kWh; (●) $0.03/kWh; (■) $0.05/kWh.

significantly lower than those of TMP, ranging from 6 to 20% of the total variable cost, while the raw material costs range from ~40 to 60% of the total variable cost.

Figure 9.6 compares the variable cost of TMP pulp to deink pulp at a constant wood chip price of $75/ADMT. The following cases are included for comparison:

Case 1: low wastepaper cost - $50/ADMT low energy cost - $0.02/kWh

Case 2: high wastepaper cost - $100/ADMT high energy cost - $0.05/kWh

Case 3: low wastepaper cost - $50/ADMT high energy cost - $0.05/kWh

Case 4: high wastepaper cost - $100/ADMT low energy cost - $0.02/kWh

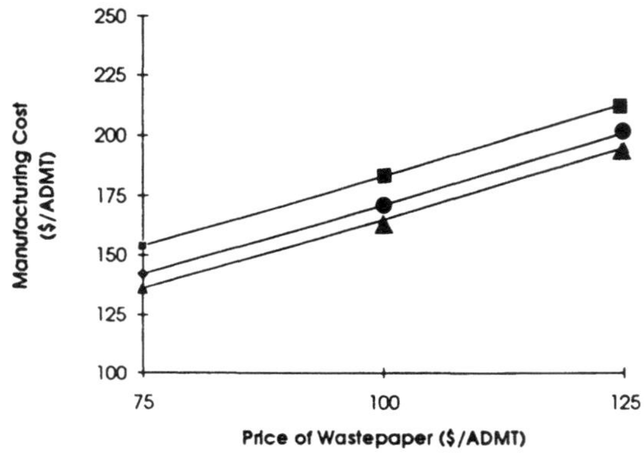

Figure 9.5 Impact of raw material and energy cost on the total variable manufacturing cost of the deink process: (▲) \$0.02/kWh; (●) \$0.03/kWh; (■) \$0.05/kWh.

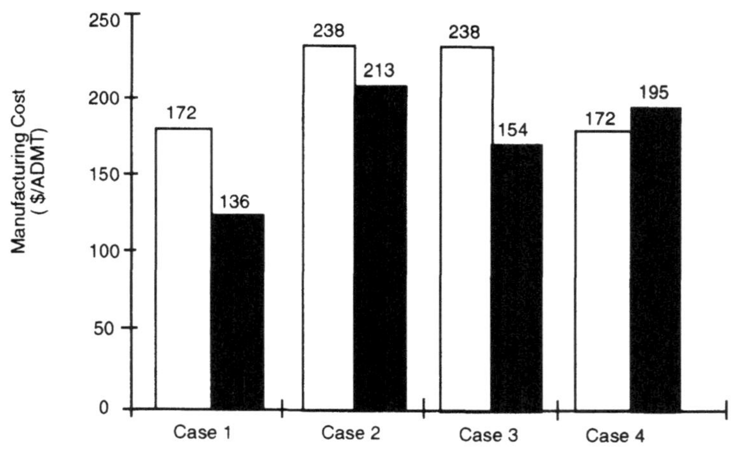

Figure 9.6 Comparison of the variable manufacturing costs of TMP (□) and deink (■) pulp at a constant wood chip price of \$75/ADMT.

The economics indicate that in most cases there is at least a slight variable cost advantage in using deink fibre. However, if an existing newsprint mill is currently integrated, and deink pulp would be replacing an existing fibre source, the cost of the capital should be included in the economic analysis.

References

1. Matussek, H., Salvesen, W. and Pearson, J. (1993) *International Fact and Price Book*, Pulp & Paper International, Miller Freeman Inc, San Francisco, p. 193.
2. *North American Wastepaper Study*, Vol. 1, Market Analysis and Forecast (Special Study Series No. 17), Resource Information Systems, Inc., (1991).
3. *The High Consistency Kneading System*, Shinhama Pump Company, Kawanoe-City, Ehime-Pref., Japan.

4. Hori, H. (1990) *Jpn Pulp Paper,* **28**(4), 27.
5. Scott, W.B. (1988) *Customer Acceptance of Waste-Based Newsprint - Quality Improvement and Trends in Consumption*, PPI Publishers Papers Conference, London.
6. McKinney, R. (1990) *TAPPI Conference Proceedings*, TAPPI Press, Toronto, p. 953.
7. Dillen, S. (September 1989) *Pulp Paper Mag.*
8. Gilkey, M. (1989) *Basics of News to News Deinking*, Black Clawson Company, Middleton, OH.
9. Rosling, M. (1990) *Pacific Paper Expo Proceedings*, Vancouver.
10. Cutts, M. (1992) *TAPPI Engineering Conference Proceedings*, TAPPI Press, Toronto, p. 147.
11. Mead, J. (1990) *PITA Annual Conference Proceedings.*
12. Rounds, D. (1992) *TAPPI Engineering Conference Proceedings*, TAPPI Press, Toronto, p. 135.
13. *Technical Constraints Limiting the Use of Recycled Fiber in Newsprint*, American Paper Institute Solid Waste Task Force, New York (1991).
14. Aspler, J. (1989) *Recycling and the Canadian Pulp and Paper Industry*, Part IV, Potential Printability Problems in Recycled Papers, PAPRICAN.
15. New York State Task Force on Recycling, Final Report, New York (December 1989).
16. Howard, R.C. (1989) *Recycling and the Canadian Pulp and Paper Industry*, Part III, Effect of Recycling on Paper Quality, PAPRICAN.
17. Abdel-Barr, D., Harris, M. and Ionides, G. (1989) *Deinked Newsprint - Market, Technology, Product Quality and Economic Issues*, Temanex Consulting.

10 The manufacture of tissue using recycled fibres

R.W.J. McKINNEY

10.1 Introduction

Tissue is unique within the pulp and paper industry in that the product which leaves an integrated converting factory is ready for use by the final consumer. Although some other paper grades are slowly becoming consumer products, they have some way to go before they become a major household product, as tissue is. This has profound implications for tissue producers, since a small number of retailers are individually major customers and competition amongst producers for retailers' business is intense. They expect tissue manufacturers to be able to respond very rapidly to changes in consumer preferences, which since the late 1980s has been for the inclusion of recycled fibres in tissue products. Consequently, pressure from retailers led to a very rapid introduction of products with high proportions of recycled fibres. Some of these had always included recycled fibres, but due to the possibility of a lower quality image, this was not always emphasised by producers or retailers. Wastepaper had been used prior to the late 1980s primarily due to significant cost advantages over virgin fibre use, with a 50% cost saving in fibre costs possible [1]. However, when using recycled fibres it is not always possible to maintain some of the attributes of high quality premium products.

Tissue consumption varies considerably, with lower consumption in less developed countries. In 1992 consumption per capita in the USA was 20 kg, whilst in Russia it was less than half a kilogram per capita. Average consumption in the EC was 10 kg/capita in 1992, but this varies within the EC; it was approximately 13 kg/capita in Germany but just over 5 kg/capita in Portugal.

10.1.1 Grade structure

In 1992, in the USA, consumer tissue accounted for about two thirds of the market and commercial and industrial (C + I) the other third [2]. In Europe, the commercial and industrial sector is not as well developed as in the USA, with consumer products representing about 75% of total tissue consumption in 1992, though this proportion varies appreciably within Europe. Within the consumer products sector, toilet (bathroom) tissue

Table 10.1 Typical wastepaper utilisation rates and grades used in toilet paper production

	USA		UK		Continental Europe	
	Wastepaper grades	(%)	Wastepaper grades	(%)	Wastepaper grades	(%)
Premium quality brands	White, unprinted	0–10	White ledger	0–50	Mixed printed [a]	50–100
Private label; quality	White ledger	0–40	White ledger	50–100	Mixed printed	50–100
Private label; budget	Mixed woodfree [b]	0–50	Mixed printed	0–100	Mixed, news magazines	50–100
Commercial and industrial	Mixed printed	100	Mixed printed	100	News and magazines	100

[a] Mixed printed - mixed woodfree and wood-containing. [b] Mixed woodfree - printed and coloured woodfree

is the most important; other products are kitchen roll and facial tissue. Overall, toilet tissue's share of the market is about 50% of total tissue demand. Each of these products is in reality a group of products, providing a variety of choice based on quality and cost, ranging from manufacturers' premium branded products to lower quality budget label products. Tissue products in continental Europe have followed a different trend from that of the USA, with much less emphasis on high quality premium products. The UK is an exception within Europe, with considerable emphasis on premium quality branded products, though even in the UK, there are fewer premium quality producers than in the USA. In very general terms, product quality and price is reflected in the quality of the fibres used, including the types of wastepaper grades used. In the commercial and industrial segment, there are few premium quality products and this sector is very price sensitive, hence a very high proportion of recycled fibres are used, with many products made from 100% recycled fibre.

These geographical differences are reflected both in wastepaper utilisation rates and the grades of wastepaper used within specific product groups. Typical grades and maximum proportions used in bathroom tissue products are given in Table 10.1

Commercial and industrial products such as hand towel or other wipes usually contain the lowest quality wastepaper grades used by the tissue industry, for example, coloured card cuttings, even some old corrugated containers.

Due to the diversity of wastepaper grades used and product qualities, recycling processes to produce tissue need to be flexible. When changing, for example, from lightly printed woodfree grades to heavily printed wood-containing grades, not only is there a change in the ink content and type

Table 10.2 Wastepaper use in total furnish (% 1988/89) [4]

Country	Wastepaper, proportion of furnish
Australia	5
Germany	37.8
Greece	8
Netherlands	70
Norway	90
Sweden	55–60
UK	40
USA	33.6

but there is also a significant change in fibre mix. This can result in slower production rates in the recycling plant, since wood-containing fibres drain more slowly than woodfree.

10.1.2 Wastepaper utilisation rates

Wastepaper has been used for a considerable time in tissue production; for example, Ponderosa Fibres started their recycled pulp business in 1965 in the USA, to supply recycled pulp primarily to a major tissue producer, Kimberly-Clark Corporation. Some companies have built their business based on 100% recycled fibre use, for example, Fort Howard Corporation in the USA. Many other tissue mills installed recycling lines, but tended to shun publicity, or not advertise their recycled fibre use, until it became acceptable to consumers in the late 1980s. Utilisation rates of individual companies vary enormously. In 1990, of the top eight tissue producers in the USA, three used 100% recycled fibre as their fibre source. The other five had utilisation rates between 0 and 20% of their total furnish [3].

There is also a wide range in wastepaper utilisation rates in individual countries, illustrated by the figures in Table 10.2 [4]. There has been a substantial increase in wastepaper use; for example, in 1992 it was approximately 60% in the UK and Germany and about 80% in Sweden.

Of the paper machines known to produce tissue, 619 (66%) were identified as using recycled fibres as raw material. In 1989 total tissue capacity was estimated to be 14.1 Mtonnes and 58% of this capacity was in recycled fibres [5]. Even though there is a high rate of wastepaper utilisation in tissue, various groups, including government organisations, are increasing the pressure on producers to use more wastepaper. In the USA, for example, in 1988 the US Environmental Protection Agency (EPA) recommended a minimum post-consumer recycled fibre content of 20% for tissue purchases on behalf of the Federal government [6]. (see Table 10.3). The following year, the Recycling Advisory Council was founded by the National Recycling Coalition, working with the EPA. Their recommendations, published in 1992, suggested that toilet tissue

Table 10.3 Recommended recycled fibre content in tissue grades [6, 7]

Grade	EPA Recommendation Federal Register (1988)	Recycling Advisory Council Recommendations (1992)	
	Post-consumer	Total recycled	Post-consumer
Toilet tissue	20	80	70
Paper towels	40	80	70
Paper napkins	30	80	70
Facial tissue	5	60	50
Industrial wipers	0	80	70

Table 10.4 Quality criteria for recycled bathroom tissue production

Quality parameter	Typical range	
	Woodfree, high quality	Wood-containing, budget
Ash (%)	1–3	3–6
Brightness (ISO)	75–80	60–65
Ink/Dirt (Tappi ppm)	30–60	30–100
Stickies Content (#/100 g bdf)	10–50	10–50
Freeness (CSF, ml)	min. 350	min. 200
Burst strength (kPa m^2 g^{-1})	min. 2.5	min. 1.5
Tear index (mN m^2 g^{-1})	min. 7.0	min. 5.0
Breaking length (km)	min. 3.5	min. 3.0

should contain 80% recycled fibres and 70% post-consumer recycled fibre [7].

Ecolabelling schemes have also increased pressures to use recycled fibres in tissue products; for example, one of the criteria for the Nordic White Swan label is 90% recycled fibre content, whilst qualification for the German Blue Angel scheme is dependent on the use of 100% recycled fibres, which must be derived from relatively low grades of wastepapers.

In a report published in 1990, Friends of the Earth, an environmental pressure group, recommended that only tissue products made from a high proportion of newsprint, magazines or mixed waste should qualify for the award of an ecolabel [8]. Thus, not only are there pressures on producers to use more recycled fibre, but also to use lower grades of wastepaper.

10.1.3 Specifications of recycled fibre for tissue production

Some specifications are related to product quality, others to tissue machine runnability and some to both. A few of the criteria vary, according to the type of product being made; for example, premium quality toilet tissue would require a minimum brightness of 75 to 80° ISO, whilst industrial toilet tissue would require a brightness of 60 to 70° ISO. Typical quality requirements are given in Table 10.4.

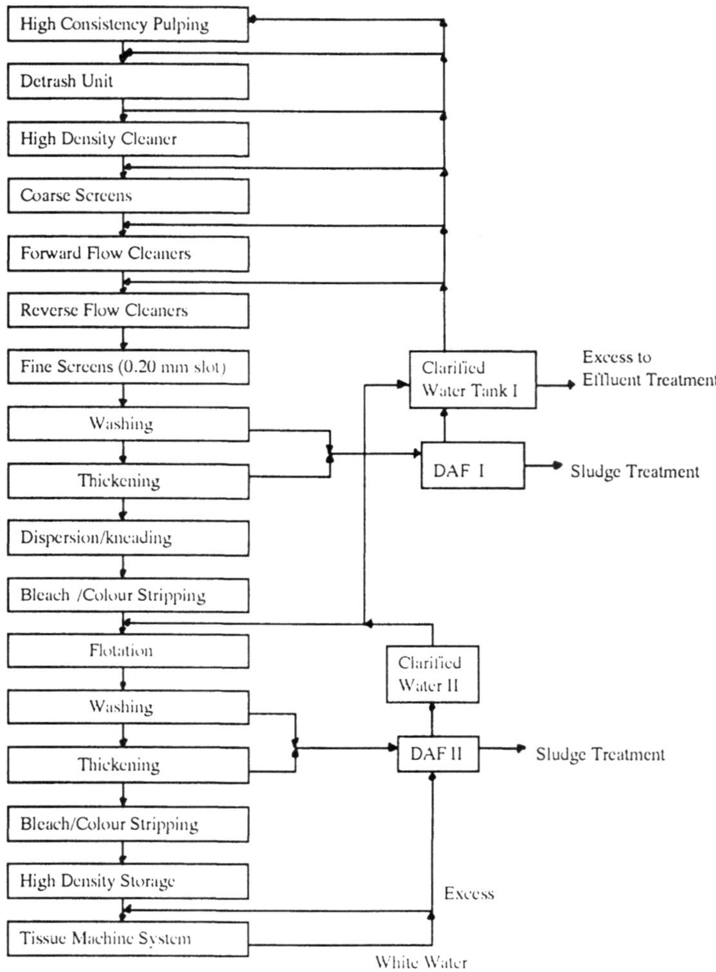

Figure 10.1 Schematic wastepaper processing system suitable for tissue.

If the characteristics of a recycled pulp for tissue production are compared with those for other grades, such as newsprint or other printing papers, in very general terms tissue needs lower ash and stickies contents, though a higher dirt count may be tolerated.

A schematic outline of a wastepaper processing system, to produce a good quality fibre from medium and low-quality wastepaper grades, is shown in Figure 10.1. Changes to this system could be based on the use of higher or lower quality wastepaper grades, or the production of lower quality tissue products. A compromise between recycled fibre quality on one hand and fibre yield on the other is always required and is usually determined by capital costs. In the sequence illustrated, two colour stripping/bleaching stages are shown; these could be based on one oxidative stage, such as hydrogen peroxide, and a further reductive stage, such as sodium hydrosulphite. With woodfree wastepapers, the most effective and economical colour stripping chemical is sodium hypochlorite, but in some countries, such as Germany, its use is considered unacceptable.

10.2 Production problems associated with recycled fibres

A variety of problems can be experienced, the majority of which are related to the criteria given in Table 10.4. The first problem is the integration of recycled fibres into the stock preparation system. Fibres from the recycling process must be blended with other fibre sources, broke, virgin fibre, etc for delivery to the tissue machine head box, after which any production and quality problems caused from the use of recycled fibres are exposed.

10.2.1 Stock preparation

In systems where tissue is produced from a blend of virgin and recycled fibres, or from only virgin fibres for some grades, there is a need to integrate recycled fibres and conventional stock preparation systems. A broke handling system has also to be considered and so the complexity of a stock preparation system is increased.

Virgin pulp may be added to enhance specific properties of individual products, for example, long fibres to increase strength, or short fibres (such as some eucalypt pulps) to improve softness. The maximum benefit from the use of these fibres is gained by separate treatment, for example, refining long-fibre (softwood) pulps alone, to develop strength, without the loss of softness or drainage from hardwood pulps which would be caused by refining.

If a multilayered head box is used, then the benefit from the use of a relatively small quantity of fibres to confer softness can be achieved. A schematic of a single layered stock preparation option is shown in Figure 10.2 and a twin layered option in Figure 10.3. In both of these, there is no refining of recycled fibres, since it is normal to use alkali in the recycling process and the action of sodium hydroxide is similar to that of refining, in that it develops strength properties. If no sodium hydroxide is used, then refining of recycled fibres may be considered, but it should be low intensity, with a low specific edge load, separate from softwood refining.

In Figures 10.2 and 10.3, no refining of hardwood is indicated, to enable softness to be developed. Water systems are integrated between the wastepaper processing line and the tissue machine, to minimise total water consumption. If no virgin pulps are used then the stock preparation system is much more simple and so lower in costs.

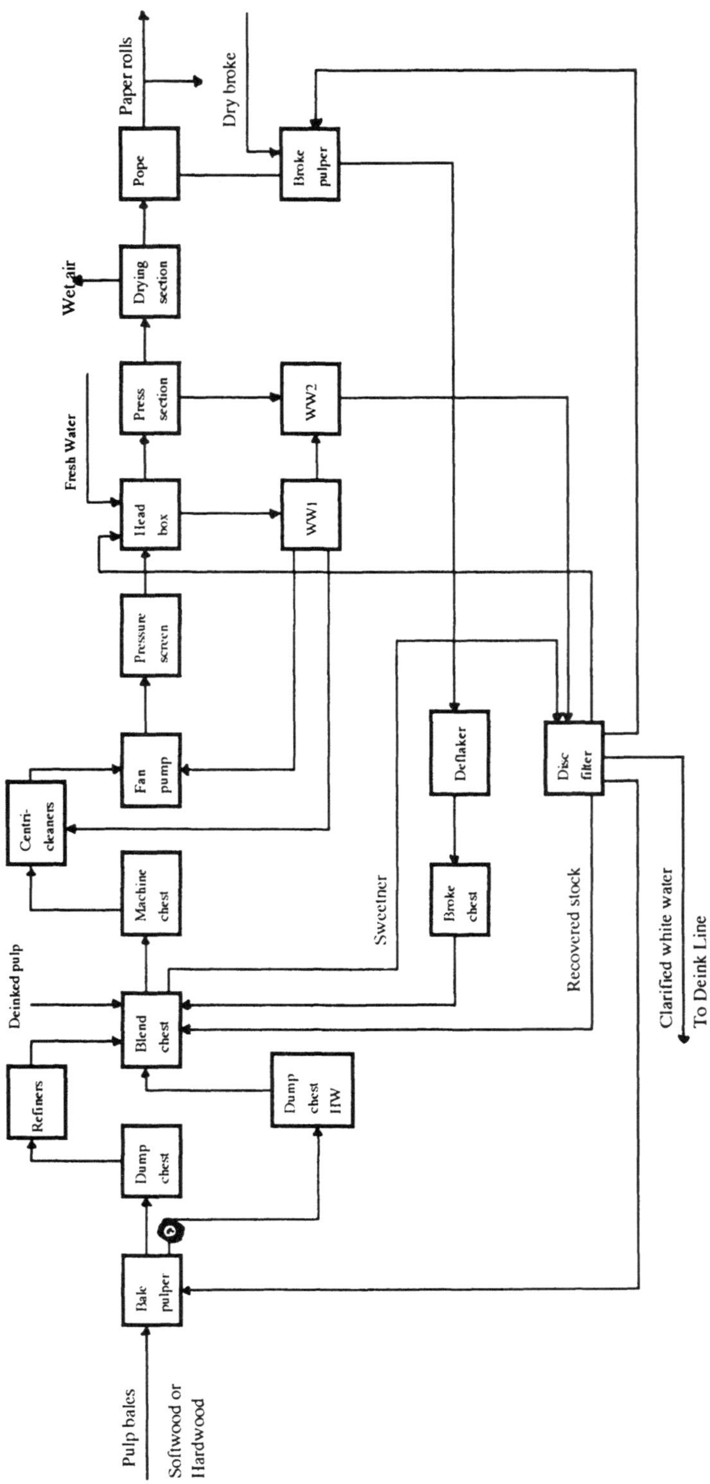

Figure 10.2 Single layer stock preparation system.

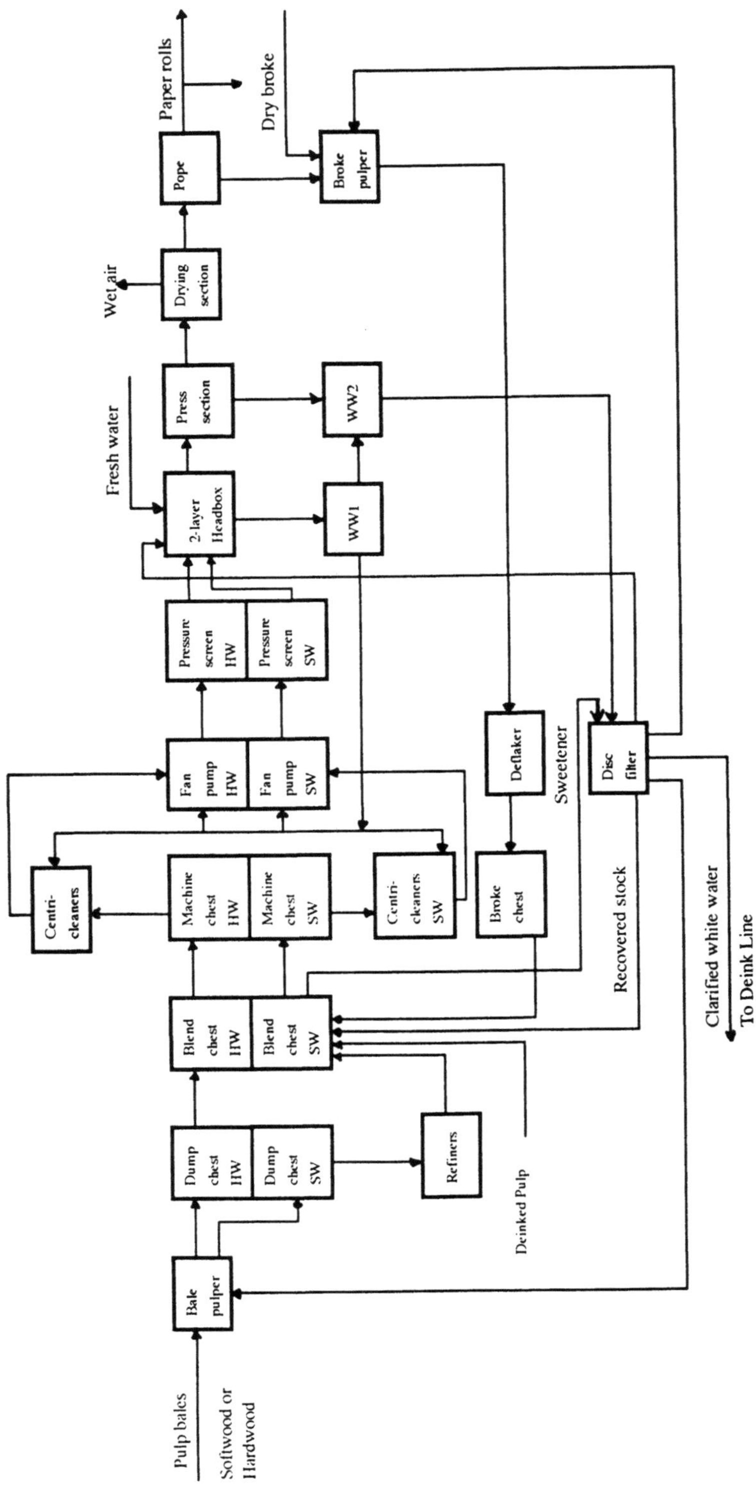

Figure 10.3 Layered option (two layers) stock preparation system.

10.2.2 Problems due to high ash

High ash has a major impact on tissue machine productivity, due to a number of interactions, including:

- reduced drainage, leading to reduced machine speed or higher drying (energy) costs. A reduction in ash content of a recycled pulp from 15% to 3% could be expected to yield about a 10% increase in tissue machine speed and a 10–15% increase in total machine production. Ash removal is achieved by efficient washing in the recycling plant, followed by treatment of the wash backwater to remove ash, prior to it being recycled for re-use.
- increased creping/(doctor) blade wear. Abrasive wear of doctor blades is increased by higher ash levels in the fibre web. This occurs even on properly coated cylinders with good doctor blades. Ash levels frequently change with the ash input from different wastepaper grades, so that the Yankee coating varies, also leading to increased abrasive wear of doctor blades. An increased rate of doctor blade wear leads to increased production losses during additional doctor blade changes. In addition, when creping blades are changed, the properties of the web change; for example, the caliper falls and returns to a maximum after about 30 min, and the tensile strength increases, falling to a minimum after about 30 min [9]. One of the impacts of non-uniform roll quality, especially caliper variations, is increased problems during converting, leading to increased waste.
- higher ash levels in tissue machine white water systems will lead to increased ash retention. This will reduce bonding between fibres and so reduce both wet and dry web strengths, leading to an increased incidence of breaks, especially at sheet transfer positions, possibly leading to increased basis weights to compensate for lower strengths.
- reduced felt and wire life. Greater ash levels lead to increased wear of wires and the tendency is for felts to become plugged, resulting in speed reductions or increased energy costs. Reduced felt and wire life not only leads to increased clothing costs, but also to increased downtime, for clothing changes.
- increased dust. Since higher ash levels in white water inevitably lead to high web ash contents and since ash and recycled fines are not bonded into the sheet, dust is created during creping and when the web is cut to size. Higher dust levels increase fire hazards arising from doctor blade sparks and can lead to an increased incidence of breaks in converting. In some countries, working atmosphere dust limits have been set; in Sweden, for example, it is 5 mg dust m^{-3} of air [10].
- fibre losses. Although the yield losses associated with efficient ash removal during wastepaper processing are high, if low-efficiency washing is used high ash and fines concentrations in the resultant recycled pulp

inevitably lead to high and increasing concentrations of suspended solids recirculating in the tissue machine white water system, since normally the only exit for solids from the white water loop is the fibre web on the tissue machine. Higher levels of ash and fines can create problems with the white water solids recovery systems, normally a disc save-all, so that solids can begin to pass the disc save-all, leading to contamination of the clear water leg. If this is used for showers, then shower nozzles can block, leading to streaks. In addition, as circulating solids levels increase in white water, the head box consistency increases, so that more dilution water is added to maintain consistency. If this is the case, then the sheet will contain more moisture and so either the machine speed has to be reduced, or drying energy consumption increases.

Other problems with high solids in white water include reduced efficiency in dye and retention aid use; increased potential for foam problems; increased corrosion of the Yankee; increased deposition problems; a buildup of organic strength leading to higher microbiological activity and problems associated with this; etc. If circulating solids increase beyond a tolerable level, then the white water system is purged, which frequently results in the uncontrolled loss of 'good' fibres circulating in the machine white water system. The balance is thus between solids losses during wastepaper processing and solids losses incurred when purging the tissue machine white water system.

10.2.3 Problems due to stickies

Stickies are small tacky particles which are derived from the many types and forms of adhesives present in wastepaper, and from many other sources [11]. Their removal and control has been extensively reviewed [12–15] and is discussed in chapter 3. Tissue production can be seriously affected by stickies, since they can affect both productivity and quality. They adhere to forming fabrics (wires) and press felts and can build up at doctor blades and on rolls, cause holes in the sheet, etc. When two plies in a jumbo roll are stuck together by stickies, there is an increase in the incidence of breaks on the rewinder (combiner) or during converting, also caused by holes in the sheet.

Other problems caused by stickies include a reduced life for machine clothing; for example, by deposition in wires and felts, by the use of high-pressure showers to remove stickies which may damage wires or felts, or the use of solvents to remove stickies which can reduce felt porosity through accelerated compaction, as well as possibly creating a health or safety hazard. Machine downtime, taken to clean wires or felts frequently results in significant production losses, as well as increased costs due to the use of chemicals to control problems. Hence stickie removal is one of the most important functions of a wastepaper processing line.

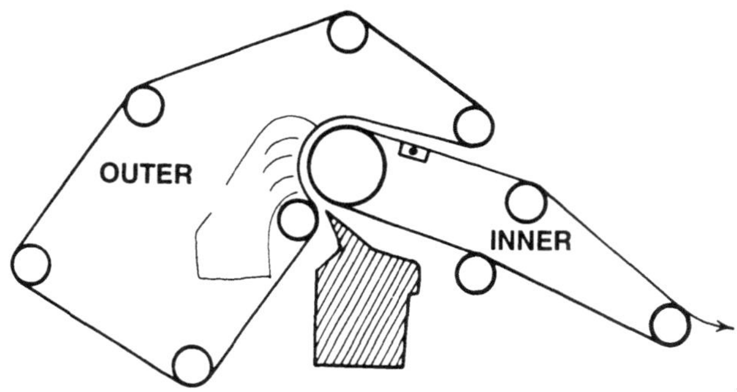

Figure 10.4 C-wrap former with solid forming roll.

Aspects of tissue machine design, especially the forming section layout, can have a substantial impact on the severity of stickies problems. Initially most tissue formers were plain or suction breast rolls in fourdrinier machines, but to gain increased speed and production, twin wire formers were introduced during the mid to late 1970s, with two forming roll arrangements, represented by a C-wrap and an S-wrap. In the mid 1980s the C-wrap former (Figure 10.4) became the primary choice, with an estimated 82 C-wrap formers (with solid forming rolls) installed up to 1990 [16]. Formers in use now include:

- crescent formers
- C-wrap twin wire, with suction or solid forming rolls
- S-wrap twin wire
- suction or plain breast rolls in horizontal or inclined fourdriniers.

In the crescent and twin wire formers, with a solid forming roll, the sheet is formed between the fabrics around the forming roll, with drainage through the outer wire in both C-wrap (Figure 10.4) and S-wrap (Figure 10.5) configurations, but in the S-wrap there is some drainage (after the forming roll) through the inner wire. In the S-wrap the drainage side is the hood side through the transfer wire, whereas in the C-wrap the drainage side is the Yankee side away from the transfer wire. With a suction forming roll there is drainage through both wires, hence the higher dewatering capacity of the C-wrap with a suction forming roll (Figure 10.6). In the crescent former the sheet is formed between a wire and a press felt around a forming roll, but in other respects it is similar to a C-wrap former; the inner fabric is a press felt.

Differences in forming are significant when sensitivity to stickies is compared. Twin wire formers are more sensitive to stickies, but there are differences between twin wire forming configurations. Mills with experience of both solid and suction C-wrap formers, using the same furnish on both, report fewer problems with fabric cleaning with the suction former [17].

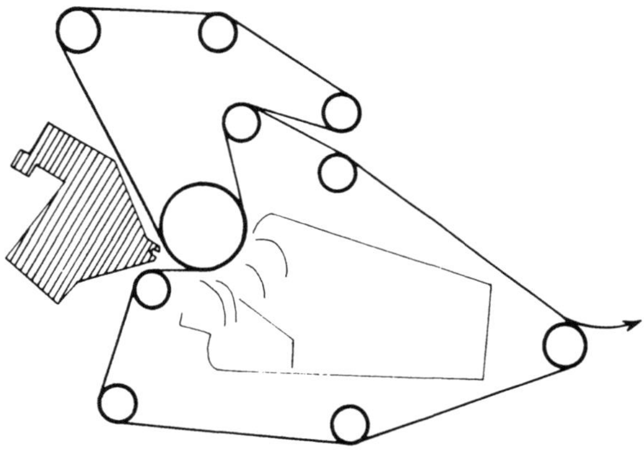

Figure 10.5 S-wrap former.

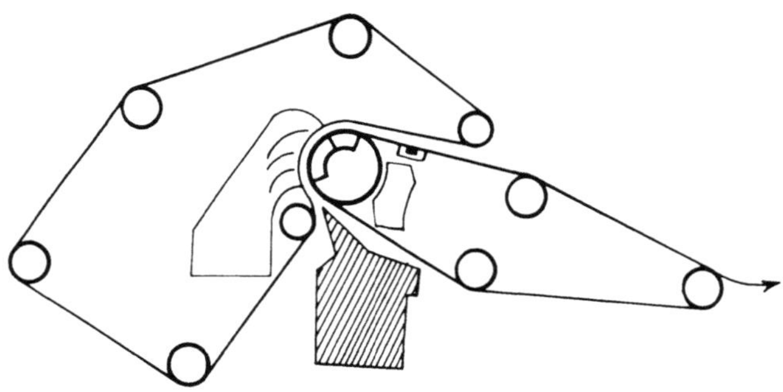

Figure 10.6 C-wrap former with vacuum forming roll.

The large flow of water through the inner fabric into the suction roll shell assists in keeping the inner fabric clean.

Studies on a pilot machine comparing S and C-wrap twin wires, with solid forming rolls, and a vacuum roll in the C-wrap, showed substantial differences in sensitivity to stickies, in terms of both holes in the sheet and the number of stickies deposited on and in the wires. Results showed sheet holes were more common and that the outer wire tended to be much more contaminated in the S-wrap configuration, but that the inner wire was much more contaminated in the C-wrap configurations, though there were fewer stickies on the inner wire with a vacuum forming roll. General conclusions were that S-wrap formers were a factor of thirty times more sensitive (in terms of sheet holes) than a C-wrap with a suction forming roll, and a C-wrap with a solid forming roll a factor of five times more sensitive than a C-wrap with a suction roll [18]. Stickies in the drainage wire are a greater problem than on the other wire, since as water drains through the wire and the sheet is formed, stickies in the wire cause drainage around the stickies, leading to holes in the sheet. Hence the greater number of holes using the S-wrap configuration. Stickies in the sheet can be picked out at a sheet transfer point, which also leads to holes.

Other design considerations include the positioning of fabric guide rolls; when located on the outside (web side) of the wire these can drive stickies (or other contaminants) into the wire, making removal more difficult. They can also pick up stickies, leading to deposits on the roll, which can break away and cause holes. Teflon coated rolls do not pick up stickies. If all the rolls are on the inside, this helps to maintain fabric cleanliness.

Location and type of showers is important. High-pressure showers help to remove stickies, but continued use can accelerate fabric or felt damage. Flood showers can wash stickies off the surface of wires and felts and flush fillers and fines to the surface, but they consume large volumes of water. If a high pressure shower is used, it should be located ahead of a flooding shower, at an angle between 15 and 30 degrees against the direction of travel of the felt for most effective stickies removal. It is important that cleaning shower drainage water is collected separately, when appropriate, and not returned to the white water system without treatment to remove stickies. Similarly, cleaning doctor blade deposits should be collected.

Chemical modification of forming wires and press felts is frequently used to minimise problems cause by stickies deposition. Some fabrics are made with 'non-tack' surface layers, but these wear off. Continuous or intermittent addition of film forming cationic polymers is a common treatment. Cationic polymers combine with surfactants and build up a coating on both wires and press felts which helps to prevent stickies from depositing, illustrating the importance of surface phenomena in stickie deposition [14]. This layer is believed to be a combination of cationic polymer and anionic materials from lignin and polysaccharides. The coating tends to discolour the wire or felt - it is usually a light tan to dark brown colour, due to the presence of lignin [19].

10.2.4 Sheet breaks

When virgin fibres are replaced by recycled fibres, it is inevitable that some aspects of machine performance will change, dependent on the type of fibre replaced and its replacement. If wood-containing wastepapers are used, for example news or magazines, to replace hardwood pulp (or woodfree recycled fibres) drainage time will increase and retention may fall. Since both wet and dry strengths of a wood-containing sheet are lower than a woodfree sheet (at equal basis weights) more breaks at sheet transfer positions in the machine, at the rewinder/combiner and in converting are probable. More dust will be created, leading to a higher fire risk and converting problems. Hence machine productivity when on lightweight products could fall significantly, though the efficiency of production of heavyweight grades, such as industrial wipes, would suffer less.

When woodfree wastepaper replaces virgin chemical pulps, machine problems are primarily related to the cleanliness of the fibre delivered to the paper machine system, in terms of contaminants such as stickies and anionic trash. The fines content of a recycled pulp is usually higher than a virgin pulp and so some reduction in drainage rate can be expected. If very efficient washing is included in the wastepaper preparation system, fines will be washed out and there will be only a small reduction in the drainage rate, which is partially compensated for by the easier drying of recycled fibres. Relative strength properties of the tissue products are dependent on the type of chemical pulp substituted by recycled fibre and the degree of refining of the chemical pulp. Normally, woodfree recycled fibres have a relatively high strength, greater than that of an unrefined hardwood pulp. One recycled pulp was shown to be equivalent or stronger than a eucalypt pulp after a power input of 50 kWh t^{-1} [20]. Hence there are few production problems attributable to strength properties when woodfree recycled fibres replace chemical hardwood pulps, or even limited replacement of softwood pulps. Since fibre selection is very important in softness development, there is an impact on quality.

10.2.5 Increased microbiological activity

Recycled fibres usually release much more organic material than virgin fibres. Efficient washing will wash out a large portion of the increased organic load, but usually the feed to the paper machine system still contains more organic material than from virgin pulps. This leads to increased biological activity in white water circuits. When water systems between the tissue machine and the recycling plant are integrated, it is possible that high levels of organic materials will reach the white water circuit and the temperature of the system will increase. Increased temperature (up to a maximum of about 55–60°C) also increases biological activity.

Increased biological activity leads to greater slime problems via increased depositions in pipework, etc., as well as odours if anaerobic conditions develop in storage chests. Routine boil outs of pipework are essential and generally, the use of slimicides and biocides has to be increased to control these problems.

10.2.6 Effect of recycled fibres on creping

Dry creping of the web from the surface of a Yankee gives tissue its characteristic feel and stretch properties. The creping action breaks some fibre bonds and makes the web softer, more bulky and absorbent, though there is a strength reduction of up to 50%. Adhesion of the sheet to the dryer is necessary so that the sheet can be creped by the doctor blade from the surface of the Yankee. Adhesion is dependent on the formation of a tacky film which accumulates on the Yankee surface. Natural adhesives are derived from pulp, mainly hemi-cellulose, but adhesive layers have also been shown to include small amounts of lignin, cellulose, extractives and dispersed fibrils [21]. Hemi-cellulose content varies with the type of pulp. Factors which influence creping include:

- Web conditions - moisture content and profile, furnish, basis weight profile, pH, hardness, chemicals in white water;
- Running conditions - press nip pressure, width and profile, refining, speed, wire and felt types;
- Coating - additives to the wet end or spray boom, including adhesives and release agents;
- Cylinder - crown and surface furnish;
- Doctor blade holder - design, fitting, strength of backing blade;
- Doctor blades - mechanical properties, edge and removal of burrs, flatness, straightness, wear and heat resistance, contact angle;
- Felt design - contact area, smoothness, condition;
- Maintenance - doctor blade grinding procedures, storage of blades;
- Sheet adhesion - hard or soft.

Using recycled fibre has an impact on a number of these factors, especially white water and thus coating chemistry. Competition from anionic trash can be introduced, which can strip organics from the cylinder, especially if creping additives are added at the wet end. Since wastepaper fibre types vary from pulper batch to batch, the concentration of important natural organics, including hemi-cellulose and lignin, also varies, so the Yankee coating from natural sources will change.

Hardness and pH are important and these vary with wastepaper, especially hardness, as the use of calcium carbonate as a mineral filler or coating in paper results in variations in hardness concentrations, with frequent high concentrations. When natural coatings are used the most favourable hardness is in the range 90–125 $mg\,l^{-1}$ as calcium carbonate [22] which is low compared to levels in recycled fibres.

Alum is still used as a sizing agent and so some wastepapers will also contain alum - at high concentrations this has a marked effect on creping via a reduction in dissolved hemi-cellulose and so a reduction in adhesion. It can also lead to increased Yankee corrosion. Hence it is clear that the concentration and type of ionic species in the wastepaper and in white water has a significant effect on the Yankee coating, and so on creping.

Negative effects can be minimised by using spray booms to add organic coating chemicals and release agents, to control adhesion to and release from the dryer. This is more efficient than wet end addition, which increases chemical use by a factor of approximately 10 [23]. Even when using recycled fibres good control of creping can be achieved, especially when adhesives and release agents are added via separate spray booms. Common adhesives include polyamides, starch, animal glues, polyvinyl alcohol, polyvinyl acetate, wet strength resins, retention aids and carboxyl-methylcellulose. Release agents include polyethylene glycols, mineral oils, surfactants, silicones and quaternary amines.

10.3 Effects of recycled-fibre use on tissue quality

Tissue is a consumer product and so product properties must match consumer demands. Until the late 1980s, most demands were focused on physical properties, such as softness, absorbency, strength and bulk, but recently consumers' environmental concerns have added questions about fibre source, so that recycled content has become a product attribute. However, many of the physical properties of tissue are gained through fibre selection and the ability to do this is reduced by the inclusion of recycled fibres, hence there is a direct conflict between some physical properties and recycled content. In the case of softness, for example, in the USA premium quality softness is achieved by the preferential use of pulps such as black and white spruce sulphite, and in Europe, by eucalypt. Strength is added by long softwood fibres. Multilayer formers allow the separation of these two furnish components, for maximal development of these properties. However, when wastepapers are used to provide fibre, selection is confined to the use of woodfree or wood-containing wastepaper grades.

If recycled fibres are used as part of the tissue product furnish, it is possible to engineer sheet properties by the selection of the other fibre component, using a multilayered approach. When recycled fibres are used as 100% of the furnish, quality attributes cannot be gained by fibre selection, and mechanical methods such as thru-drying, or embossing are the only practical methods to achieve desired attributes. These techniques do not restore properties to the levels achieved when selected virgin fibres are used; for example, a 100% recycled thru-dried sheet does not have the softness and bulk of a thru-dried sheet produced from selected virgin fibres. Differences are due in part to changes in fibre properties with recycling, but are primarily because of the use of different species of fibre.

10.3.1 Effect on softness

Premium quality softness can be achieved by a combination of means, which include fibre selection, multilayering, chemical treatment, double creping, thru-air drying, embossing and calendering. The use of recycled fibre has an effect on these in a variety of ways.

The property of 'softness' is usually divided into two components, which are bulk softness and surface softness. Bulk softness is the perception of softness when a sheet is crumpled in the hand, whereas surface softness is the velvet feel when the fingertips are brushed over the surface. Both can be correlated to sheet properties and can be described mathematically [24]. There is a strong correlation between bulk softness and some physical properties, such as tensile stiffness, bending stiffness and bulk; a high bulk and low bending stiffness is necessary for a high-bulk softness. High bulk is gained by using stiff fibres with a large diameter and a low degree of bonding. However, stiff thick-walled fibres which give good bulk do not give a good bulk softness, this is achieved by using thin and flexible fibres. Some sulphite softwoods are thin and flexible and their large diameter gives a higher bulk than hardwoods alone. A low degree of bonding is achieved through minimal refining, reduced pressing, chemical treatment of the fibres, etc.

Surface softness is favoured by a fine crepe and flexible fibres at the surface and is sensitive to protruding fibres and irregularities in the surface. Thus long stiff fibres, which will give good bulk will not give a good surface softness. Shorter fibres have a smaller fibre diameter and so are less stiff and will give a good surface softness, hence hardwoods are used, especially short fibres such as eucalypt. Thick-walled species can be used, to give good bulk as well as surface softness [25].

These considerations illustrate how important fibre selection is in developing softness, and this is lost when using recycled fibres. Recycled fibres, due to hornification, are relatively stiff fibres and, more importantly, selection of fibre type to give maximum flexibility at the surface is not possible. In addition, recycled fibres have been refined on their first cycle and so they have a higher bonding affinity than unrefined eucalypt [20] which reduces bulk.

Overall, the use of recycled fibres, even without consideration of increased ash in the web, has a major adverse impact on softness. If used as 100% of furnish, only chemical treatment, using chemical softeners, thru-air drying, embossing etc. can be used, but super soft quality cannot be achieved. C-wrap or crescent formers would be appropriate for this furnish, to take advantage of their stickie handling ability.

If recycled fibres are used as a portion of the fibre furnish, then multilayering may be used. As well as allowing fibre selection this technique allows other benefits to be gained, including higher caliper and strength, and higher retention. In the case of a single ply product a three-layered head box allows recycled fibres to be buried between two virgin fibre layers and these can be selected to provide maximum softness. Broke has to be used in the centre ply and if reinforcement by softwood is necessary, these restrict the amount of recycled fibre which can be used. Since single ply products have heavier basis weights a C-wrap former with a vacuum forming roll would be appropriate.

If a twin ply product is produced, a two-layer head box can be used to allow fibre selection. Maximum softness would be gained (and higher retention) by having unrefined virgin hardwood (eucalypt) next to the (outer) drainage wire with recycled fibre on top of this layer, next to the inner wire. An S-wrap former would be required to allow the virgin fibre layer to adhere to the Yankee and hence be the side creped from the Yankee, to achieve maximum softness. However, the S-wrap former is most sensitive to stickies though the virgin fibre next to the outer wire should reduce problems. Hence this approach could be used and problems minimised if special attention were given to stickies removal.

When chemical softeners are used, anionic substances can interfere with the softener. Recycled fibre can have a high carry-over of anionic trash, which is minimised by having separate water loops on washing stages, especially when high-efficiency washers are used. Spray addition of softener onto the web just before the Yankee improves surface softness but was shown to have no effect on bulk softness [26].

When thru-dried or embossing is used to enhance softness, recycled fibres bond more than unrefined pulp and so caliper is lower, hence bulk softness is lower than with a virgin sheet. However these techniques do produce recycled products with good softness properties.

10.3.2 Effect on appearance

Inevitably tissue product appearance is adversely affected by recycled fibre. Since ink removal is <100% efficient, specks are normally present in the final sheet, with the speck count dependent on the quality of the wastepapers used and the recycling process.

A large proportion of consumers have accepted the recycled signature of a specky product. It is also usual for the brightness of a recycled product to be lower, especially if wood-containing wastepapers have been used. Again, this is acceptable to a large proportion of consumers, as is an increase in the shade variation of coloured products. Holes in the sheet are an indication of stickies problems, which have a serious effect on machine productivity. Holes in a sheet can be tolerated only to a limited degree, hence stickies can cause serious quality problems.

10.4 Future use of recycled fibres in tissue

A major problem facing tissue producers is the rising cost of sludge disposal, which in some countries eliminates cost advantages from using recycled fibres. However, consumer and other pressures will ensure that the proportion of recycled fibres used increases. Nevertheless, some consumers will continue to expect premium quality products, with high softness, and so there will continue to be a place for virgin fibre in tissue products, unless substantial cost penalties against virgin fibre use are introduced via government taxes or other penalties. Techniques for improving softness by increased debonding, such as double creping or thru drying will be used more, to overcome some of the inherent disadvantages, in terms of softness, from using recycled fibres, as will embossing, chemical softeners, etc. Multilayering may be used in high-quality tissue products, to allow the introduction of some recycled fibres, whilst maintaining premium level softness. Techniques such as fractionation may be used to provide long and short fibre-enriched fractions for multilayering of 100% recycled fibre furnishes [27].

As more wastepaper is used, recovery systems will come under more pressure and quality levels will probably fall. Hence there will be pressure on recycling systems to eliminate the additional burden from a wider variety and higher concentration of contaminants. Nevertheless, recycled fibres will grow in importance in tissue production.

References

1. Siewert, H.W. (1989) The use of wastepaper in tissue production, *Tappi J.*, **72**(12).
2. Anon (1993) Grade profile, *Pulp and Paper*, **67**(2), 13.
3. Goodman, J. Secondary fibre to the tissue industry, In *Tissue Issues*, produced by Niagra Lockport. Undated.
4. Anon (1992) *Environmental Issues Waste Paper*, OECD Report OCDE/GD Paris.
5. Steffner, S. (1991) Deinking - process concepts, *Valmet Tissue Making Seminar*, Karlstad, Sweden.
6. Anon (June 22 1988) Federal Register, Vol 53, No 120.
7. Miller, E. (1992) Final report on recycled paper definitions, standards, measurement and labelling guidelines, *Proceedings of TAPPI Finishing and Converting Conference*, TAPPI Press.
8. Walker, P. (ed.) (1990) *Market Barriers to Paper Recycling*, Friends of the Earth, London.
9. Klerelid, I. (1991) New possibilities in creping and coating control, *Valmet Tissue Making Seminar*, Karlstad, Sweden.
10. Glifberg, B. (1989) Dust removal systems, *Valmet Tissue Making Seminar*, Karlstad, Sweden.
11. Moreland, R.D. (1986) Stickies control by detackification, *1986 TAPPI Pulping Conference*, TAPPI Press, Vol 1, pp 193.
12. McKinney, R.W.J. (1987) Stickies removal, *PITA Paper Week Conference*, York.
13. McKinney, R.W.J. (1989) Colloidal interactions and their important in stickies removal and control, *Asian Pacific Wastepaper Conference*, Taipei.
14. McKinney, R.W.J. (1989) A review of stickie control methods, including the role of surface phenomena in control, *TAPPI Recycling Seminar*, TAPPI Press, Madison.
15. Fogarty, T.J. (1992) Cost effective, common sense approach to stickies control, *TAPPI Pulping Conference*, TAPPI Press, Vol 2, pp 429–437.
16. Thomas, J. (1991) Forming fabric application on tissue and towel twin wire C-wrap formers, with solid and suction forming rolls, *Valmet Tissue Symposium*, Karlstad, Sweden.
17. Erikson, J. (1989) Twin wire tissue and towel forming with suction forming roll, *Valmet Tissue Making Symposium*, Karlstad, Sweden.
18. New, N.L. (1989) Paper presented at the *TAPPI Recycling Seminar*, TAPPI Press, Madison.

19. Hassler, T. (1991) A new approach to controlling pitch and stickies deposits in tissue manufacture, *Valmet Tissue Making Seminar*, Karlstad, Sweden.
20. McKinney, R.W.J. (1989) Recovery of high quality wastepaper from offices, *EUCEPA Symposium*, Ljubljana, Vol 1, p. 29.
21. Fuxelius, K. (1967) *Svensk. Papperstid* **70**, 164.
22. Oliver, J.F. (1980) Dry-creping of tissue papers - a review of basic factors, *Tappi J.*, **63**(12).
23. Marzullo, G. (1987) Practical methods for the control of creping, *Valmet Tissue Making Seminar*, Karlstad, Sweden.
24. Hollmark, H. (1983) Evaluation of Tissue Paper Softness, *Tappi J.*, **66**(2).
25. Hollmark, H. (1987) Pulp influence on sheet quality, *Valmet Tissue Making Seminar*, Karlstad, Sweden.
26. Andersson, A. (1991) A new concept for tissue softening, *Valmet Tissue Making Seminar*, Karlstad, Sweden.
27. McKinney, R.W.J. (1993) Tissue product trends and market developments, *Valmet Tissue Making Seminar*, Karlstad, Sweden.

11 Manufacturing of printing and writing papers using recycled fibres

J.B. MORRISON

11.1 Introduction

The successful manufacture of printing and writing papers from recycled fibres has resulted from the evolution of processing technology. Although it is a technology beset with challenges, an ever-growing number of manufacturers have mastered the demanding attributes of recycled fibre in the production of printing and writing papers.

The fundamentals of this recycling technology are steeped in the history of paper making and have developed throughout the twentieth century. Today it stands as a viable fibering alternative for the manufacture of papers. The application of this technology requires not only the development of paper making expertise to deal with the unique properties of recycled fibres but perhaps more importantly, the preparation of the fibres. Not unlike virgin fibres, recycled fibres must be processed correctly and in a consistent manner if quality grades of papers are to be manufactured.

11.2 Printing and writing grades

Printing grades include a broad classification of paper products that have traditionally been referred to as fine paper. Such products encompass a whole spectrum of paper grades that are utilized for the universal communication and dissemination of information and knowledge. Whether for written publications or the advertising printing media, many varieties of paper grades are manufactured displaying functional and aesthetic characteristics that contribute to communications.

The general classification of printing and writing papers is:

- Uncoated groundwood grades
- Coated groundwood grades
- Uncoated free sheet grades
- Coated free sheet grades
- Speciality grades

Within each classification are a multitude of paper grades that possess unique properties, especially the largest segment, the uncoated free sheet

Table 11.1 Paper and paper board capacity by major area [1]

	Capacity (1000's mt)			Average annual growth (%)	
	1986	1990	1994	1986–1990	1990–1994
United States	69 544	77 104	84 535	2.6	2.3
Canada	16 158	18 185	20 344	3.0	2.8
Nordic countries	18 014	20 725	24 058	3.6	3.8
European Community	35 207	42 320	52 894	4.7	5.7
Other western Europe	4988	6034	6961	4.9	3.6
Japan	24 617	30 728	36 680	5.7	4.5
Asia and Oceania	21 645	29 369	32 482	7.9	2.5
Latin America	12 483	13 528	16 631	2.0	5.3
Africa	2885	2884	3081	-	1.7
USSR/eastern Europe	20 685	19 619	18 411	nm [a]	−1.6
Total	226 226	260 496	296 077	3.6	3.3

Source: FAO Pulp and Paper Capacities Survey, 1987 and 1991. [a] nm = not measured.

grades. It is this segment where recycled fibres have the greatest utilization rate and potential for future growth.

Grades including reprographic bond and writing, ledger, forms bond, carbonless, tablet, envelope, offset, premium text and cover, commercial printing, book paper and technical specialities can be commercially produced from recycled fibre. The manufacture of the remaining printing and writing grades from varying levels of recycle fibre is technologically feasible but limited due to a questionable quality perception.

Worldwide, the combined annual capacity growth of printing and writing grades is greater than all other areas of paper and paper board manufacturing. The capacity growth and projection through the period 1986 to 1994 for printing and writing papers by the major manufacturing areas of the world is shown in Table 11.1 [1]. Of the total world capacity of printing and writing papers, the United States has the largest manufacturing capability (Table 11.1) at 21 475 000 tonnes (1990) with the European Community next at 15 250 000 tonnes (1990).

In the United States, until recently, the utilization of recycled fibre in the production of many printing and writing grades was limited and declining. For many years, the growth in new printing and writing manufacturing capacity has been vested mainly in integrated virgin-fibre facilities of a world class design, producing primarily commodity grades. The existing mills employing recycled fibre were characterized as older and marginal non-integrated and semi-integrated facilities that manufactured a narrow range of printing and writing grades. It was inevitable that given the absence of the development of new recycling technology to offset increasing cost and changes in the recovered waste material markets, the future growth and expansion of these facilities was not practical or realistic, with limited future viability.

Compounding these difficulties was the growing enactment of environmental laws and manufacturing constraints in many nations of the world that placed unacceptable capital cost demands on recycling mills. Consequently, many of the recycling facilities, especially in the United States, were curtailed and virgin fibre (as market pulp) was substituted for recycled fibres. Nevertheless, the utilization of recycled fibres endured in European countries, specifically in non-integrated mills, driven mainly by economic factors.

However, the development of the next generation of recycling technology required effectively to process the emerging new and more difficult printing systems and graphics arts medias diminished, due to the lack of an economic incentive. Further, continued research into the fundamental principles and concepts of recycling technology as well as the paper making characteristics of recycled fibre narrowed and was assigned a minor priority by the industry. During this technology hibernation, many mills that had previously processed recycled fibre for paper making commenced revising stock preparation areas and wet end of their machines to accommodate virgin fibres.

In the late 1980s, environmental issues in the United States and the world again dramatically altered the manufacturing philosophies of printing and writing paper mills. The demand for recycled paper products saw a rapid resurgence not only from the legislative action of governmental agencies, but more significantly from an environmentally concerned private marketplace. Since the paper industry, especially in the United States, lacked sufficient installed capacity to meet this abnormal recycled paper demand, it has attempted to modify existing virgin-fibre machines and mills to increase the utilization of recycled fibre. Manufacturing's reaction to the demands from the marketplace has encountered marginal success, not due to the unavailability of techniques and knowledge but by the complexity of the issue and the lack of economic incentives.

Recycling is a technology that has traditionally been an accepted and practical manufacturing concept, albeit limited in application, but it is now being seriously challenged by the growing demand for recycled paper. The paper industry has quietly acknowledged the technological reality of recycling. A crucial factor is whether the industry will react rapidly to the growing demand for recycled-fibre utilization. Many of the major paper manufacturers are doing so, even though the response has required the quick restructuring of many stoic manufacturing perceptions.

11.3 Recovered wastepaper grades

It is generally acknowledged that except for specialized technical products, every grade of printing and writing paper can be manufactured using recycled fibre, to one degree or another.

In defining what grades of recovered wastepaper can be employed in the manufacture of printing and writing grades, the process technology, as well as the market acceptance of the product determines the viability of the material. Historically, the specific type of recovered waste material utilized for recycling depended primarily upon the quality definitions for the grade of printing and writing papers to be produced. Until the early 1990s, the identification of individual grades of recovered wastepaper had been well defined and provided a material base for managing the process, especially grades that require deinking. Recently, however, market factors in conjunction with legislative guidelines, have governed the classification and source of the recovered material and the type of wastepaper no longer is the specifying factor. Since there is no universally accepted definition as to what constitutes recycled paper, the raw material base has varied considerably, but within fibre type limitations, and is influenced by the available processing equipment and technology.

Currently, in the United States, recycled paper definitions and specifications are embroiled in a legislative quagmire. The definition of recovered wastepaper that can be utilized in the manufacture of printing and writing papers is now falling within guidelines established by the United States Environmental Protection Agency. These guidelines for the "Federal Procurement of Paper and Paper Products Containing Recovered Materials" were developed as part of the Resource Conservation and Recovery Act of 1976, Section 6002 (RCRA) to encourage governmental agencies to purchase products containing material recovered from solid waste [2]. Recovered material includes both post-consumer materials and manufacturing, forest residues and other wastes.

Post-consumer material as defined by RCRA, "are items which have passed through their end-usage as a consumer item and would include old newspapers, magazines, used corrugated containers and office waste paper." [3]. The Act was amended in 1984 to require that, "in the case of paper, the guidelines would maximize the use of post-consumer recovered materials" [3].

The other part of recovered paper materials as defined by RCRA, "manufacturing, forest residues and other waste" - are pre-consumer wastes. This would include manufacturing waste such as paper and paper board waste, bag, box and carton waste, printed paper which has never reached the consumer and obsolete inventories [3].

Wastepaper is a general RCRA category that includes both the post-consumer and the pre-consumer classifications. In this category, the EPA has determined that, "mill broke is specifically excluded from the definition of recovered materials because it is waste generated before completion of the paper making process" [4].

Included in the RCRA wastepaper definition is a pre-consumer category that is commonly referred to by the industry as 'pulp substitutes.' This is material that can be processed into recycled fibre with a minimum of treatment and consists of, "dry paper and paper board waste generated after completion of the paper making process including envelope cuttings, bindery trimmings and other paper and paper board waste resulting from printing, cutting, forming and other converting operations: bag, box, and carton manufacturing wastes and butt rolls, mill wrappers and rejected unused stock" [5].

The RCRA guidelines are by *de facto* establishing recovered materials grades that are causing considerable processing difficulties for the paper industry, especially the printing and writing sector. In many instances, acceptable grades of recovered wastepaper are being contaminated through a comingling of printing medias, such as impact printing and non-impact printing. Until the technology for the effective processing of such material is commercially developed, the recovered materials utilized will continue to be the more traditional grades.

The Institute of Scrap Recycling Industries specification circular defines over fifty recovered grades of wastepaper [6]. The recovered materials that can be utilized in the manufacture of many of the printing and writing grades have been pre-consumer wastepaper, generated by convertors and printers. Grades such as hard white envelope cuttings, manila and coloured tabulating cards, sorted white and coloured ledgers, manifold white and coloured ledgers, computer print-outs and various grades of unprinted and printed bleached sulphate have all been successfully utilized in the manufacture of recycled fibre for printing and writing grades. However, these established and well-defined grade definitions are being severely questioned by the pressing need to increase the consumption of wastepaper that is usually utilized in the lower quality paper board grades.

An important and key definition in the United States is the generic term, free sheet papers, both coated and uncoated. This has limited the use of a large quantity of groundwood-based waste material in the manufacture of recycled printing and writing papers. European, as well as Japanese paper manufacturers have not necessarily followed this restriction, and in many instances are using groundwood-containing waste material to produce acceptable printing and writing papers for the local marketplace.

However, the obvious effect of utilizing mechanical fibres from recovered groundwood-based wastepaper on the functional, aesthetic and archival characteristics of recycled printing and writing papers is not acceptable in the United States marketplace. The quality of recycled printing and writing papers produced for the United States market is much more critical. For all practical purposes, the quality of recycled grades produced in the United States is expected by the marketplace to be equal to virgin fibre except for slightly lower brightness levels. Consequently, the selection of the usable grades, in conjunction with a stringent inspection of the incoming recovered waste material, is the first important step in the management of the recycling process, especially for deinking.

11.4 Recycled-fibre manufacturing

By definition, deinking is simply the dispersion and separation of print from the fibre substrate. By virtue of the separation phase, the unusable fibre fraction of the deinked material, as well as the inert fillers, are removed with the dispersed inks and other contaminants.

This proper preparation of the wastepaper grades, via the deinking process, is an essential step in the manufacture of printing and writing papers. Further, it is necessary to employ the correct processing technology in order to provide a commercially acceptable fibre furnish for paper making. By design, the deinking process should be structured to ensure the required dispersion and separation of the print as well as the removal of the non-fibrous substances from the recovered wastepaper grades. Process management is an inherent part of a successful deinking plant and is required to support the selection and inspection of the waste material. Process control functions must be detailed and consistently applied to the operation of the systems to ensure consistent quality output from seemingly variable raw material. Modern day distributed control process computers have added greatly to the consistent manufacture of recycled-fibre furnish for printing and writing paper making.

11.5 Properties of recycled fibres

It is well known and reported in the literature, that recycled fibres demonstrate a noticeable loss in strength properties [7–9]. Further, the paper making characteristics of recycled fibre are assumed by their very nature to be inferior and unpredictable. Consequently, there has been an overall opinion that recycled material is unsuitable for the manufacture of quality grades of printing and writing papers.

This is a paradoxical statement when contrasted with the successful commercial production of a variety of printing and writing grades. Many paper manufacturers have developed an understanding of, and applied the correct processing technology to utilize the unique characteristics of, recycled fibre, made possible through the application of a fibre preparation technology that compensates for the morphological characteristics of recycled fibre.

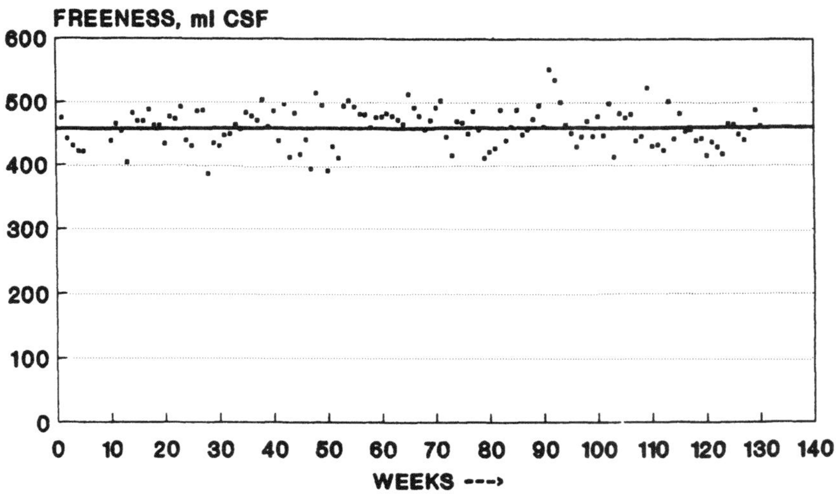

Figure 11.1 Recycled-fibre deinked pulp characteristic trend freeness.

Figures 11.1–11.7 illustrate the properties of recycled fibres produced from a commercially operating 200 short tons per day deinking facility. These trend charts are the result of weekly composite samples and represent in excess of two years of production experience and demonstrate the paper making potential of recycled fibre when processed correctly. The test results are from in-house evaluations that follow the TAPPI methods of preparation and testing procedures [10].

Traditionally, it is assumed that the softwood to hardwood fibre mixture is within the 20/80% range normally utilized in the manufacture of printing and writing grades of paper, but fibre analysis shows the recycled-fibre mixture is 23% softwood to 77% hardwood. Apparently, the separation process removes unusable portions of the hardwood fibre fraction in an unequal proportion to softwood.

Figure 11.1 shows the weekly freeness level of the processed recycled fibre, with a running average of 460 Canadian standard freeness (cfs) with a nominal range of between 400 and 500 cfs. Although there are weekly variations, the trend line for this period is basically flat. Both the burst strength trends in Figure 11.2, and the caliper trends in Figure 11.3, confirm this consistent freeness trend in values and show similar flat curves. For the period involved, the burst strength has averaged 64 psig within a fairly narrow range. The caliper averaged 0.0171 inches for four sheets and exhibited very little variation.

The trend line for tear strength in Figure 11.4, is atypical of the freeness, burst and caliper trend lines and demonstrates an overall increase. Tear strength variations from week to week could be due to a combination of factors but in all probability are caused by changes in the recycling raw material. The long-term upward trend of the tear strength can be more

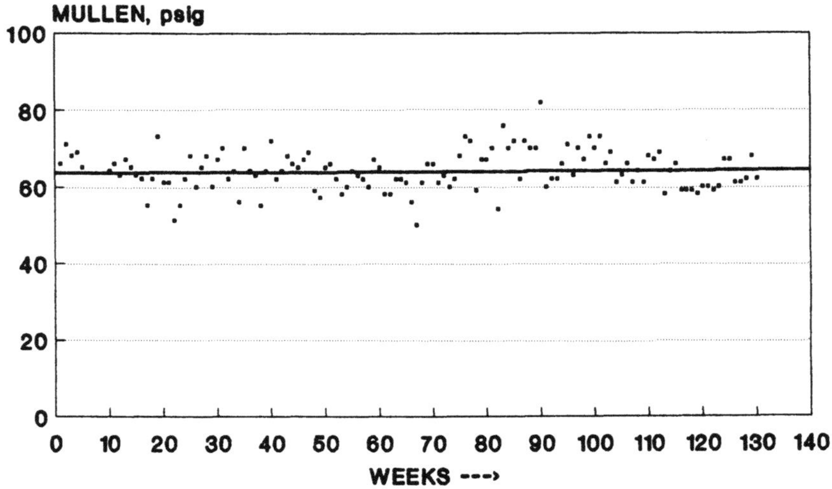

Figure 11.2 Recycled-fibre deinked pulp characteristic trend muller strength.

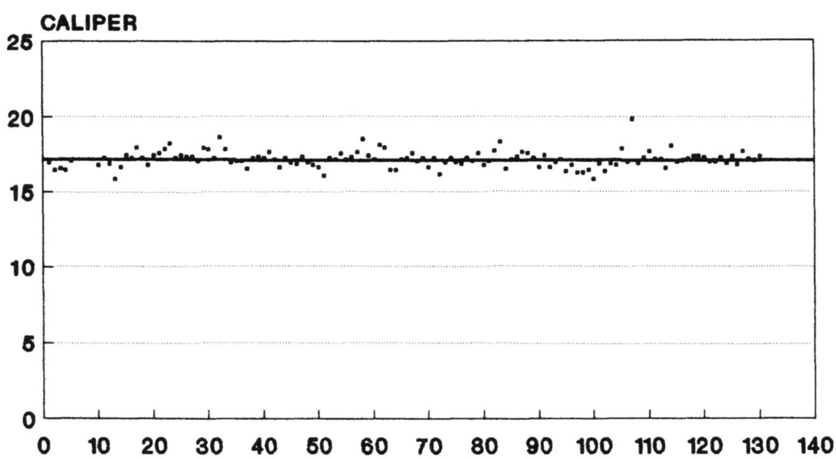

Figure 11.3 Recycled-fibre deinked pulp characteristic trend caliper.

clearly understood when compared to the percent ash level in Figure 11.5. The ash level contained in the finished stock is a good measurement of the separation efficiency of the deinking process. The lower ash levels reflect the improved removal not only of dispersed inks and other contaminants, but perhaps more importantly the removal of the fines, the unusable portions of the fibre fraction. Such improved fines removal could result in a minor reduction in the hardwood fraction of the recycle furnish. Thus, the higher ratio of softwood fibre could produce a corresponding gain

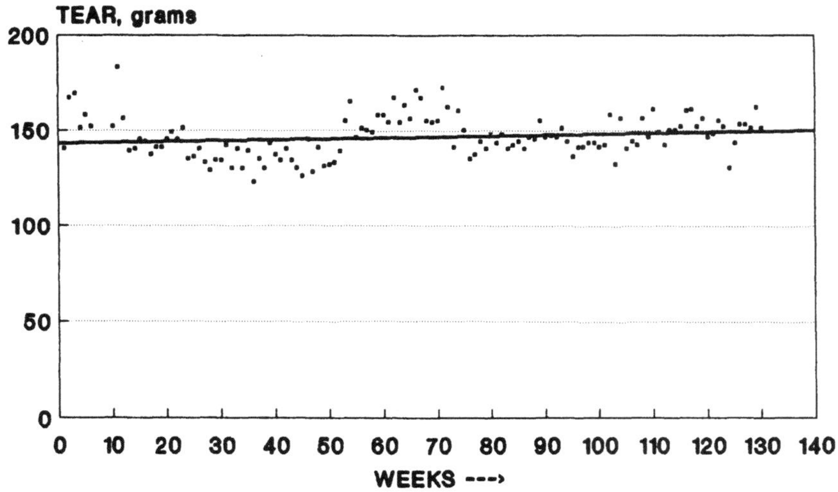

Figure 11.4 Recycled-fibre deinked pulp characteristic trend tear strength.

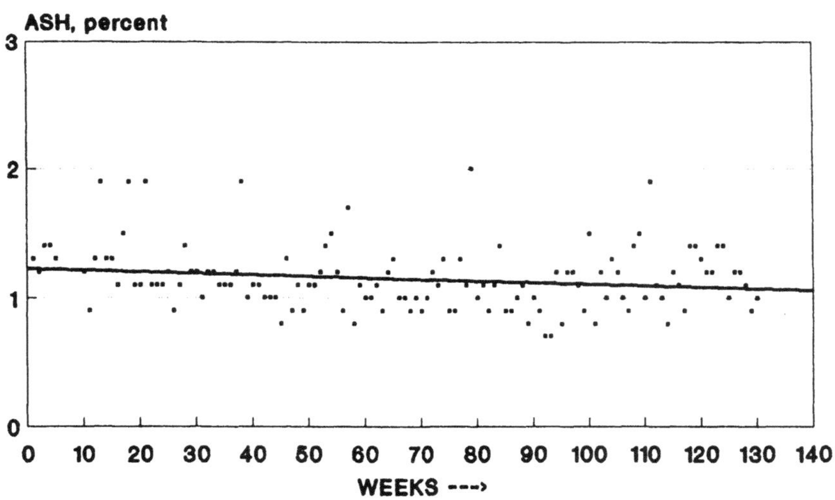

Figure 11.5 Recycled-fibre deinked pulp characteristic trend ash level.

in tear values. Since the time span of this change in fibre properties is over a long period of time, without a weekly fibre composition analysis of changes in the fibre mix, this conclusion cannot be confirmed and is only conjecture.

In the manufacture of printing and writing grades, the aesthetic values of the recycled-fibre furnish are as important as the strength values. Figure 11.6 shows the brightness levels, which display a slight upward trend, from the 78% range to the 80% range. The corresponding opacity values, in

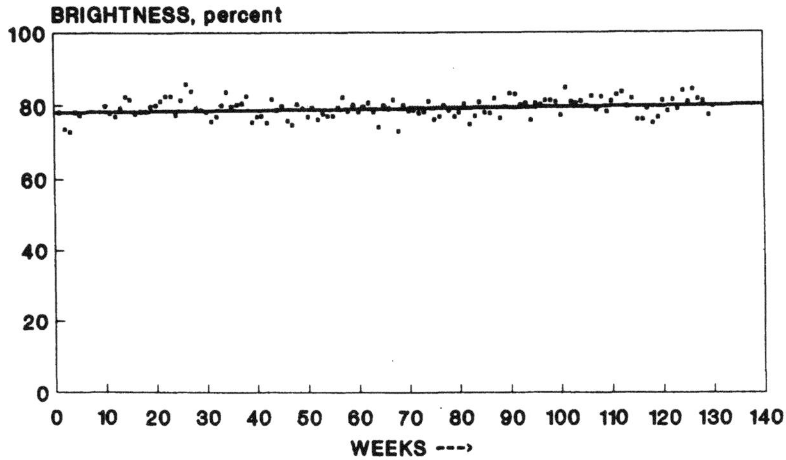

Figure 11.6 Recycled-fibre deinked pulp characteristic trend brightness.

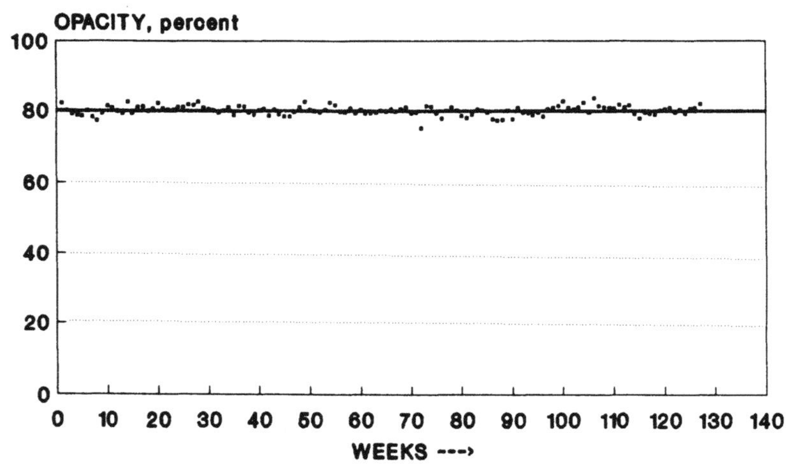

Figure 11.7 Recycled-fibre deinked pulp characteristic trend opacity.

Figure 11.7, reflect the opposite, with a slight downward trend over the same time frame. Both of these trends can be directly related to the trend line exhibited by the ash levels (Figure 11.5) and clearly demonstrate the effect of improved separation efficiency. The lower ash levels indicate improved removal of the inks and fillers in the recycled furnish with the resulting improvement in brightness. As to be expected, the opacity results are also affected as the residual inks and fillers are separated from the recycled-fibre furnish.

11.6 Recycled-fibre stock preparation systems

The paper maker understands implicitly that the recycled furnish is composed of pre-developed fibres, and is aware of the techniques that can be utilized to adjust for a variety of paper grades and stock ratios, from fully recycled fibre to a varied mixture with virgin fibre. Currently, a limited number of paper manufacturers are producing printing and writing grades from 100% recycled fibre. The majority are employing, however, a noticeable amount of virgin fibre as a component of the stock mixture.

The stock preparation and processing technology for recycled/virgin fibre mixtures is more critical than for recycled furnish alone. Separate refining of the various fibre streams prior to blending ensures the successful production of quality paper.

The key element in the paper making process is the proper sequencing of the stock preparation equipment for the pre-developed fibres to be adapted to meet grade specifications. A well grounded understanding of the morphological characteristics of recycled fibres is paramount in the effective utilization of recovered wastepaper. In recent years, recycling technology has grown, but without integrating the fibre preparation and utilization functions of the process into a total system concept, a myriad of problems can occur.

Through employing the correct process management, including the utilization of the proper grades of recovered wastepaper, a consistent recycled fibre furnish can be produced. As to be expected, the overall paper making characteristics of a recycled-fibre furnish will vary more than a comparable virgin-fibre furnish, but over a range that can be tolerated and compensated for by knowledgeable paper makers. The stock preparation equipment, as well as the wet end of the paper machine, can be structured to provide the paper maker with the ability to readily adjust, if needed, for the variation in the furnish. In reality, if the recycled furnish is processed correctly in the deinking operations such variations are long term, not short term, minimizing the variations in recycled fibre.

11.6.1 Non-integrated printing and writing mills

The proper preparation of the recycled-fibre stock is a troublesome area for numerous non-integrated printing and writing mills that are now producing recycled papers. These smaller, more specialized facilities, have stock preparation systems that are designed primarily for virgin-fibre processing. Many do not separately refine the hardwood and softwood pulp, but usually combine the pulp for processing.

There are innate strength development dilemmas associated with this technique. However, when recycled fibre such as deinked market pulp is blended with virgin pulp via a pulper, the problems are compounded. The development of consistent drainage and strength characteristics is essentially impossible.

Although there is no supporting research to confirm this condition, practical paper making has demonstrated that when refined together, the pre-developed recycled fibre will preferentially continue to develop prior to the development of the unrefined hardwood or softwood virgin fibre. The resulting fibre furnish is characterized as being comparable to a mixture of 'sticks and straw.' Consequently, the strength levels gained through the development process of the virgin fibre, especially the softwood, is marginal. Further, the underdeveloped softwood in the stock, at a given drainage level, has a tendency to form poorly, resulting in excess fibre flocs.

In the small non-integrated printing and writing mills, this problem is complicated by the difficulty in making the process transition from batch pulping to continuous stock feed at the wet end of the paper machine. There is an inherent change in development levels as the refined stock furnish is recirculated, via pressure controls on the refiners or stock box level controls, to the unrefined stock chest. Each time a pulper of blended stock is dropped to the beater chest, the degree of fibre development will vary with a corresponding change in strength and drainage characteristic on the paper machine.

This is a common fault with most stock preparation systems of non-integrated printing and writing mills and is further exacerbated by pre-developed recycled fibre. Statements that recycled fibre forms poorly are not due to the nature of the fibre, but more to the design and management of the stock preparation system.

11.6.2 Semi-integrated printing and writing mills

The limited number of printing and writing mills that operate an integrated deinking facility fibre have an improved opportunity to reduce the developed-fibre syndrome. These semi-integrated mills still utilize varying amounts of virgin fibre in the stock furnish, but usually at much lower levels. As detailed in the section 11.4, typical drainage and strength levels of commercially produced deinked pulp from an integrated facility are consistent and within an acceptable variation range.

Ideally, a recycled-fibre/virgin fibre stock preparation system would be a distributed ratio controlled (on a dry weight basis) stock blending system. In such a system, the virgin fibre is processed separately to a pre-determined drainage level. For many recycled printing and writing grades, the virgin fibre is usually either softwood or hardwood, which reduces the difficulty of refining the blended virgin fibres.

The pre-developed recycled fibre and developed virgin fibre are then combined in the stock blending stage, providing a more consistent fibre furnish to the machine systems. A 'tickler' refining system at the machine provides the paper makers with the ability to develop further the drainage and strength levels. Although not perfect, this tickling refining process for mixtures of developed recycled and virgin fibres is a more effective means of maintaining functional quality specifications consistently for printing and writing grades.

11.6.3 Integrated printing and writing mills

A fully integrated printing and writing mill that can effectively utilize 100% recycled fibre would require a much simpler stock preparation system than a non-integrated or a semi-integrated mill.

By knowing the developed characteristics of the recycled fibre, freeness and strength levels, the stock preparation system would be designed and operated quite like an integrated virgin-fibre facility.

11.7 Paper making characteristics of recycled fibres

Paper mills that utilize recycled fibres in the production of printing and writing grades view such material as just another source of paper making fibres. The development of the manufacturing expertise needed to utilize effectively recycled fibre depends upon many factors. Each stage of the process, the stock preparation systems, the wet-end configuration of the machine, the chemistry of the wet-end, etc. all influence the successful production of recycled-fibre-containing paper. Obviously, the grade of printing and writing paper to be produced, as well as the functional qualities and specification, also have a measurable impact on the technology.

Many of the paper making techniques and skills developed are obtained via a long and, in many instances, difficult learning curve. The knowledge gained through this learning process is more practical and based on intuitive paper making abilities rather than through a defined technology. The paper making characteristics of recycled fibre, however, are not unknown or unique if considered in the proper context.

The demand for practical and meaningful information about the paper making attributes of recycled fibre has been answered more from in-house trial and error research than from the academic world. Such research has been driven by the economic necessity to develop a feasible fibering source for printing and writing mills. When considered as a total system, the selection of the recovered waste material, the processing of material, the preparation of the fibre furnish and the application of the paper making skills, recycled fibres can perform equally to virgin fibres.

The impression that recycling degrades fibre strength is well known and widely reported in the technical journals [7, 8, 10]. The reduced swelling

Table 11.2 Recycled versus deinked fibre characteristics[a]

Property	Zero recycle	Recycle-1	Deink-1
Bulk factor	0.76	0.78	0.78
Opacity	72.1	74.9	74.8
Tensile index	86.61	72.83	79.70
Breaking length	8.834	7.368	8.127
Tear index	8.835	8.857	8.871
Scott bond	250+	245	250+
MIT folds	641	601	667
Final freeness	445	445	440
Initial freeness	720	510	630
PFI revolutions	3000	800	1800

[a] Sheet properties of the first stage recycle and deinked hand sheets compared to that of zero recycle virgin pulp at similar freeness. Both the recycled and deinked pulps were refined in a PFI mill prior to forming the hand sheets.

capacity of recycled fibres caused by irreversible hornification is the accepted morphological reason for this strength loss.

Commercial manufacturers of printing and writing grades from recycled fibres are aware of this phenomenon and have developed the technology to compensate for this deficiency. In part, this paper making technology is based on the apparent fact that the deinking process is contributing much more to the successful utilization of recovered waste material than just a separation process. The alkaline medium, pH 10–10.5, employed in the dispersion stage as well as the elevated process temperature, 120–140°F, appears to reverse the hornification of the recycled fibres to a degree.

Table 11.2 shows the results of a study conducted to measure the properties of recycled fibres prepared by (1) recycling of formed hand sheets and (2) deinking the same material under laboratory conditions [11]. The initially formed hand sheet stock, the recycled stock and the deinked stock were refined in a Papirindustriens Forskningsinstatt (PFI) mill to equivalent freeness levels.

The strength characteristics of the deinked stock is noticeably greater than the recycled stock and comparable to the original stock strength values (Table 11.2). The freeness level prior to the PFI refining was higher for the deinked stock than for the recycled stock which confirms the removal of the unusable fibre fines during the separation phase of the process.

This apparent reversing of hornification provides the paper maker with a fibre source that has development potential. In commercial application, the strength of the deinked recycled fibres can be developed to a level comparable to virgin fibre but probably at a lower drainage value.

The development potential of recycled fibres provides not only the paper maker with operating flexibility but also the process engineer with a clearer understanding of its paper making characteristics. In a study conducted at a commercial recycling printing and writing mill, several conclusions were drawn that supported the paper making potential of recycled (deinked) fibre [12]. Based on the results of this study:

- Recycled pulp has a similar strength development profile, but at a lower freeness level, to virgin pulp blended at the same softwood to hardwood ratio.
- The intrinsic fibre strength of recycled pulp is comparable to virgin pulp blends at the same fibre ratio as measured by the zero-span tensile test. Page *et al.* proposed [13] that the applied stress during zero-span tensile testing is borne by the cellulose fibres and that the bonding between the fibres plays an insignificant role in the process. Thus fibre strength is constant as long as the cellulose has not been degraded. Hence it can be inferred that the zero-span tensile strength is a measure of intrinsic fibre strength.
- Both the recycled pulp and the virgin-pulp blend exhibited similar tear strength development profiles, after the initial characteristic hump in the curve for the virgin blend. This is to be expected, since the recycled fibres have been through the paper making process and have a lower initial freeness value.
- Scott internal bond development profile for the recycled pulp and the virgin-pulp blend demonstrates similar fibre properties. However, the strength values for the virgin-pulp blend are higher at a given freeness. It is evident that a given strength value can be obtained with recycled pulp through refining to a lower level than the virgin-pulp blend. Alternatively, the strength difference can be narrowed by improvements in the separation stage of the deinking process. It has been reported that the bonding strength is affected by the amount of contaminants and anionic material which can interfere with fibre-to-fibre bonding [14].

This study was limited in scope, but has confirmed the paper making characteristics of recycled fibres that are commercially utilized in the manufacture of printing and writing papers. Recognizing and understanding the development potential of recycled fibre is extremely important to the paper maker.

11.8 Paper machine manufacturing technology

Changes to paper machine technology, for the manufacture of printing and writing papers from recovered materials, are needed to counteract the demanding and variable characteristics of recycled fibre. It is unrealistic to assume that recycled fibre furnishes, especially for printing and writing papers, are contaminant free. To compensate for the potential carry-over of contaminant such as process chemicals and non-fibrous materials from the deinking operations requires both the modification and in many instances the addition of wet end cleaning equipment.

Obviously, the efficiency of the deinking separation process will have an effect on the level of contaminant experienced in paper manufacturing. For the most part, knowledgeable paper manufacturers will as a preventative measure install the necessary equipment as insurance against process upsets and product quality problems. The type of cleaning equipment and its correct application is well known in the industry and depends to a great extent on the design and configuration of the wet end of the machine. Further, the reduced drainage of pre-developed recycled fibre compared to virgin fibre, at equivalent strength levels, requires modifications in the equipment as well as the operating techniques of the forming section of the wet end of the machine. For commodity grades of printing and writing papers, changes in the design of the forming fabric, additional dewatering elements, wet end chemistry, etc., could be required to maintain productivity levels. In the case of speciality printing and writing grades, which are not necessarily production driven, the modifications to the wet end operations will be influenced more by quality factors.

The operation of paper machines designed specifically for recycled fibre is within the realm of present day technology. The apparent difficulties that many manufacturers encounter are related more to the conversion of existing machines to recycled-fibre furnishes. Without proper planning and knowledge of the potential problems that can be experienced, the utilization of recycled fibre in the manufacture of printing and writing papers can be a very difficult and demanding experience for the mill.

11.9 Quality and performance of recycled printing and writing papers

There is a general perception that the manufacture of printing and writing papers using recycled fibres is limited to slower machines producing more specialized grades. Like many perceptions, this has been perceived as reality and is based on the impression that recycled fibres are impractical and unsuitable as a long-term fibering source due to extreme variations in quality, especially as a furnish for world class machines.

11.9.1 Recycled-fibre quality myths

This is a conclusion not confirmed by fact but by the many myths that have been circulated about the paper making characteristics of recycled fibre. Dispelling the myths about recycled fibres is no less an important task for the industry than meeting the technological challenges of the recycling effort.

Of the many myths, the most prevalent is that printing and writing papers produced from recycled fibres are lower in overall product quality, both functional and aesthetic. This mistaken image may have originated in the early 1970s when many paper companies, reacting to economic and environmental concerns, literally began recycling overnight. Unfortunately, much poor quality printing and writing paper entered the marketplace during that time.

That first impression remains today with many paper makers as well as printers. The paper makers that attempted to produce equivalent printing and writing papers from improperly processed recycled fibres suffered the consequences of lower manufacturing efficiency and poor product quality. For the printers, many still believe that printing and writing grades produced from recycled fibre have poor printability, marginal runnability and are aesthetically inferior to virgin-fibre papers.

Today, however, quality problems with printing and writing grades should not be attributed to the use of recycled fibres *per se*, but to the inability of paper makers to adapt to the somewhat demanding characteristics of recycled fibres. This problem is further compounded by the lack of adequate processing technology to prepare the recovered wastepaper properly for paper making. Without the correct equipment and sufficient knowledge of recycled-fibre utilization, it is very difficult to meet the quality demands of the marketplace. If processed correctly, recycled fibres have the advantage of slightly higher bulk, good dimensional stability and a consistent level of strength development.

For many paper makers, the use of recycled fibre is a new experience and problems occur with existing stock preparation systems, especially when processing a mixture of recycled fibres and virgin fibres. As discussed in section 11.5, the strength development techniques usually employed can result in an improperly processed machine furnish. Printing and writing papers manufactured from such furnishes often have poor formation with large fibre flocs and variable strength levels. Further, the improper separation of the soluble anionic substances, commonly called 'trash' in the deinking process will have an impact on the efficiency of the paper machine's wet end chemistry. This can result in paper of marginal and variable internal sizing, especially in acid-free alkaline grades. Variations in internal sizing will also affect the surface sizing characteristics, which will result in printing problems.

11.9.2 Manufacturing efficiency for recycled printing and writing papers

The manufacture of recycled printing and writing grades on world class machines is a feasible technology. There is no technological or paper making limitation that would prevent a commodity printing and writing paper mill from producing uncoated offset and reprographic grades efficiently with recycled fibre as a part of the machine furnish. The successful manufacture of 100% deinked newsprint on many world class machines has confirmed the paper making feasibility of this concept.

The desirable quality parameters for commodity products, good printability surface, moisture stability, bulk, stiffness, consistent strength levels and aesthetic quality are all obtainable from recycled fibre either as a part of the virgin furnish or even at 100% usage. Obviously, at the higher machine speeds, consistent quality of the recycled-fibre pulp will be a much more critical requirement to maintain efficient machine performance. Modern day technology is available that will produce consistent and viable recycled fibres from usable recovered wastepaper.

11.9.3 Recycled alkaline printing and writing papers

The question whether or not alkaline printing and writing papers can be produced from recycled fibres is not a technological issue. There is no valid reason why the quality characteristics of alkaline grades can not be met with recycled fibre. In many commercial printing grades, the use of calcium carbonates as filler as well as coating pigments enhances the optical characteristics of recycled fibres. High-opacity alkaline grades can be produced from 100% recycled fibres that have brightness values in the high 80s with no apparent functional differences to a virgin-fibre sheet.

In a recent senior research project [15], an investigation was conducted into the strength characteristics of recycled alkaline and acid hand sheets. This study showed that under repeated recycling, all of the strength tests were greater for the alkaline hand sheets than for the acid. The study suggested that alkaline papers were more resistant to strength losses associated with recycling.

For commodity printing and writing papers, the caveat however, is that the use of recycled fibres in such a manufacturing environment is not a defined science and although it is a feasible technology, it will undoubtedly be a long and difficult learning curve for many paper makers.

11.10 Economics of recovered wastepaper usage

The view that the manufacture of printing and writing papers from recycled fibres is an uneconomical technology is not always totally incorrect. The economics of recycled-fibre usage depends, to a great extent, on the products manufactured and the markets served. With printing and writing manufacturers, the economic question has more to do with the productivity level of the mill than to relative costs of recycled and virgin fibres. Even with an installed deinking plant, a speciality printing and writing paper mill cannot compete in commodity grades against a fully integrated world class facility.

It is a correct assumption that the cost of slushed recycled fibre delivered to the paper machine is comparable to the cost of virgin fibre. Thus, the economics of recycled printing and writing papers are affected more by the over-the-machine cost than by the fibre cost. The revenue generated by any specific grade is based on the productivity level of the particular segment of the printing and writing market that it serves. For commodity grades, the productivity level is very high whereas, for speciality grades, it is at low to medium levels. The operating margins produced by the various printing and writing grades are proportional to the value added in manufacturing.

There are short-term economic factors that are affecting the profitability of recycled printing and writing papers but in the final analysis the marketplace will determine the value to be derived from the production of recycled papers.

11.11 Future for recycled printing and writing papers

In recent months, the paper industry worldwide has been seriously challenged by the growing demand for recycled printing and writing products. In recent history, no other event has so dramatically changed the industry's course of direction so quickly than this unprecedented demand for recycled paper products. The paper industry is slowly accepting the technological reality of recycling. However, accepting and understanding the need to respond in a credible manner is the crucial factor in meeting the recycling challenges of the future.

This increased demand for recycled printing and writing paper has promoted an accelerated search for a practical and cost effective manufacturing technology. For the paper industry, which has in the past regarded recycling as an unsuitable long-term fibering source, finding an adequate technology has been difficult and demanding, in the absence of fundamental knowledge of the process. There is much to be learned and accepting the need to define basic principles and concepts for recycling is the first step in answering the challenge.

To meet the demands of this technology challenge will require a dedicated research commitment by the industry to define the fundamentals of the recycling process, for the industry knows very little about the chemistry or hydrodynamics involved in the technology of dispersion and separation of the print from the fibre substrate. What is known is mainly an extension of existing process technology.

To meet the recycling expectation of the printing and writing marketplace requires an in-depth analysis of the various process parameters that affect the efficiency of the technology. This will require a long-term commitment from the paper industry.

The paper industry must avail itself of the opportunity to expand and develop the technology needed to utilize many of the grades of paper that are currently unusable in the manufacture of printing and writing papers. The industry has a limited window of opportunity to develop the technology required to utilize effectively these more difficult waste materials, especially office waste. Once the fundamentals of the technology are defined and understood, then the next generation of equipment and processes will be developed that will provide the operating margins needed to improve the quality of printing and writing papers.

References

1. Food and Agriculture Organisation's (FAO) Survey, *Pulp and Paper Capacities, 1991-1995*, American Paper Institute, World Paper and Paperboard Capacities, 1991-1994, (August 1991).
2. Federal Register, Vol. 53, No. 120., Rules and Regulations, US Government, Washington DC, June 22, 1988, p. 23546.
3. Federal Register, Vol. 53, No. 120., Rules and Regulations, US Government, Washington DC, June 22, 1988, p. 23548.
4. Federal Register, Vol. 53, No. 120., Rules and Regulations, US Government, Washington DC, June 22, 1988, p. 23551.
5. Federal Register, Vol. 53, No. 120., Rules and Regulations, US Government, Washington DC, June 22, 1988, p. 23531.
6. Institute of Scrap Recycling Industries, Scrap Specification Circular (1991), *Guidelines for Paper Stock*, 1627 K Street, N.W., Washington, D.C. 2226.
7. McKinney, R.W.J. (1989) Changes in fibre properties with recycling, *Pira Seminar Use and Abuse of Recycled Fibres*, Ref: 31/912/CM/102.
8. Howard, R.C. (1990) The effects of recycling on paper quality, *CPPA Annual Meeting*, B337-B346.
9. Ferguson, L.D. (1992) Effects of recycling on strength properties, *Paper Technol.*, **33**, (No. 10), 14-20.
10. *TAPPI Test Methods 1991*, (1990) (2 Volumes) TAPPI Press, Atlanta.
11. Khambadkone, G.G., *Effect of Recycling Process on the Papermaking Characteristics of Cellulose Fibres*, Unpublished research work at the Miami Mill of Cross Pointe Paper Corporation, West Carrollton, Ohio, USA.
12. Khambadkone, G.G., *Virgin Versus Deinked Pulp - A Comparison*, A research project conducted at the Miami Mill of Cross Pointe Paper Corporation, West Carrollton, Ohio, USA.
13. Page, D.H., Seth, R.S. and El-Hosseinny, F. (Sept 1985) Strength and chemical composition of wood pulp fibres, *Trans. Eight Fundamental Research Symposium*, British Paper and Board Makers' Association, Oxford, England, 77-91.
14. Klugness, J.H. (1974) *Tappi*, **49**(7) 71.
15. Fernstrum, P. (December 1992) Effect of acid and alkali on strength properties of paper, *Paper Sci. Eng. Schof*, Western Michigan University.

12 Printing trends - impact on paper recycling

B. THOYER

12.1 Introduction

The first aim of printing is to form a coherent ink layer on a print medium, which has to be bound to the paper to have enough resistance to carry the printed message to the audience. For this purpose, sophisticated ink formulations are associated with different paper grades to produce a variety of printed products obtained with a wide range of printing processes.

In conjunction with these developments, an increasing environmental awareness has made paper recycling a topical subject. Whereas in the past an ink was only developed to have good performance on the printing machine and to give a final printed product of good quality, the ink maker now has to take into account recycling. For certain products, emphasis will be on producing inks which are easily deinkable. The share of ink in a wastepaper stock is about 2% (maximum) of the total weight. When the recycling process needs the wastepaper to be deinked, the difficulty is to remove this small quantity of ink out of the pulp produced from the wastepaper. Dependent on the printing process used, the paper and the ink are bound together in different ways. In the future, the relative development of printing processes will influence paper recycling.

Ink chemistry is a very complicated science and each ink maker has their own formulations. For one printing process and for a particular application, there are dozens of possible ink recipes. When the printed products are recycled, the wastepaper stock is a blend of many different papers and inks which go through the deinking process. The composition of the wastepaper stock depends on the relative importance of the different printing processes. Consequently, the future developments of these processes have to be taken into account. If some difficult-to-deink grades should develop in the future, the industry has to be ready. This chapter will try to show the influence of the printing processes on the further recyclability of wastepaper, with an overview of difficult-to-deink grades.

12.2 The influence of paper grade and paper properties

12.2.1 Non-coated papers

Non-coated papers can be considered as three-dimensional fibre networks. The spaces between the paper fibres (constituted of air) occupy about 50% of the total volume. With non-coated papers, printing is achieved directly on the paper fibres. The paper grades concerned are, for example, newsprint (printed with the letterpress, offset or flexographic processes), laser papers and electrophotographic papers. With these, the ink is directly bound to the paper fibres, so to deink these it is necessary to detach the ink from the fibres.

12.2.2 Coated papers

Printing can also be made on coated paper, especially when high-quality reproduction is required, which is mostly the case with magazines, for example, offset and gravure processes. In this case, the ink is deposited on the layer of coating material, so that the deinking process requires a separation of the inked layer from the fibres.

These two examples show the influence of the paper grade on recycling. It will be easier to deink an ink that was printed on a layer of coating material than an ink that was directly deposited on the paper fibres. A number of trials have shown that printed products made with coated papers are easier to deink than printed products made from uncoated papers [1].

12.2.3 Influence of smoothness and absorbency of paper

The smoothness and the absorbency of a paper can also play a role. If the paper is very rough or porous, the ink film is entangled in its surface and the adhesion is very good. A greater surface area of paper fibres will be in contact with the ink layer. Even if the substrate is non-porous, roughness will help adhesion by increasing the area over which adhesive forces will operate, without increasing the amount of film which has to be held by these forces.

12.3 General composition of a printing ink

Generally, a printing ink is composed of three or more distinct ingredients: a pigment, a vehicle and additives.

12.3.1 The pigment

The pigment is the colouring agent which is dispersed in a very fine grain form. Pigments are almost completely insoluble in the medium in which they are used.

Due to its special properties, the most important black pigment for the printing ink industry is carbon black. Carbon black is produced by the incomplete combustion of hydrocarbons (mostly mineral oils). The primary size of carbon black particles is in the range of 10–50 nm, though these are structured into large agglomerates during ink production. The size of these aggregates determines the colouring power of the pigment.

The pigments for coloured inks originate almost exclusively from organic chemical synthesis of complex organic molecules. There are hundreds of different types of pigment produced; for example, blue and green pigments can be made with phthalocyanine pigments (a molecule which contains copper). Heavy metals have almost been eliminated from most printing inks due to environmental considerations [2].

In the production process, the pigments go through a surface treatment to give them specific properties. During this treatment, colophane resins, amines or synthetic polymers can be added. The aim is to modify the dispersion ability, the rheological behaviour and the size of the pigment particles.

12.3.2 The vehicle

The function of the vehicle is to ensure the transportation of the pigment during printing and to give the ink its final aspect after drying. The vehicle is formed with different components of varying viscosity, depending on which ink it is destined for and which pigment is used. During drying, physical and chemical properties of the vehicle will change to give the printed sheet a definitive aspect; for example, the layer can harden by chemical modification, evaporation of volatile products or absorption of specific products by the paper sheet.

In most cases, the vehicle is composed of a hard resin, a solvent and a diluent. The *hard resin* is so called because it is solid at normal temperatures. This resin can have a natural (animal or vegetable) or synthetic origin. For offset and letterpress, a modified rosin-based resin is often used. In addition, hydrocarbon resins and, for black inks, a natural fossil asphalt can also be used. For rotogravure and flexographic inks, the vehicle can include rosin-based resin, acrylic hydrosoluble resin, chlorinated rubber, vinyl chloride acetate, polyamide, etc. For the production of vehicles, a combination of different hard resins is used. The resin gives cohesion, gloss and the right mechanical properties to the ink. The hard resin is the most significant component influencing the further deinkability of the printed product [3].

The *diluent* and the *solvent* are difficult to differentiate. Some products can be both solvent and diluent. The role of a diluent is to adjust the viscosity. These oils evaporate very slowly so that there is no drying of the ink on the rollers of the printing machine. In the cases of gravure and flexographic processes, the solvent evaporates to allow the ink to dry. Spirit, toluene, water, hydrocarbon fractions (petrol) or esters (ethyl acetates) can be used as a solvent for these inks.

12.3.3 Additives

Several additives can be added to the ink formulation to achieve a particular performance or to meet a specific quality requirement. They may be used to adjust flow, set-off characteristics, print-through, ink transfer, etc. These additives are finely powdered filling material, oil compatible gel formers, waxes or wetting agents. As these additives are often very different from one ink to another, and as they can substantially influence the properties of the dry ink film, the additives can also affect the future deinkability of a particular ink.

12.4 Ink drying

Drying means that the ink will be transformed from a viscous liquid to a solid. The ink will have very quickly to lose all its initial flow properties. This is made possible by the chemical or physical modification of the ink vehicle. When the ink film has reached a condition in which it has an acceptable resistance to deformation, it is said to be dry. The final aspect of the ink has also to be suitable for further handling. Consequently, it is important that the ink manufacturer takes into account further treatment undergone by the printed product, for example, the resistance to fold. The final form of the ink layer deposited on a particular printed product plays an important role in the ability of this product to be deinked.

Four main drying methods can be described; chemical change, penetration into the substrate, solvent evaporation and precipitation. For certain inks, two or more mechanisms can be involved in drying. The best example is the combination of penetration and oxidation to give the quick-setting inks that dominate letterpress and litho formulations.

An important aspect for recycling is the speed of drying of an ink. Moreover, it is important to know if ageing will change the ink characteristics of a specific wastepaper grade. Air supply, temperature and humidity have a major influence on drying speed. The characteristics (mainly smoothness, absorbency and acidity) of a paper or board can also play a role. Modification of the formulations of inks have constantly improved the speed, or it can be achieved by providing energy from an external source. Energy-assisted drying is heat assistance to evaporation drying of flexo, gravure, screen and web-offset inks. Infrared and ultraviolet drying technologies have also developed. The use of radiofrequency (microwaves) for water-based inks and even electron beam for special inks could develop in the future.

12.4.1 Penetration and absorption drying

Penetration drying represents the simplest form of drying. The ink (or some of the ink) penetrates into the inner part of the paper. It is generally explained as capillary components of the attraction of the liquid into the spaces between the fibres. The ease with which a liquid will wet paper fibres depends on both the liquid and the treatment the paper has undergone. Cellulose fibres are very easily wetted by water-based systems which explains the almost immediate drying of water-based flexo inks on newsprint, kraft papers or corrugated boards. Sizing of the paper with rosin-based materials reduces the ability of water to wet the paper fibres but increases the affinity for oil-based inks.

The inks that dry exclusively by penetration are rotary letterpress news inks, some other letterpress inks for uncoated papers, many cold set web offset inks and some small offset press inks. Absorption also plays a very important role in the drying of most other inks for all papers and boards, except for the least absorbing grades. Consequently, with respect to deinking of these inks, there is no physical separation between the ink layer and the paper fibres. The two layers are intermixed.

12.4.2 Oxidation drying

Just after printing, the 'liquid' ink undergoes a chemical change causing solidification. The normal requirement is that oxidation should take place as quickly as possible under normal ambient conditions, but that the ink should be as stable as possible during storage and use on the printing machine.

Linseed oil was one of the first ink vehicles used. This oil dries very slowly on contact with atmospheric oxygen. Linseed oil is a substantially neutral medium showing little reactivity towards pigments and substrates but, when unaided by catalysts or elevated temperature, the drying processes are slow. This is an example of printed products where ageing plays an important role in deinkability: an aged ink containing an oxidised oil is less deinkable than a fresh one [4, 5]. However, the mechanisms of oxidation are not very well known but it is possible that, after a sufficient period of time, the ink could react with the paper fibres to form new stable bindings.

Oxidation drying is the traditional drying process for litho and letterpress ink and early screen inks. Drying oils and resin systems derived from them can have excellent press performance characteristics and the properties of the dried film obtained include good gloss and good rub resistance. In the search for quick drying inks, many chemical modifications of drying oils and many completely synthetic media have been developed. Some specific requirements have been added to this objective, like the resistance to various chemicals, vapours, light and heat. These requirements are often in contradiction with the manufacture of an easily recyclable product since the more the ink layer is bound to the paper, the more difficult it will be to deink.

12.4.3 Precipitation and neutralisation

The precipitation drying process can be explained by the use of an example: traditional moisture set inks are based on high acid value resins, dissolved in a high boiling point polar solvent, which is hygroscopic (readily absorbs water). The resin remains soluble in the solvent until a sufficient amount of moisture is absorbed from paper or air, when it precipitates out. These inks have not been successful because of poor stability on the press.

Traditional water-based flexo inks are good examples of neutralisation drying. The acidic resin is solubilised in water by addition of a base. During and just after the transfer onto the paper, the base is mostly removed by absorption into the paper but also by evaporation. The resin becomes water insoluble because of the removal of the amines and the ink film becomes rub- and water-resistant very quickly. In the meantime, a lot of water is absorbed by the paper contributing to the formation of a very resistant ink film. The more modern water-based flexo inks are based on resin emulsions rather than solutions (or a combination of both). The same drying speed is retained with inks made with resin emulsions.

12.4.4 Evaporation

Solvent evaporation is a drying method used in most major printing processes. It is of greatest importance in flexo, gravure, screen and web offset printing. The evaporation of the light elements (solvent or diluent) can be facilitated if the interaction with the other elements (resin, additives, pigments) is low. In gravure, flexographic, screen and heat set printing, whenever solvent evaporation is used to achieve ink drying, formulations are needed which evaporate as little as possible in the ink duct and on the applying rollers, but which dry as rapidly as possible after printing.

The properties of a dry film resulting from evaporation can depend on the solvents used in the ink. This is due to the presence in the dry film of traces of solvent, which can act as a plasticiser, making the ink film more flexible and therefore increasing its apparent adhesion. That can be misleading because films which are flexible can become brittle within hours, days or perhaps weeks. This possible modification of the printed product after ageing is also an important factor for recycling.

12.4.5 UV curing

To meet the demand of faster printing presses, UV curing inks have been developed. Just after printing, the printed product passes under an UV lamp. Drying of the ink is caused by the action of ultraviolet radiation, which acts as an external energy source. The drying reaction is a free radical mechanism. Within the ink vehicle are compounds which can undergo polymerisation - monomers. The UV radiation of appropriate wavelengths causes the photoinitiator (also present in the ink formulation) to decompose to create free radicals, which transfer their reactivity to the monomer, and the subsequent chain propagation creates stable polymers. This drying occurs almost instantaneously. When recycled, UV inks can create large specks which are not removed easily by conventional deinking methods and which are visible on the recycled paper. These specks are chemically resistant because of the stability given by the polymer. For example, cover pages of magazines are often UV varnished, which can create problems during recycling. Deinking of these has been researched, but a satisfactory solution has not yet been found.

12.4.6 Other drying processes

12.4.6.1 Infrared drying. Infrared drying differs from UV curing because special inks are not required. It is an improvement to the printing press by the addition of a stage to generate heat and thus accelerate the normal drying process. At higher temperatures, oxidation, penetration and evaporation mechanisms are all accelerated.

12.4.6.2 Microwave or radiofrequency drying. Water and other polar solvents are readily heated by microwave energy. Polar solvent molecules behave like a tiny magnet. When exposed to an electric field, the molecules line up with the field. Because the microwave field is alternating rapidly, the solvent molecules are oscillating, trying to line up with the field. The movement of the molecules then generates heat. Equipment costs are very high and this process will only be used where it offers particular benefits. The most active solvent, by far, towards microwave energy is water, so microwave drying could be a possible alternative for water-based flexo inks or it could be a possibility for the future if water-based gravure inks are developed, since water-based inks create a drying problem caused by the low evaporation rate of water.

12.4.6.3 Electron beam curing. It has been suggested that electron beam curing would be a more efficient alternative than UV curing because of the higher energy levels involved. This enables simpler inks (free of photoinitiator) to be formulated and should also allow thicker ink films to be cured. Conventional oxidation drying inks could also be used without any special additives. However, this process requires a high level of shielding to protect operators. Also, the process is fully effective in a vacuum (or near vacuum) and the cost of achieving these conditions is considerable. The result is that the equipment is very expensive and the future development of this process is linked to the evolution of the associated costs.

12.5 Impact of different printing processes and recycling

The main printing processes are reviewed in this section and their impact on recycling is assessed. When a 'deinking coefficient' [6, 7] is calculated, the printed samples are tested for deinking by a method based on three test procedures:

A The printed samples are pulped in water and treated with chemicals in the flotation cell. Hand sheets are produced on a laboratory former.
B Unprinted paper is pulped in water and treated with chemicals in the flotation cell. Hand sheets are produced on a laboratory former.
C The printed samples are pulped in water without chemicals and hand sheets are produced on a laboratory former.

After deinking and the preparation of the hand sheets, the brightness of the samples is measured as a *Y*-value at 557 nm. On the basis of the measurements of the brightness of the samples A, B and C, a deinking coefficient DEM is calculated as follows:

$$\text{DEM (\%)} = 100 \times \frac{\text{brightness(A)} - \text{brightness(C)}}{\text{brightness(B)} - \text{brightness(C)}}$$

A deinking coefficient near 100% represents a perfectly deinkable sample, whereas a deinking coefficient near 0% means very poor deinkability.

The values of deinking coefficients presented in this chapter are only to give examples and should not be taken as absolute values, and are included only to give an idea of the relative deinkability of different printed products. Rather than give absolute values, trends are illustrated in Figures 12.1 to 12.3.

12.5.1 Letterpress

12.5.1.1 Introduction to the letterpress printing process. This process was the first printing process to be developed and up to 1975 was the most important printing process in use within the industry. Since then letterpress has constantly declined, being replaced by other processes such as lithography and flexo. However, letterpress is still in use for newspaper printing, even if it is still declining and is much used for book printing.

Printing is achieved directly from relief plates. Lead plates were used in the past; photopolymer plates are now more common. For letterpress printing, a high viscosity ink is needed and the ink is applied with a relatively high pressure on the paper, enhancing its penetration into the fibres.

A typical letterpress rotary letterpress black ink is composed of:

- Pigment 10–14%
- Mineral oil 80–88%
- Resin 0–3%
- Solvent 0–3%
- Additives 2–5%

As no water comes in contact with the ink for this printing process, it is necessary only to coat the pigment on the paper. Mineral oils are capable of this, with only a very small addition of hydrocarbon resin varnish being necessary on occasion. The letterpress process is more direct than the offset process: the number of times the ink is split between inking rollers and paper is less than with offset. That means that the ink film weight which is applied is higher and therefore the pigment content to achieve the same density will be lower.

12.5.1.2 Deinking of letterpress printed products. This type of ink dries mainly by absorption of ink components into the paper surface. After a sufficient period of time, the components also dry by oxidation. The deinkability of this type of ink is influenced by the quantity of hydrocarbon resin present in the formulation. Mineral oil is rather neutral for the deinking process. However, after a certain period of ageing (3 to 6 months), the mineral oil on the printed product can undergo chemical changes which can reduce ease of deinkability. This ageing time limit is mainly dependent on the fraction of oxidative drying components used in the ink formulation.

Nowadays, newspapers represent most of the tonnage produced by letterpress printing. Most of the time, only black inks are used for newsprint letterpress printing. Colour printing is usually with the lithographic process. Pre-prints in newspapers are made in gravure or in heat set offset. Colour in newspapers using letterpress is limited to the use of spot colours. Black news inks are generally based on the cheapest available materials (lowest

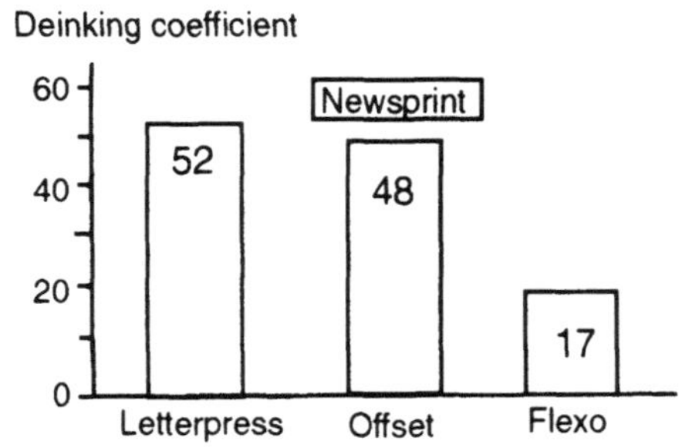

Figure 12.1 Deinkability of different newspapers collected in Germany and Austria.

cost carbon black and mineral oils). The share of letterpress in newspaper printing is very variable between different countries. In Germany the amount of letterpress printed newspapers is about 1%, whereas the share is >40% in the United Kingdom. In Europe, in a wastepaper stock collected for newsprint production, the amount of newspapers is generally 50–70%. Consequently, in some regions, the amount of letterpress printing products in a wastepaper stock can be relatively high and have an influence on the recycling process. A substantial quantity of books are also letterpress printed and can also be present in the wastepaper stocks. The ease of deinking is compared in Figure 12.1, which shows that in general terms, letterpress printed newsprint is easier to deink. This helps to explain different deinking results in different geographic areas; for example, in the UK very good newsprint deinking results are achieved, compared with areas such as Germany and Japan, which have a much lower proportion of letterpress printed newsprints.

12.5.2 Offset lithography

12.5.2.1 Introduction to the offset printing process. The basic principle of offset lithography relies on mechanical damping of a printing plate which consists of an oleophilic (or hydrophobic) image area and a hydrophilic non-image area. The plate is then inked and put into contact with a rubber blanket which, under pressure, transfers the image to the printing substrate. On the printing plate the image areas are covered with inks and non-image areas are covered with water. Intimate contact between ink and water must lead to an emulsion being formed between the two. The formation of oil-in-water and water-in-oil emulsions and the balance between these two systems is a complicated problem and directly influences the quality of the printed result.

The weight of the deposited ink film depends on the paper grade, the printing pressure and the blanket compressibility. The substrate contributes a lot to the transfer of the ink. It may vary from newsprint, which is extremely porous and absorbent, through to smooth calendered coated paper, which has a closed structure and is much less absorbent. The formulation of an ink and whether it will dry by penetration, quicksetting or oxidation depends on the paper grade being printed.

Sheet-fed offset inks. When the printed product is stacked just after printing, it is still wet. To prevent this wet ink from setting-off, spray powders are used to separate the sheets. The development of rapid setting inks in recent years has led to a general reduction in the quantity of powder to be used to avoid set-off. For modern quickset inks, the vehicle contains a thin and a thick phase. The thin phase (a printing ink distillate) is able to be quickly absorbed by the surface of the paper to render the ink 'touch-dry' in a matter of minutes. As the thin phase is absorbed, the thick phase of pigment, resin and oxidisable material is left on the paper surface. Oxidation converts the touch-dry film to a chemically and physically resistant film over a period of some hours.

Web offset inks. The web offset market is still growing, representing important paper tonnages. The two most common examples are the printing of newspapers (generally web offset cold set) and the printing of magazines and quality printed products (generally heat set web offset). Heat set and the cold set inks are very different and have to be studied individually.

Cold set web offset. Uncoated papers with an open structure (such as newsprint) are generally printed with this process which has become the most common method for the printing of newspapers and therefore represents a very important tonnage. It is also used in the continuous stationery and business forms market. The ink drying method is by penetration. It is essential that rapid penetration takes place so as to avoid marking on turner bars or in the folder of the rotary press. The ink formulation has also to take into account the potential for further rub-off, off the newspaper copy.

Heat set web offset. The web offset heat set market is typically centred around publication printing and competes with sheet-fed offset, web and sheet-fed letterpress, and gravure for the magazine, catalogue and advertising brochure market. Most heat set work is four-colour printing. The process involves the passage of the paper web through a dryer after printing before the folder, so that the ink is dry before the end of the press.

The most common dryers are hot air and infrared dryers; for example, the web is heated by hot air to enable the distillate in the ink to evaporate. In this stage, the drying process is mainly by evaporation. The web temperature to evaporate the distillate is approximately 120–175°C. The proportion of distillate is around 30% of the total ink weight and it is

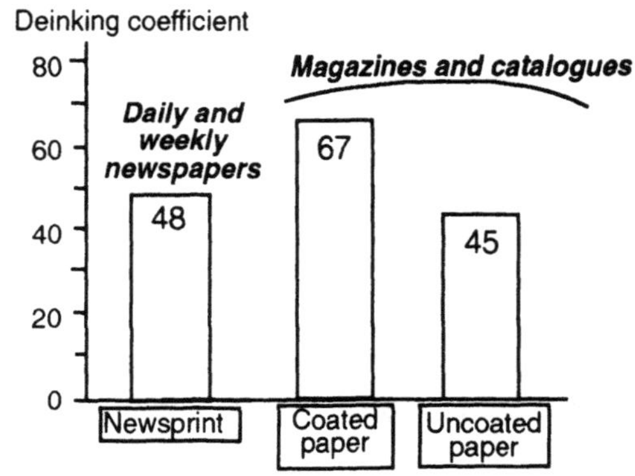

Figure 12.2 Deinkability of different offset pointed products from Germany and Autria (source: PTS, Munich).

reduced to 5–10% after the dryer, the ink being then fixed onto the paper.

12.5.2.2 Deinking of offset printed products. When the offset process started to develop, deinkers had major problems with their conventional recycling process. Early versions of these inks contained oxidisable alkyde resins (soft resins) as binding agents and drying oils (like linseed oils). After a certain period of ageing (4 to 6 weeks), during which the binding agent solidified by oxidation, printed products containing these components became very difficult to deink because the ink film could only be partially removed from the surface of the fibres. After research, more suitable ink formulations and better deinking chemicals were developed and deinking problems were overcome. Today, deinking processes have been optimised for the treatment of offset printed products.

Deinking of various types of offset printed products is shown in Figure 12.2. The deinking coefficients were averaged from 30 daily and weekly newspapers and 38 magazines and catalogues collected in Germany and Austria in 1991. Newspapers were printed by the cold set web offset process, whereas the magazines and catalogues were mostly printed with the heat set web offset process. Magazines printed on coated papers are the most easily deinked offset products. On average (coated and uncoated papers), it can be seen that magazines printed by offset are more deinkable than newspapers, using conventional flotation deinking processes. Generally speaking, no substantial deinking problems are caused with offset printed products, though ageing (after several months) can have an adverse effect on deinkability. The deinking coefficient for offset printed newspapers is compared with other newsprints in Figure 12.1.

Very recently, *vegetable oil-based inks* have started to be developed, especially for printing newspapers, including soybean, linseed, tall and rapeseed oils, etc. When such samples are deinked immediately after printing no particular problems have been reported. However, some problems with deinkability can appear after a certain period of ageing and overall yield of the process can fall. Experience has shown that these types of inks can cause problems if the ageing period exceeds about 6 months [4].

12.5.3 Flexography

12.5.3.1 Introduction to the flexographic printing process. The flexo printing process has developed rapidly. One reason for the growth of flexo is the boom in its biggest market, flexible packaging. It is also widely used for packaging printing in general and for printing labels. With respect to printing papers, flexo has developed primarily in printing newspapers in the United States and in Italy (where it represents approximately 20% of the newspaper market), and also to a smaller extent in the United Kingdom.

A flexo press consists of an anilox roller and a cylinder onto which a photopolymer plate containing the image is mounted. The ink is transferred direct from an ink duct to the anilox roller, to the photopolymer plate and then to the paper. This very short ink train means that the ink must distribute rapidly and thus must be of much lower viscosity than conventional oil-based offset or letterpress systems. A flexographic ink consists of three basic components: a solvent, a resin binder and the colouring agent. The choice of these components depends mainly on the speed on the press and the printed substrate. Inks have to be formulated to dry quickly. It is essential that all the solvent is removed from the ink layer before the end of the printing press.

The inks used in flexo printing are either solvent or water-based. Organic solvents are used for non-absorbent substrates. This type of ink will dry by evaporation. For the flexo printing of newspapers, water-based inks are mostly used.

12.5.3.2 Deinking of water-based flexographic inks. Water-based flexo inks contain 60–75% water with 12–20% pigments. They are also based on approximately 10% of acidic resins such as acrylic polymers. These resins are in solution or dispersed in water, when neutralised with organic bases, such as amines. During and just after the transfer of the ink onto the newsprint, the amines are removed by absorption into the newsprint and by evaporation. The resin becomes water-insoluble and the ink film solidifies. At the same time, a lot of water is absorbed by the paper.

The deinking coefficients of the three printing processes used in newspaper printing including flexographic are compared and illustrated in Figure 12.1.

Coefficients were calculated and averaged by the ‘Papiertechnische

Stiftung' (PTS) in Munich from 30 different newspapers collected in Germany and Austria [8]. The deinking coefficient for flexo printed newspapers has been averaged from three newspapers from Italy. The deinking method used for these tests was based on flotation. As illustrated in Figure 12.1, flexo printed newspapers are poorly deinked using flotation and the same deinking conditions used for letterpress or offset newspapers. The introduction of a small share (>10–15%) of flexo printed newspapers in the wastepaper stock disturbs the deinking process and makes the stock unsuitable for the production of conventional newsprint because of its low brightness [9, 10]. The pulp is discoloured by the water-based flexo inks, which produce very small ink particles and so lower brightness. In Europe, where the flotation deinking process is mostly used, flexographic printed papers pose deinking problems, and even the use of the wash deinking process does not completely resolve problems, since ink removed by washing must be removed from the wash water before this can be successfully reused. Flexo ink particles are mainly between 0.5 and 2 μm, whereas the optimum particle size for efficient flotation deinking is between 5 and 60 μm. This size range is ideal for removal by washing, but particles are so numerous, it is very difficult to achieve total removal [9].

Some chemicals are quite successful in deinking flexographic (water-based) inks, even with the flotation process, but other inks are not removed as efficiently as under normal conditions. When wastepaper is deinked, the temperature, pH and the chemical conditions have to be fixed and these conditions cannot be perfect for all wastepaper grades. A solution would be selective collection and processing of the different wastepaper grades, which is unrealistic.

Two options are possible to solve this problem - the development of a new deinking process which is able to process both conventional and water-based inks, or to develop water-based inks deinkable with the conventional flotation deinking method. New several stage deinking processes have been developed, for example, a process developed by the Centre Technique du Papier in Grenoble (France) [11]. In this, a non-alkaline first flotation stage is used to eliminate the flexo inks and a second alkaline stage used to eliminate the conventional inks. With this process, the results obtained with water-based inks are comparable with those obtained with conventional inks. Some ink manufacturers are now working to optimise water-based ink formulations to be deinkable with conventional deinking methods and chemicals. Some of these manufacturers have already developed such inks and are making full-scale printing and deinking tests [12]. It is probable that these inks will be available in the marketplace within a few years.

In conclusion, a number of problems are still created by the use of water-based flexo inks for the deinking of printed products. Even if new processes and new inks are successfully developed, some new problems may appear, especially because of the effect of ageing on the printed product [4, 9].

It is necessary to be aware of potential deinking problems in the development of these inks.

12.5.4 Rotogravure

12.5.4.1 Introduction to the gravure printing process. With this printing process, the image is etched or mechanically engraved into the surface of the printing cylinder which is partly submerged in ink and the excess ink is removed with a blade, leaving ink in the recessed image. The ink is transferred from the small cells of the printing cylinder by placing the paper in direct contact with it under pressure.

A high percentage of illustrated weekly and monthly magazines are printed by gravure as are some mail order catalogues and types of packaging. Generally speaking, the gravure process is used when long runs have to be printed since for long runs, it is economically better to use gravure than offset.

Today, toluene is mostly used as a solvent for gravure inks, though with packaging, alcohol-based inks can be used. Solvent represents approximately 50% of the ink weight. Gravure inks are considerably lower in viscosity compared to offset inks. After printing, the ink transferred to the paper dries mostly by evaporation of the solvent but also by absorption into the paper. Heated air dryers are the usual means of increasing the rate of evaporation of the solvent which is usually around 80%. Devices for solvent recovery have to be installed to limit emissions into the atmosphere.

12.5.4.2 Deinking of gravure printed products. Results obtained by PTS in Munich are illustrated in Figure 12.3, which illustrates average deinking coefficients for magazines and catalogues, printed by either offset or gravure, on uncoated and coated papers.

These deinking coefficients are averages from 38 different magazines and catalogues collected in Germany and Austria. Generally speaking, it can be seen that gravure printed products are more deinkable than offset printed products (using conventional deinking methods) with highest deinking coefficients obtained with magazines and catalogues, gravure printed on coated paper. With these examples, the influence of the paper type (coated or uncoated) is also shown - higher deinking coefficients are obtained with coated papers.

A lot of research work is being conducted to find water-based gravure inks [13] but a problem with these inks is the drying time, which is longer than conventional solvent-based inks. The energy necessary for drying water-based gravure inks is six times higher than the energy needed for conventional inks. Currently, these inks are used on lower speed presses and give lower quality, especially in colour printing. However, new water-based

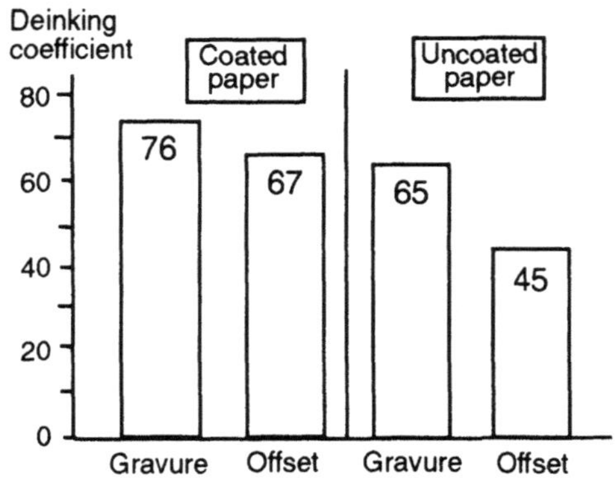

Figure 12.3 Deinkability of different magazines and catalogues from Germany and Austria (source: PTS, Munich).

inks are increasingly available on the market and, with emphasis on environmental awareness, it can be forecast that this type of ink will be further developed. To deink these water-based gravure inks, the paper maker will be faced by the same problems as for water-based flexo inks.

12.5.5 Non-impact printing

12.5.5.1 Presentation of non-impact printing techniques. Non-impact printing is printing with no striking of the print substrate. There are several mechanisms by which the ink (or toner) can be deposited on the paper such as electrophotographic, ink jet and electrostatic printers.

Electrophotographic is the most common process, which is used in most photocopiers and laser printers. Electrophotographic printers have a rotating drum with a photoconductive surface, which can be made of selenium or an organic compound. This surface is insulating in the dark and semi-conducting when exposed to a light of a certain wavelength. Initially the drum is charged up to a particular potential. The light source (photocopier) or a laser beam (laser printer) then writes the image onto the drum by discharging parts of the surface. This may be a positive or negative image. In the developing stage, the discharged area may attract (positive image) or repel (negative image) toner. The end result is a positive image which is transferred to paper by either cold or hot fusing. In hot fusing the toner is heated to approximately 100°C where it melts onto the paper. In cold fusing, the toner is heated to 50°C and pressure is applied.

Ink-jet is also a developing non-impact printing technique. A liquid ink is accelerated in the form of droplets which are guided by an electric field and projected onto the paper.

Thermal printing is becoming more common; an example is for fax machines. The paper is coloured in black when it is heated over a specified temperature.

12.5.5.2 Deinking of non-impact printed papers. Toner printed products are difficult to deink as, when pulped, large ink particles are produced. Some of these are too large to be removed by conventional deinking processes. When a large proportion of toner printed products is present in the wastepaper stock and deinked with conventional methods, the final sheet of recycled paper usually shows black specks. This can be overcome by the inclusion of deinking after a dispersion or kneading stage, by adjusting the quantity and/or types of deinking chemicals, or by optimising the dispersion stage after deinking, etc. Much progress has been made recently to overcome these problems, but more research is still necessary. The manufacturers of deinking equipment have developed new systems to assist in processing these paper grades.

Thermal paper for fax machines has been reported to create problems sometimes. When heated, a small quantity of this type of paper gives an overall colour to the wastepaper pulp. The colour can appear at the drying stage on the paper machine. As long as fax paper represents only a very small percentage of wastepaper, no particular problems appear.

12.5.5.3 The future of non-impact printing technologies. With the development of new technologies, it can be forecast that non-impact printing applications will grow to include on-demand short run and one-off books, and short-run colour work. They will also include other niche markets and work now handled by monochrome small offset. Increased integration with other printing and manufacturing can be anticipated. Non-impact printing output speed depends on data generation rate which is limited by computer power. Within a few years, significantly more powerful computers, available at lower costs, will greatly increase non-impact printing capabilities. Currently, offset presses are much faster. Toner systems are limited in speed by toner fusing. Ink jet printing is limited by jet velocity, ink chamber pressure, drop interaction in flight and by mechanics. The rate of development of new technologies and the price of the equipment will condition the share taken by non-impact printing in the future relative to the other printing processes.

12.6 Future development trends of different printing processes

A number of studies have been published on the future of different printing processes. It is very difficult to forecast what the relative shares of these will be in 10 years. Many unknown factors will influence the development of printing processes, with the results that different studies that have been

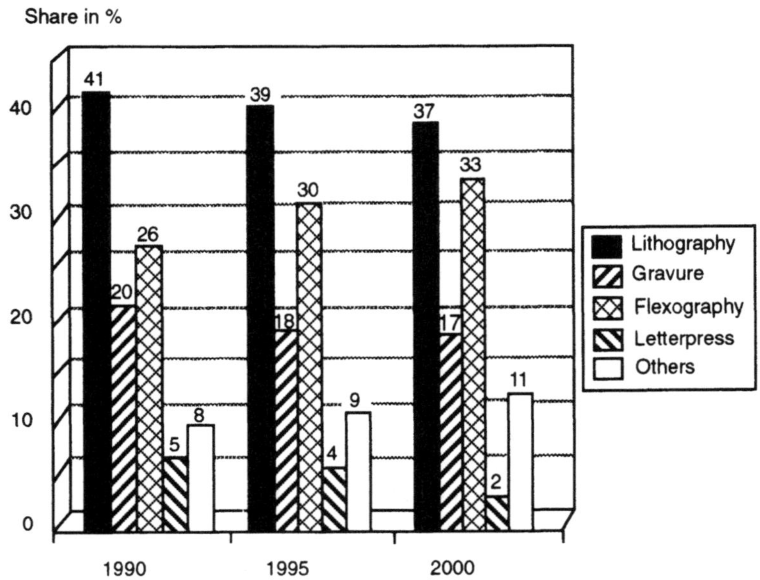

Figure 12.4 Future development of different printing processes in Europe.

published are contradictory. Two studies (one for Europe and one for America) which we think could reflect the situation in the future are given below and illustrated in Figures 12.4 and 12.5. The first of these, Figure 12.4 gives an idea of the future development trends of printing processes in Europe.

It can be seen that the two processes forecast to develop in the future are flexo and non-impact printing techniques. This trend is reinforced by the second study which concerns the United States, illustrated in Figure 12.5 [14].

Here too, it is forecast that flexography and non-impact printing techniques will develop in the future. The author forecasts a very substantial development of non-impact printing in the future.

Over a similar time period the most problematic areas of developments in printing processes that may have an impact on the recycling and deinking processes could be:

- water-based flexo inks;
- water-based gravure inks;
- non-impact printing technologies;
- to a lesser extent, vegetable oil-based inks for newspaper printing.

For the paper recycler an interest in the development of printing processes is very important because this will condition the composition of the

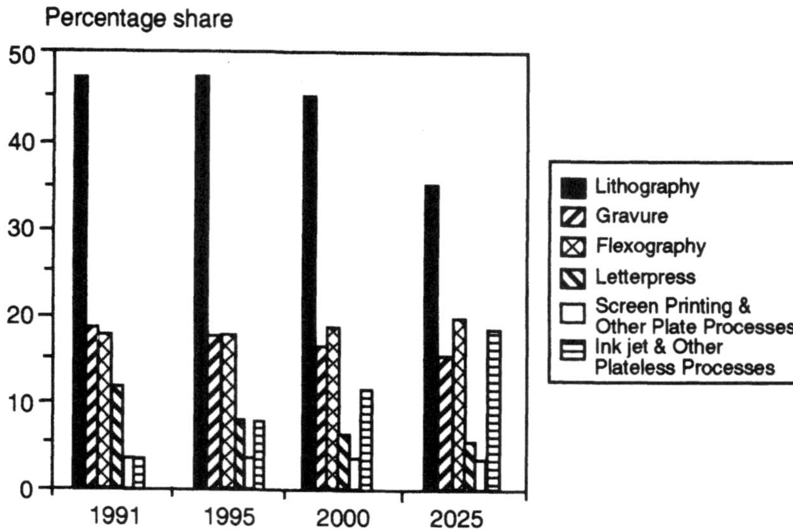

Figure 12.5 Future development of different printing processes in the United States [14].

wastepaper stock and the future possible problems for the recycling process. Other factors, such as the binding and finishing methods used, can also play a role; for example, the type and quantity of glues or the products used for certain inserts in magazines have been reported to create problems during recycling.

However, the evolution of the recycling technology and the modernisation of installations will allow problems to be minimised. Nowadays, the quality of recycled papers produced is constantly increasing. Many parties are involved in the recycling cycle: the papermaker, the printer, the inkmaker, the deinking chemicals manufacturer, the deinking plant manufacturer etc. The key to success in avoiding all the problems is cooperation between these different parties.

References

1. Neue Forschungsergebnisse zur Druckfarbenentfernung durch Flotation (New Research Findings on Flotation Deinking). Paper presented by Dr. E. Hanecker at the 5th Deinking Symposium, Munich, May 6, 1992. *Proceedings of the PTS-Deinking-Symposium 1992*. Deutscher Fachverlag GmbH, Frankfurt am Main.
2. Anon (1992) Druckfarben sind keine Belastung für die Umwelt, *Druckwelt*, **14**, 33–36.
3. Kübler, R. (1988) Deinking aus der Sicht der Druckfarbe, *Wochenblatt für Papierfabrikation*, Nr **10**, 386–392.
4. Göttsching, L., Putz, H.-J. and Renner, K. (April 1992) Bericht über das

Alterungsverhalten verschiedener Offset - Zeitungsdrücke im Auftrag von IFRA. Institut für Papierfabrikation Technische Hochschule, Darmstadt.
5. Tuovinen, J. (1992) Welche Einflüße haben Druckfarbenkomponenten auf das Deinkenergebnis? *PTS/PTI Deinking Symposium*, München.
6. PTS - Method PTS-RH 010/87. *Testing of Wastepaper. Identification of the flotation deinkability of printed wastepaper* (January 1987).
7. PTS - Method PTS-RH 010/88. *Testing of Wastepaper. Identification of the wash deinkability of printed wastepaper* (August 1988).
8. IFRA special report, *Deinkability of Flexo Printed Newspapers*, IFRA, Darmstadt (October 1987).
9. Cathie, K., McKinney, R.W.J. and Biddlecombe, J. (August 1988) *Deinking of Flexoprinted Newspapers*, Progress Report, Pira, Leatherhead.
10. Thoyer, B. (1992) Newsprint from paper with water-based flexo inks - a survey, *Newspaper Techniques,* **7**, 50–53.
11. Galland, G. and Vernac, Y. (1991) Deinking of wastepaper containing water-based flexoprinted newsprint, *1st Research Forum on Recycling*, CPPA.
12. Hornfeck, K., Liphard, M. and Schreck, B. (1991) Interfacial Studies and Application Tests on the Flotation of Water-Based Printing Inks, Henkel Publication, Ref. No. 27, 90–97.
13. Williams, C. (24 July 1991) Suppliers get an inkling, *Lithoweek*, 20–21.
14. Bruno, M.H. (1991) *Status of Printing in the USA - 1991 Update*, IFRA, Darmstadt.

Bibliography

Putz, H.J., Török, I. and Göttsching, L. (1988) *Deinking von Holzhaltigem Altpapier*, Institut für Papierfabrikation Darmstadt.

Inca-Fiej Research Association (IFRA) (1982–1991) *Newsprint and Newsink Guide.*

Thoyer, B. (May 1991) IFRA special report 1.6, *Paper Recycling and Deinking - What is their Importance for the Newspaper Industry?*

McKinney, R., Cathie, K., Staves, J. and Biddlecombe, J. (February 1989) Deinking efficiency improvement, *PIRA*.

Sirost, J.C. (Mai 1990) L'encre et le vernis face au désencrage, *Lorilleux International.*

Williams, C. (24 July 1991) Suppliers get an inkling, *Lithoweek*, 20–21.

Norgate, C.D. (1990) Recent developments in ink for newspaper printing, *Professional Printer* **34** (5), 8–11.

Sirost, J.C. (1992) L'encre d'imprimerie, *Nouvelles Graphiques*, No. 6–12.

Frank, E. (August 1990) Einfluß verschiedener Druckfarbenzusammensetzungen auf die Deinkbarkeit, *Coating 8/90*, 315–318.

Schneider, J. (1990) Non-impact printing: Technologie-Entwicklungen, Einsatzbereiche, Markttrends, *Deutscher Drucker* Nr. 4, 1–2.

White, W. (1988) Non-impact printing: a budding technology, *American Printer*, 10.

Guiroy, M. (1988) Impression sans impact, *Nouvelles Graphiques,* **3/88**, 18–16.

13 Environmental impacts of paper recycling

R.W.J. McKINNEY

13.1 Introduction

Recycling paper has both negative and positive environmental impacts, but is environmentally efficient provided that environmental costs do not exceed benefits. Environmental costs arise from the use of resources in wastepaper sorting, collection, transport and reprocessing, the production of emissions, the possible concentration of potentially hazardous materials in sludges, and the disposal of sludges and other residues arising from the recycling process. Benefits from recycling include the extension of yield from forests and the avoidance of disposal of paper products and related costs, such as methane and leachate production from landfills, etc.

This environmental model is equivalent to a conventional economic model in which low-cost producers are successful. Up to the late 1980s, economics drove up recycling levels. In each country where recycling developed for economic reasons, costs and product prices favoured locally produced products, made from wastepaper, over imported equivalents, usually produced from wood. Other than setting import quotas and tariffs, the role of governments in developing recycling was limited to economic assistance, for example, the provision of grants and soft loans, and in this recycling was not usually favoured above other industrial activities. However, since the late 1980s, governments have adopted more environmentally interventionist policies, for example, in arbitrarily setting targets for recycling levels or in the case of paper, recycled-fibre content - in the USA, Germany, etc. Generally, these targets ignore market forces, as well as the concept that there is an environmentally optimal level of recycling, beyond which there are net environmental costs, rather than benefits. Environmentally optimal levels of recycling are dependent on geographic factors such as wastepaper availability, environmental costs of alternatives to recycling, etc. Recycling has thus become a political issue and has lost the certainties imposed by market forces, that is, when recycling was economically efficient it increased, and it continued, provided it remained economically viable. On the other hand, politically inspired recycling carries no certainties and if political patronage is withdrawn, it is unlikely to survive if economics are against continued recycling.

Other problems associated with setting arbitrary recycling and recycled-fibre content levels include:

- they can restrict imports and so serve as an invisible trade barrier;
- where recovery targets are set these can lead to high levels of exports, so that a solid waste disposal problem is exported to economically weaker countries, rather than being solved at source. The prime example of this is the German Packaging Ordinance;
- national environmental effects are considered, rather than global pollution loads;
- environmental and public health cost benefit relationships are ignored; for example, in terms of public health, diseases from pets are a much greater public health hazard than pollution from the paper industry. In terms of global environmental health, more substantial benefits would be gained by targeting pollution control assistance to developing countries rather than introducing more stringent controls on emissions in developed countries.

In introducing recycling and other environment related targets, many governments have reacted to the successful publicity campaigns of pressure groups, which is not necessarily the best way to develop government strategies. Another, more rational approach to set public health and environmental priorities would include:

- an examination of national public health and environmental issues leading to public health and environmental priorities;
- an examination of these in the context of global environmental and public health issues;
- development of a programme, reviewed at intervals as new data is generated, which provides, at least cost, the maximum beneficial impact, which could be via national and/or international projects;
- a review of company taxation systems to provide incentives to investments meeting targets set by the adopted programme, national or international. In the case of the paper industry, for example, every dollar spent on environmental projects in the former USSR would provide a better environmental return than many dollars spent in developed countries.

Weightings which allow the interpretation of related and unrelated environmental effects also need to be developed; for example, the significance of the generation of greenhouse gases compared to chemical oxygen demand (COD) discharges, and would be based on priorities adopted. If, for example, limits on greenhouse gas emissions had a higher priority than COD discharges, then the weighting system would ensure more resources were given to greenhouse gas emission control.

Since global resources (capital and expertise) essential to limit and reverse environmental problems and improve public health are limited, it is necessary to set priorities and establish goals to ensure resources are consumed in a fashion which brings maximal benefits. Should the West, for

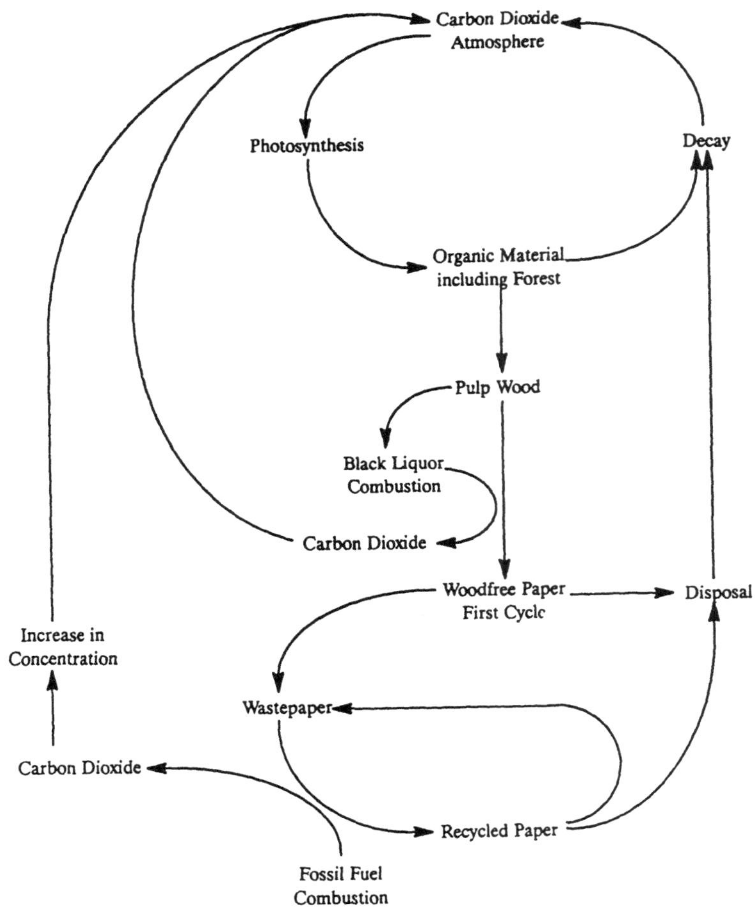

Figure 13.1 The carbon dioxide cycle and woodfree paper production.

example, reduce COD discharges, whilst continuing to ignore the transmission of water borne diseases in the third world? Rational analysis has not occurred in any country which has set recycling targets and current policies will inevitably result in the inefficient use of resources, both funds (raised via taxes) and time.

Life cycle analysis has frequently been presented as a means by which environmentally preferable options can be selected. This is an illusion. There is no mechanism by which different environmental effects can be compared, for example, a comparison between paper recycling and virgin paper production. Recycling results in a contribution to greenhouse gases from the generation of carbon dioxide from fossil fuels, whereas virgin paper production is largely independent of fossil fuel use and so, when operated with forest renewal, is not a net contributor to increased carbon dioxide concentrations and hence to the greenhouse effect (see Figure 13.1). If the greenhouse effect were considered to be the greatest environmental concern facing the world, then virgin paper production, combined with incineration of wastepaper, would probably be the preferred environmental option, despite other environmental differences. Since various environmental challenges have not been assigned weightings, life cycle analysis has become another tool for proving the views of those undertaking the analysis. In other words, until internationally accepted environmental weightings are established, environmental analysis is open to abuse, and is entirely dependent on the assumptions made by those making the analysis. At best, life cycle analysis provides comparative data when similar environmental effects are being compared for specific cases, but with no methodology to interpret the data when different effects are involved. In general, international comparisons are even less practical, since data varies from country to country, as well as case to case.

This chapter makes no judgements on the relative merits of virgin or recycled paper production. Recycling paper is totally dependent on the virgin paper industry. It is thus not realistic to isolate the environmental impacts of virgin paper production from those of recycling paper. A healthy recycling industry depends on the availability of wastepaper, with a substantial input of first cycle paper. If wastepaper is not readily available, then at some point the economic and environmental costs of collection outweigh advantages gained from recycling. Background information is given so that the impacts arising from recycled paper production can be placed in the context of environmental challenges.

13.1.1 Public health hazards

During the last century there has been a remarkable improvement in the health and life expectancy of populations in developed countries. This is primarily due to a recognition of the role pollution pays in public health, following, for example, in the UK, reports from several Royal Commission investigations into pollution. The first Commission was appointed in 1898 and a series of reports on sewage disposal, published in the period 1901 to 1915, firmly established the link between pollution from domestic sewage and the spread of water borne diseases, such as typhoid, cholera and dysentery [1]. This led to both sewage treatment and the treatment by chlorination of drinking water supplies. The effects of these measures are illustrated by the reduction in death rates, given in Table 13.1. In view of the campaign against the industrial use of chlorine, it is ironic to consider that the gains illustrated in Table 13.1 were achieved by the use of chlorine, in effect, due to the population of the UK drinking dilute solutions of chlorine. Over the same period, average life expectancy increased from about 53 to 74 years, an astonishing increase.

Table 13.1 UK death rates from enteric fever

Deaths per million of population	Year
371	1871-1875
47	1914
0.3	1955
0.05	1991

Even though there was considerable pollution from industrial sources, it was not regarded as a major threat to public health and so it was not until much later that restrictions began to be imposed on industrial effluents, discharged to inland waterways, and it was not until 1951 that this was enshrined in law [2]. Also in the UK, the London fogs of the early 1950s were shown to result in thousands of deaths, due to air pollution and in 1956, the Clean Air Act [3] was the first legislation arising from the recognition of the effect of air pollution on public health.

Looking towards a second century of pollution control, it is unlikely that the same remarkable results as those of the first will be achieved. Legislation which sets increasingly more stringent standards yields progressively smaller improvements in public health. Possibly as a consequence of this, in developed countries, public health has ceased to be the prime motivation for change. Instead, the more open and difficult to define concept of 'the environment' has replaced public health as an indicator of pollution. On some occasions the terms 'the environment' and public health are used as if they are related, but this is rarely the case. One of the benefits of being able to assess the effects of pollution on public health is that this is measurable and quantifiable, whereas effects on the environment is not. Hence, when discussing the effects of pollution on the environment discussions are not normally based on statistically valid scientific information, which leads to widely differing interpretations of available information.

It should also be noted that a majority of the world's population have not benefited from what in developed countries is regarded as basic public health provisions. Safe effluent disposal practices and clean drinking water are not available and millions die each year from drinking polluted water, so that the death rate from enteric fever given in Table 13.1, for 1871-1875, is still typical of most the earth's population.

13.1.2 Contribution of the paper industry to pollution

During the 17th and 18th centuries, a paper mill was as unwelcome a neighbour as a nuclear power plant is today. Rags, which were the raw material, frequently carried fleas, which in turn sometimes carried the bacillus responsible for the bubonic plague, the Black Death. Although the presence of the bacillus was not known, the higher incidence of bubonic plague outbreaks around paper mills resulted in their poor reputation.

Despite much adverse publicity which suggests that present day pulp or paper mills carry similar risks, the risk to public health from a pulp or paper mill is so low that it is insignificant. Even when chlorine gas is used in lignin extraction during the production of bleached chemical pulp, full flow effluent treatment, using biological treatment, eliminates any recognisable risk to public health. There is no scientific evidence which would support a pulp or paper mill being classified as a public health hazard. Despite this, militant groups have picketed pulp and paper mills, whilst ignoring serious public health hazards. In the UK, for example, on average, more than 100 children a year either are totally blinded or have their sight seriously impaired from toxocariasis, caused by *Toxocara canis*, a parasitic round worm sometimes found in dog faeces. The public health risk from a pulp or paper mill is probably less than that of one dog infected with *Toxocara canis*, and there are about 7.5 million dogs in the UK alone, most of which are infected at some time.

Pulp and paper production is however, not without an environmental impact. Environmental impacts arise from air emissions, solid wastes and liquid effluents. A majority of pulp and paper mills in developed countries have restrictions imposed on emissions, especially the quality of water they can discharge and so most have full effluent treatment. However, in developing countries and in eastern Europe, many mills discharge untreated effluent. Those mills which discharge untreated effluents have the greatest environmental impact, discussed in chapter 7. The installation of recovery boilers (when technically feasible) with emission controls and full effluent treatment at all pulp and paper mills would have a much greater effect in reducing the global pollution load from the pulp and paper industry than more stringent standards in developed countries. The imposition of very restrictive discharge standards, on air or water emissions from pulp and paper mills in the developed world, will result in relatively minor reductions in the global pollution load compared with what would be obtained if all mills had up-to-date emission control, including effluent treatment.

When safeguards are applied, the pulp and paper industry poses no serious environmental hazard, in other words, there is little or no environmental degradation arising from the use of paper products produced in most of the world's large pulp and paper mills. Smaller pulp and paper mills, primarily located in developing countries, may have little or no control on their emissions and these can have a considerable negative impact on their local environment, predominantly via water pollution, but also through air pollution. Environmental groups and national governments concerned with the impact of paper use on the environment should promote policies to encourage assistance to this type of mill, which would have a much greater impact in reducing pollution from paper use than applying more stringent discharge conditions to mills already with strict controls on emissions. In developing countries, no alternatives to paper use exist; paper is essential to transmit knowledge in education, in journals, etc. The availability of paper enables the ready transmission of political philosophies, which contribute to the liberalisation of society so that the availability and use of paper is a cornerstone of a free, democratic society. Education and political pluralism would be impossible without communication papers. Better food packaging in developing countries would help to reduce their high levels of food waste. Hence increased paper use in developing countries would assist appreciably in improving public health and living standards and assistance should be to achieve these goals, whilst limiting any adverse environmental effects arising from existing or increased paper production.

13.2 Relationships between paper recycling and current environmental issues

Set against the age of the earth, *Homo sapiens* is a very new newcomer. Despite this we have had a marked impact on the world, which with respect to our continued presence on earth, has largely been a negative impact. However, it is supremely arrogant to believe that if the earth were changed to the extent that it would not support human life, it will die - this is simply not the case. Life on earth has survived many cataclysms, including asteroid collisions, extreme climate changes (ice ages) and the drowning of continents. The diversity of life is too wide for all forms of life to become extinct through our activities, even though it appears possible, for a variety of reasons, that there may be a time limit on human dominance of the earth. Other life forms would replace humans, much as dinosaurs were replaced millions of years ago.

However, there is a widespread recognition that our activities are having profound effects on the earth's ecosystems, though the outcomes from these changes are not fully understood. Differing interpretations of changes can result in considerable controversy; for example, many groups suggest that the buildup of greenhouse gases, such as carbon dioxide, has already resulted in climatic changes, but some researchers suggest there has not been any change due to an increased greenhouse effect. Unfortunately, in many cases, science gives way to ideology.

Paper production, virgin and recycled, contributes to changes in ecosystems, and it is clear that emissions must be controlled, for example, to eliminate any acute toxicity to aquatic life forms from pollution, and in almost all developed countries effective controls have been implemented for many years. Rather than compare recycled paper production with virgin paper production, some of the pressing global concerns are reviewed below and the effects of paper recycling briefly reviewed. Despite many comparisons between virgin and recycled paper production, the symbiotic relationship which exists between these make this type of comparison meaningless. It is similar to a comparison between different environmental effects of males and females. Recycled paper production could not exist without virgin paper production and if virgin paper was not recycled, the demand for wood would be such that paper would become much more costly. In addition, there are differing environmental effects from the production of the two types of paper and in the absence of science-based weightings given to these different impacts, it is not possible to say which is preferable. Life cycle analyses which attempt to do this have no credibility.

13.2.1 Ozone depletion

At the edge of the earth's atmosphere, there is a thin layer of ozone gas, the stratospheric ozone layer. A large hole was discovered in this layer over the South Pole in 1987 and significant ozone depletion found over the North Pole in 1989. Depletion of the ozone layer could lead to increases in skin cancer, suppression of the immune system, lower crop yields, etc. through increased solar radiation. Stratospheric ozone destruction is complex, but refrigerants such as chlorofluorocarbons (CFCs) and halon are believed to be responsible.

In 1983 the earliest moves to control the use of CFCs as aerosol propellants were made by Sweden, Finland, Norway and the USA, which led to the Montreal Protocol in 1987, signed by 50 countries. In 1992, the total had increased to 93 countries, agreeing to phase out some of the chemicals responsible for ozone layer destruction [4]. Consumer campaigns, spearheaded by environmental groups, were effective in convincing retailers not to stock aerosols containing CFCs, which led to many manufacturers phasing out CFC aerosols considerably in advance of internationally agreed dates. This was one of the first indications of the power consumers exercise through selective purchases.

There are only minor steps that recycled paper producers can take to limit the effect of ozone depletion, since the use of ozone depleting chemicals is small. Steps include:

- use of alternatives to halon in fire extinguishers and the controlled disposal of halon-containing extinguishers;
- substitution of ozone depleting solvents – (CFC 1,1,3 trichlorotrifluoroethane) and methyl chloroform (1,1,1 trichloroethane) with alternatives;
- use of CFC alternatives in air conditioning and cooling systems, with controlled disposal of CFC refrigerants;
- use of alternative insulation in buildings to CFC blown foam, and controlled disposal of existing foam during building renovations.

13.2.2 Global warming

The effects of global warming would be felt across the globe, through changes in weather patterns and increased sea levels in many regions. An increase of only 1–2°C would be sufficient to have a significant effect. Some researchers predict that by 2030 the earth's average temperature could increase by up to 5°C. This is contested by others, who are not convinced that any increase will occur, with still more others predicting increases of 1–2°C. Carbon dioxide levels have been higher than they are currently; for example, 130 000 years ago the carbon dioxide concentration in the atmosphere was more than 10% higher than today. A reduction in concentration of about 15% coincided with the last ice age, which ended about 100 000 years ago [5]. However, cause and effect are not known. Even though the effects of increased global temperatures would be widespread, life on earth is not threatened, but without the greenhouse effect the earth would be frozen and lifeless, similar to Mars; it is essential to warm the earth's climate.

A variety of gases contribute to the greenhouse effect but the most important is carbon dioxide, produced from burning fossil fuel and biomass. Greenhouse gases are listed in Table 13.2.

Paper recycling is dependent on purchased electricity and so on fossil fuel combustion, hence recycling paper makes a contribution to increased carbon dioxide concentrations, with the quantity dependent on the blend of fuels used in a specific geographic region to generate electricity. Emissions of sulphur dioxide and oxides of nitrogen (NO_x) may also occur when electricity is generated, dependent on the stringency of emission controls, the type of fuel and boiler, etc.

Collection and transport of wastepapers also makes some contribution to greenhouse gases, via vehicle emissions. Normally this would be relatively minor when compared to fossil fuel combustion. Emissions vary for each mill, dependent on wastepaper collection systems and transport.

Since the paper industry as a whole is heavily involved in forestry, with most companies planting more trees than are felled to make paper, and since forests convert carbon dioxide into biomass, the net contribution to carbon dioxide buildup by paper use is probably small. With high levels of tree replanting and the slow release of carbon dioxide stored in paper buried in landfills, paper use probably results in a reduction in carbon levels despite the release of carbon dioxide from fossil fuels with recycling. Ways to reduce carbon dioxide emissions from recycling include:

- increased energy efficiency, for example, through a combined heat and power system;
- a switch to a renewable fuel;
- sludge incineration, with energy recovery, rather than landfill, which could produce methane, though methane could be collected and used as a fuel;

Table 13.2 Greenhouse gases and their sources [5]

Gas	Global increase (% per year)	Relative strength	Contribution to greenhouse effect (%) (1980s)	Life (years)	Sources
Carbon dioxide	0.5	1	50	7	Burning of fossil fuels and biomass (forests, etc.)
Methane	1.0	30	18	10	Anaerobic bacterial breakdown of biomass, natural gas
CFCs and halons	6.0	20 000	14	110	Coolants, solvents, fire extinguishers
Nitrous oxide	0.4	150	6	170	Vehicle emissions, fossil fuel and biomass burning
Tropospheric ozone	2.0	2000	12	Renewed	Photochemical oxidation of vehicle emissions, smog

- anaerobic effluent treatment, rather than aerobic alone, with collection and use of methane gas as a fuel. Normally anaerobic and aerobic need to be combined for adequate treatment, but anaerobic is not always technically feasible.

13.2.3 Forest depletion

A frequently quoted environmental benefit of increased recycling is that this saves trees. There is some truth in this, but a substantial proportion of virgin paper is produced from wood residues (from saw mills) and forest thinnings, so recycling paper does not always save trees. In addition, most wood used for pulp and paper production is from commercial forests, so recycling paper has a very small impact on saving virgin forest resources. Nevertheless, paper recycling lessens the pressure on forest resources and without the existing level of recycling, forest resources in Western Europe would be stretched to maintain the level of paper and board consumption currently enjoyed. In developing countries in which wood is used extensively as a domestic fuel, forest resources could not provide papers needed for education, packaging, etc. so recycling is essential in these areas.

Recycling has a beneficial economic benefit, in that it extends the ability of an area to provide paper products, without an equivalent increase in the demand for wood. Hence the economic effect of recycling is to keep paper product prices down, since without recycling the increased demand for wood would result in higher prices.

A detailed study of the effects of recycling included an examination of the impact on wood balances in Western Europe [6]. In this study three theoretical cases were evaluated, which were:

1. A maximum recycling scenario for all paper and board qualities. In this, 90% of available wastepaper was collected with an average 56% used across all paper and board grades.
2. A selective recycling scenario, based on the use of wastepaper collected but without a requirement for long-distance transport, with an overall utilisation rate of 35% - similar to the rate in 1993.
3. A zero recycling scenario, with maximum energy recovery from fibres through the use of wastepaper as a fuel and 100% wastepaper disposal by incineration, with energy recovery.

The study indicated that moving from selective recycling (35% overall utilisation rate) to a maximum recycling (56% utilisation rate) results in the consumption of about 9% less wood. However, as detailed in the analysis, this would not result in 9% less tree felling, but would lead to an increase in the waste of wood resources, since wood residues and thinnings would be left to rot in the forest, or burned.

Inter-relationships between economic and environmental results are complex. Since the pulp and paper industry provides a substantial income for commercial forestry operations, a reduction in this cash flow would have significant results on forestry operations. Wood prices would have to increase to cover all the costs of commercial forestry, leading to substantial increases in the price of wood product - construction timber, furniture, fencing, etc. Resources for forest silviculture would probably fall, leading to increased degradation of forest resources [6].

That this outcome is probable is illustrated by changes in forest management in the UK. Government forestry policies were altered at the end of the 1980s, which led to very substantially reduced planting rates in the UK. A low new planting rate resulted in reduced income for forestry management and related companies, many of whom went bankrupt or withdrew from forestry. One forestry management company with an exemplary record in the management of forest ecosystems withdrew [7] and successor companies devote less resources to the overall management of forest ecosystems. This has already resulted in some degeneration in affected forests in the UK.

13.2.4 Utilisation of non-renewable resources

Non-renewable resources are materials such as oil, coal, metal ores, etc. In the case of paper recycling, the most important are fuels, since recycling and paper making are energy intensive. In most recycling mills, energy is supplied as purchased electricity with steam generated at the mill. Hence recycling involves the use of non-renewable resources to generate electricity and as a boiler fuel. Since electricity is also generated by nuclear and hydroelectric plants the actual use of non-renewable resources, fossil fuels, for energy production will vary from region to region, dependent on the fuels used within a specific region to generate power.

In a few cases, wastepaper surpluses and processing residues are used as fuels, to provide steam and generate electricity. This reduces the dependence of recycling on fossil fuel use. Nevertheless, increased recycling increases fossil fuel use; in one study it was estimated that total fossil fuel use would increase by about 7%, if wastepaper use increased from 35 to 56% in Western Europe [6]. Authors of this study assumed that with paper recycling there is a substantial energy saving, which is not always correct; for example, a comparison between wastepaper-based test liner and virgin pulp-based kraft liner showed energy consumption was the same [8], so that at equivalent performance levels, recycling actually increases energy consumption. Even with this incorrect assumption, the authors showed fossil fuel use increased with increased recycling, though the consumption of renewable resources (wood) falls [6].

Another comparison suggested there are considerable energy savings when producing test liner rather than kraft liner, which illustrates the care necessary when making energy consumption comparisons, since the variation from mill to mill is very large. With lightweight coated and uncoated free sheet papers this comparison suggested there are very small energy savings in making their recycled equivalents. In the case of newsprint, a substantial energy saving was indicated [9].

Since the use of non-renewable resources is linked to energy use and global warming, ways to limit the use of non-renewable resources are similar to those given previously, to reduce emissions contributing to global warming.

13.2.5 Acidification

Acidification is primarily due to oxides of nitrogen and sulphur discharged as airborne pollutants, from combustion, predominantly combustion of fossil fuels. A minor secondary cause, in some conifer forests, is the effect of acidification from the decay of conifer needles. Although these gases have been discharged for more than a century in industrialised countries, problems have been recognised only relatively recently - initially public health problems, followed by environmental. One study estimated the total deposition of sulphur over the period 1880–1991, and in the most affected areas of Europe, it exceeds 6000 kg of sulphur per hectare [10].

Air pollution was recognised as a problem many years ago; in 1273 a law was passed in England to try and control atmospheric pollution. However, it was not until after the second world war that air pollution was finally connected with public health. From Friday 5th December to the morning of Tuesday 9th December, 1952, London was enclosed in still, cold air. The result was a dense fog in which air pollution reached an unusually high level. It was later estimated that this was responsible for the deaths of 3500 to 4000 people, either during the fog, or soon afterwards [11]. Although this was not the first episode which connected increased mortality rate and air pollution, its severity led directly to the Clean Air Act of 1956 [3], which controlled smoke (particulates) but not sulphur dioxide emissions.

Problems due to sulphur dioxide and oxides of nitrogen became obvious with the discovery in the 1960s of acidified lakes and streams and forest damage, initially in Scandinavia, but later shown to be widespread across industrialised nations. Problems are most acute in soils which have a low buffer capacity, so that pH falls more rapidly.

Sources of sulphur dioxide are primarily from the combustion of coal and oil, whilst the largest source of nitrogen oxides is road traffic. Since the 1970s, when some countries introduced controls on sulphur emissions, sulphur dioxide levels have been falling, but the emission of nitrogen oxides is still growing, due to increased road traffic.

Since recycling paper is energy intensive and much of the energy used is produced by combustion of fossil fuels, in common with all industries recycling makes a contribution to emissions of sulphur dioxide and nitrogen oxides, through fossil fuel combustion at mill sites, electricity production and from transportation of wastepaper. Contributions vary, according to type of fuel used, distances wastepaper is transported, wastepaper collection systems, emission controls, etc. In countries such as Sweden, which has strict controls on sulphur emissions, recycling produces less sulphur emission than in the UK which has fewer controls.

Wastepaper is a dispersed raw material and collection and transport inevitably result in emissions of nitrogen oxides. In Japan and the USA, where vehicle exhausts have catalytic converters fitted, transport associated with recycling produces less nitrogen oxides than in most of Europe, where standards are lower. Progress on the control of nitrogen oxides is much slower than sulphur emissions. These can be minimised by:

- using vehicles fitted with catalytic converters;
- organisation of an efficient wastepaper collection system and the avoidance of special vehicular journeys to deposit small quantities of wastepaper;
- where possible, the use of rail, rather than road transport;
- construction of recycling mills close to wastepaper sources, to avoid long transport distances;
- switch to cleaner, renewable fuels such as ethanol, methanol, or natural gas, etc.

Collection of wastepaper by door to door services is more efficient than the use of paper banks, unless individual journeys by car to a paper bank are combined with other reasons for making the journey. A round trip of 10 km to deliver 10 kg of newsprint would consume about 30 MJ of petrol (dependent on the car size, traffic conditions, speed, etc.) equivalent to 3000 MJ t^{-1}. In the UK, the largest door to door collection system, in Milton Keynes, averages about 130 MJ t^{-1} of wastepaper collected [12].

Pre-consumer wastepaper collection is the most efficient collection system, since large quantities of wastepaper are collected from each site. Collection of post-consumer wastes, such as packaging, from supermarket sites is also efficient. It is clear than in term of emissions of nitrogen oxides, the collection of relatively large sources is more preferable than collection from dispersed sources. In urban areas with low air quality it is essential to develop efficient methods for the collection of dispersed post-consumer wastepapers, to avoid contributing to lower air quality. In the design of new buildings, especially office complexes, more attention needs to be given to separation, storage and collection of recyclable materials.

Other means which recycling mills can adopt to reduce sulphur and nitrogen oxide emissions include:

- use of low sulphur fuels;
- use of low NO_x burners;
- flue gas cleaning - to remove particulates; for desulphurisation, denitrification;
- more efficient energy use.

Acidification is linked very strongly to air pollution, other aspects of which are covered in the next section.

13.2.6 Water and air emissions

Air emissions are more difficult to control than emissions to water, and air pollution is due partially to actual emissions and partly to interactions between them and sunlight. As highlighted above, most of the contributions to air pollution are created from energy use. Emissions other than sulphur dioxide and nitrogen oxides include carbon dioxide (see global warming, section 13.2.2) and volatile organic compounds (VOCs).

In recycling, chemicals used can contribute to emissions, especially of VOCs. In bleaching, if high levels of hypochlorite are used, with a mixture of woodfree and wood-containing papers, chloroform may be formed, most of which will escape to the atmosphere. Other VOCs are released from recycling mills if solvents are used in deinking formulations or to clean stickies from machine clothing, etc. These releases are small, but could be significant if they contribute to the development of a photochemical smog. Various photochemical oxidants (including ozone) can be formed from nitrogen oxides (from vehicle emissions) and hydrocarbons, in the presence of sunlight, leading to a photochemical smog, experienced in many major conurbations. A combination of airborne pollutants, such as ozone, nitrogen oxides and sulphur dioxide, is more potent than these components individually, so that environmental damage will result at lower levels when all are present than would occur if each were acting separately. Ozone is being introduced as a bleaching agent, but any escape of ozone could have a substantial impact so that great care will have to be associated with ozone use.

A variety of international treaties with the objective of reducing emissions have been signed by different governments, including:

- SO_2 Protocol – reduce sulphur dioxide levels by 30% by 1993, relative to 1980 levels, to be followed by steps to reduce NO_x, to be implemented by 1996;
- VOC Protocol – the target is to reduce the episodes of high ozone concentrations, with a commitment to reduce these by 30% by 1999, using 1988 as the base year;
- NO_x Protocol – limit NO_x emissions so that after 1994, levels are not to exceed 1987 levels, to be followed by steps to reduce NO_x, to be implemented by 1996.
- CO_2 Emissions – a reduction in CO_2 emissions by 20% by the year 2005, based on 1988 levels.

Pressure to meet these targets is passed on to industry by governments either by direct command and control approaches (legislation to restrict emissions) or by indirect measures, such as ecolabelling systems or pollution taxes. Progress towards these targets, whether forced or voluntary, would form an important part of a recycling mill's environmental programme, possibly for inclusion in an annual audit.

Water emissions from paper recycling mills have been examined in detail [13–18]. Waste waters produced have high organic strengths relative to domestic sewage. However, the effluent, although aesthetically unappealing, has no major public health risks and is easy to treat, to reduce the organic strength and remove suspended solids. Provided secondary (biological) treatment is installed, water pollution from a recycling mill is minimal, with little or no long-term environmental effects on receiving waters. Untreated effluent would have a greater impact, dependent on factors such as the degree of dilution, receiving water quality, recycle mill grade, etc. Some potential impacts are discussed in chapter 7, and others are discussed below.

13.2.6.1 Dioxins. Very low quantities of dioxins and furans have been found in effluents from recycling paper mills [13–15]. In one study, concentrations found were in the range 1–5 $ng\,kg^{-1}$, with one sample of about

$20\,ng\,kg^{-1}$ in a waste-based printing and writing paper [13]. Since recycled papers with known histories had not been chlorine bleached, the presence of dioxins and furans was unexpected. The dioxin and furan 'fingerprint' was different from that found due to paper bleaching. Possible sources suggested were carbon black used in printing inks, chlorination of mill water supplies or the original paper, though this was discounted due to the different fingerprint [14].

Further work showed that of the dioxins and furans present, about 75% of the dioxin coming in with the recycled paper exits with the flotation sludge, which suggested ink as its source. Only a small percentage is present in paper and the remainder appears in the effluent, but is removed during effluent treatment, possibly in sludge.

A later study found dioxins and furans in grades which had never been chlorine bleached, such as packaging (OCC) grades. Sources explored were:

- chlorine-based bleaching;
- chlorophenol content of wood chips;
- printing ink;
- ambient environment.

The conclusion, which was based on the fingerprints of dioxins and furans, was that they were from the ambient environment. Recycled papers and final (treated) mill effluents were shown to have low concentrations in comparison with sludges [15], confirming the apparent removal of dioxins and furans during treatment and wastepaper processing.

Since dioxins and furans were shown to originate from the ambient environment, it is clear that levels in recycled papers present no increased public health or environmental risk and that concentrations found can be reduced only as ambient concentrations are reduced. Incineration of sludges, which under controlled conditions will destroy dioxins, is clearly a preferred option.

Little is known of the fate of ambient dioxins, so it is advisable to institute monitoring programmes to measure routinely the concentration when sludge is applied to land. Since levels in deinking mill sludge are likely to be similar to other sludges, for example, from domestic sewage works, which have been applied to land for many years, problems are unlikely.

13.2.6.2 Polychlorinated biphenyls (PCBs). Contamination of recycling mill effluents with polychlorinated biphenyls, which are carcinogens, occurred in the 1960s and 1970s. They were used in the production of carbonless copy papers and in some inks, but when their carcinogenic properties became known, their use was discontinued, for example, in the USA in 1971. They could still enter a mill effluent system if very old wastepapers were used, but this is very unusual and levels are very low. Another possible contamination route is from electrical transformers, but normally the

Table 13.3 Concentration of PCBs in recycled papers [16]

Date	PCB concentration (ppm)
1969	1.82
1972	1.18
1975	0.23
1980	0.08

removal and disposal of PCBs from electrical transformers is rigorously controlled.

The reduction of PCBs in recycled papers is illustrated by the concentration in recycled papers, in Table 13.3 [16]. In mill effluents concentrations are usually below detection limits, approximately 0.5 parts per billion, whilst in sludges concentrations are typically <3 parts per million (ppm) on dry solids and are often below detection limits [17].

13.2.6.3 Effluent toxicity. Bioassays of deinking mill effluents show that acute toxicity is normally low. In some acute toxicity tests in 1990 on effluents from waste-based packaging and fine paper mills, without secondary biological treatment, the concentration meeting the LC50 was 80% or more of mill effluent, for both trout and daphnia (LC50 is the concentration which allows a survival rate of 50% of the organism specified, over a given period of time, usually 96 hours). Other mills which had lower LC50 concentrations reported problems due to residual chlorine from a chlorine or hypochlorite bleaching stage [18].

Bioassays on deinking mill effluents (tissue, fine paper and newsprint) following secondary treatment showed low acute toxicities; in general, LC50 values could not be estimated, due to survival rates in 100% effluent. However, when secondary biological treatment plants were not operating satisfactorily, higher acute toxicities were found [17], which emphasises the need for continuous monitoring of effluent treatment plant performance, especially when dilution of the effluent by the receiving water is low.

Seven day chronic toxicity bioassays of deinked mill effluents using *C. dubia* and fat head minnows were estimated. The chronic value (ChV) is the geometric mean of the no observed effect concentration (NOEC) and the lowest observed effect concentration (LOEC). Respectively, these are the concentrations used in the test which showed no observable effect on the survival, growth or reproduction of the test organisms and the lowest concentrations which do cause an effect. The average of the ChVs ranged from 6 to 87% when *C. dubia* was the test organism and from 17 to 100% with fat head minnows [17].

High concentrations of dissolved salts in effluents result from a policy of low water consumption and a high recycle rate. The resultant high salinity can have a negative impact on bioassay test results. Dilution of salts by receiving waters eliminates problems. This illustrates difficulties inherent in interpreting bioassay test results. Poor results can be a consequence of high salinity, or natural resins and acids released from wastepapers during the recycling process. The latter are normally removed by biological effluent treatment. One of the major advantages of a biological effluent treatment system is that it acts as an indicator of potential problems; for example, if there is residual chlorine the biological stage will be adversely affected. This provides an indication that remedial action is necessary and prevents the discharge of an effluent containing, for example, residual chlorine.

13.2.6.4 Other chemicals used in recycling. Direct controls on the emissions of some chemicals have been introduced. Slimicides and biocides are controlled; for example in Sweden only a limited range is allowed to be used, and in the USA, levels of trichlorophenols and pentachlorophenol which are contained in some slimicides must be within given limits and routinely monitored.

Indirect controls have also been imposed, for example, by the use of ecolabels. In Germany, the Blue Angel ecolabel cannot be awarded if chelating agents have been used, for example, to stabilise hydrogen peroxide. This is due to fears that excess chelating agent will remove metals necessary as micronutrients from receiving waters. Evidence for such an effect is limited. Other ecolabelling systems also apply restrictions; for example, under the Nordic White Swan criteria, all chemicals used must be biodegradable and the use of nonylphenolethoxylate surfactants is not allowed, nor is the use of carbonless copy wastepapers (in kitchen and toilet paper), optical brightening agents or chlorine-containing bleaching agents [19].

13.2.6.5 Nutrient Enrichment/Eutrophication. Macronutrient (nitrogen and phosphorus) enrichment is a problem with discharges from effluent plants treating domestic sewage. These contribute to problems from algal blooms in receiving waters, both freshwater and saline. Problems caused by algal blooms vary; in freshwater they are unsightly and can lead to oxygen depletion as algae die off, resulting in fish kills. At sea, algal blooms cause more problems; as well as being unsightly, some algae produce toxins, which result in shellfish contamination, making them unfit for human consumption. Oxygen depletion also can result in fish kills.

Effluents from recycled mills are usually deficient in both nitrogen and phosphorus and these have to be added for effective biological treatment. Added nutrients are incorporated into cellular mass and removed with the surplus micrograrnisms. Provided nutrient addition is controlled, there is no

Table 13.4 Solids losses and volume of residues from recycling

Wastepaper type and product grade	Typical losses (%)	Volume at 35% solids[a] (m^{-3}) (per tonne input)
OCC; packaging	8–15	0.2–0.5
News and magazines; news	10–15	0.3–0.5
Coated paper; tissue	25–35	0.8–1.2

[a]Assumes sludge crumb has a density of 850 kg m^{-3}

macronutrient addition to receiving waters. Ideally, recycled mill raw effluent would be treated with domestic sewage, so that no additional nutrients would be required. This would reduce the nutrient load discharged from domestic effluent treatment systems.

13.2.6.6 Solid waste disposal. Increased recycling of post-consumer wastepapers has been strongly encouraged, especially in the USA, as a method to reduce the volume of solid wastes being disposed of by landfill. Provided post-consumer wastepapers are used in addition to other waste-papers and not simply to replace them, increased recycling will result in lower volumes of solid waste. However, the actual reduction is dependent on the type of wastepaper being used, the paper produced and the treatment of residues and sludges from the recycling mill. In some cases the volume of residues produced by a recycling mill can exceed the volume of waste-paper being recycled, for example, when coated papers are used to produce tissue and simple dewatering is carried out on residues before landfill.

Typical solids losses from recycling are given in Table 13.4 and it is clear from this that when sludge is disposed of by landfill the actual volume reduction due to recycling can be small.

Sludge volumes can be reduced by increased dewatering, drying and incineration. Wet sludge is most difficult to landfill, due to handling problems. Within a landfill, wet sludge is more likely to result in gas and leachate production. Studies have shown that paper in a landfill, properly covered and capped, remains dry and hence the rate of biodegradation is extremely low [20]. With wet sludge, biodegradation will be relatively rapid. However, in properly planned and operated landfills, gas collection and leachate treatment will prevent these and gas can be used as a fuel. Problems with sludge disposal have been recognised within the industry and several beneficial uses of sludge have been developed, for example, re-use by some board mills, energy recovery, land application, use in cement production, etc.

Recycled mill sludges analysis has not revealed any specific reasons for concern. Heavy metals content is lower than in sludge from domestic sewage plants (cf. chapter 7) which has been successfully used as a soil conditioner for many years. Even when domestic sewage sludges have been applied to land for many years, for example on 'sewage farms', no specific problems have been found, even when some heavy metal concentrations are higher than recommended under UK guidelines [21]. Control of pH is important in preventing mobilisation of heavy metals.

Analyses of ash residues from incinerated sludges indicate that provided temperatures are maintained below 1050°C, particles in the incinerator do not fuse or sinter. Samples analysed were predominantly clay (as calcined clay) with smaller proportions of calcium oxide (from $CaCO_3$) and smaller amounts of magnesium oxide and titanium dioxide, from talc and rutile/anatase respectively [22]. Since some of these materials have a high value as paper mill additives, it may be economically possible to recover and reuse them.

Other solid wastes from recycling mills also require management and careful disposal. These wastes include:

- empty chemical containers, oil drums, etc;
- pallets;
- unacceptable wastepapers, wind blown litter;
- surplus supplies;
- bale wires, other metal scrap;
- office wastes;
- food residues.

Although volumes for disposal are small relative to that of sludge, they can be further reduced by reuse and recycling where possible; for example, suppliers should accept responsibility for collection of empty chemical and oil drums, with preference given to suppliers using reusable containers. Ferrous metal in rejects can be minimised through the use of a magnetic separator. Not all scrap metal dealers will accept some hi-tensile steels used and appropriate baling wires should be specified in wastepaper specifications. The use of standard pallets should be encouraged, to ensure that pallet waste can be kept to a minimum. Segregation of office wastes can allow collection by wastepaper merchants of papers unsuitable for recycling in the mill, together with rejected wastepapers.

13.2.7 Summary of environmental impacts

Overall the effects of increased recycling have three parts - those which have positive or those which have negative environmental outcomes and those which are not clear or are area- or site-dependent.
Negative impacts include:

- increased fossil fuel use;
- increased emissions (fossil fuels are dirtier than wood) of SO_2 and NO_x, in urban areas;
- increased net carbon dioxide emissions, from increased fossil fuel use;
- increased emissions in urban areas, due to the switch of transport

from forested areas to urban areas, leading to increased urban air pollution;
- reduced cash flow for forestry management activities.

Positive impacts include:

- reduced volume for disposal in landfills;
- increased yield from forest resources.

Impacts which have both positive and negative effects include:

- energy savings with some grades, though this is offset to an extent by increased emissions of air pollutants, such as SO_2, NO_x and the greenhouse gas CO_2;
- reductions in chemical oxygen demand (COD) discharges, though since COD discharge originates from atmospheric CO_2 the net effect is not clear, especially since the environmental effect of COD associated with a relatively low biological oxygen demand (BOD) is obscure.

Overall, it is clear that there is a balance between environmental costs and benefits and that at some level of recycling, the net result will be increased environmental costs. The level of recycling which results in net increased environmental costs will vary, since the inputs into the equation vary, with paper grade and geographic area. A blind belief in recycling, with no consideration of this equation, bodes ill for the environment. Better knowledge of optimal levels of recycling, with some consideration of the broad effects of increased recycling, would allow more environmentally efficient use of fibre resources. This could lead to, for example, the use of targets which allows the use of both virgin and recycled paper grades, avoiding the potential for environmental error implicit in both recycled content and wastepaper recovery targets.

13.3 Environmental management systems

In different geographic areas a variety of environmental management systems are being introduced, which include annual environmental audits, environmental management systems, environmental reviews etc. Environmental audits are reviews of actual performance in meeting emission limits, whether these are legislative or adopted voluntarily. Results may be published, similar to the familiar annual financial report, as an environmental statement. In the European Communities Ecoaudit system, these must be validated by independent environmental auditors and released to the public. This allows the use of the European Union (EU) Ecoaudit logo.

Environmental management systems have been introduced in the UK and

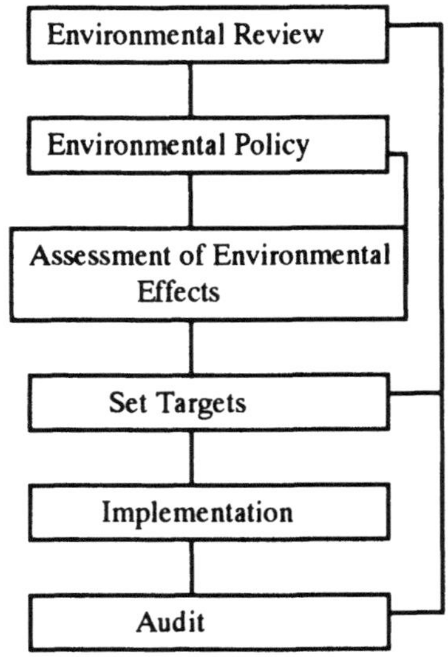

Figure 13.2 Implementation of an environmental management system - a schematic illustration.

Canada. In the UK, they are related to the ISO 9000 quality standards. In the UK, the equivalent British Standard to ISO 9000 is BS 5750 and the related environmental management standard is BS 7750. Under this system, it is essential to have a detailed and documented environmental management system, covering a variety of related topics, including definition of personnel responsibilities, employee training programmes, detailed environmental policy with targets, an environmental management manual, reviews of the system, etc. [23,24]. Implementation of BS 7750 is illustrated in Figure 13.2.

Although many of these environmental initiatives will encourage paper recycling, paper recycling mills will have to respond to them in a similar fashion to other industries, to demonstrate that paper recycling does not pose any unacceptable environmental risks, and in many cases is an environmentally (and economically) desirable use of fibre resources.

References

1. Anon, *Royal Commission on Sewage Disposal, 1901–1915*, Nine reports with Appendices and a Final Report in 1915, London, HMSO.
2. *Rivers (Prevention of Pollution) Act* (1951) London, HMSO.
3. *Clean Air Act* (1956) London, HMSO.
4. MacKenzie, D. (5th December 1992) Agreement reduces damage to ozone layer, *New Scientist*, p 10.
5. Financial Times Survey (March 16th, 1990) *Industry and the Environment.*

6. Virtanen, V. and Nilsson, S. (1993) *Environmental Impacts of Wastepaper Recycling*, International Institute for Applied Systems Analysis.
7. McKinney, R.W.J. (1991) Environmental issues in tissue production, *Valmet Tissue Making Symposium*, Valmet-Karlstad, Karlstad, Sweden.
8. Anon, Energy used in testliner and kraftliner, SCA Environmental Facts, No. 9, October 1990.
9. Pesonen, K.V. (1991) Recycled vs virgin - energy and manufacturing cost differentials: four hypothetical case studies, *Proceedings of Focus 95+ Symposium*, TAPPI Press.
10. Mylona, S., Trends of Sulphur Dioxide Emissions Air Concentrations and Depositions of Sulphur in Europe since 1880, Report 2/93 European Monitoring and Evaluation Programme (EMEP) MSc-W, Norway.
11. Anon (1970) *Air Pollution and Health*, Committee of the Royal College of Physicians of London, Pitman.
12. Flood, M. (Nov 1992) The resource cost of moving materials, *WARMER Bulletin*, No. 35.
13. Svensson, T. and Solyom, P. *Chlorine Bleached Pulp and Paper Products*, Kemikalieninspektionen, Stockholm, Sweden, Dur 40-380-38.
14. Magnusson, H. (1990) Dioxin and waste pulp, *Proceedings of PITA Conference*, PITA.
15. Nguyen, X.T., Shariff, A. and Jean, M. (1991) Impact of paper recycling on the environment and quality of paper and board products, *Proceedings of 1st Research Forum on Recycling* CPPA, Montreal.
16. Badar, A.T. (1993) Environmental impact of recycling in the paper industry, *Prog. Paper Recycling*, **2** (3), 42-52.
17. Miner, R., Somehwar, A., Weigand, P., Fischer, R., Berger, H., Borton, D. and Unwin, J. (1993) Characterisation of wastes and emissions from mills using recovered fibres, *Secondary Fibre Recycling*, Ed. Spangenberg, R.J., TAPPI Press.
18. McCubbin, N. *et al.* (1992) *Best Available Technology for the Ontario Pulp and Paper Industry*, Ontario Ministry of the Environment, Appendix A.
19. McKinney, R.W.J. (Feb 1993) Environmental labelling of paper products, *Prog. Paper Recycling*, **2**(2), 15-23.
20. Rathje, W. (Jan 1991) Dig with a difference, *WARMER Bulletin*, No 28, p. 4.
21. Rundle, H., Calcroft, M. and Holt, C. (1982) Agricultural disposal of sludges on a historic sludge disposal site, *Water Pollution Control*, **81**(5), 619-632.
22. Latva-Somppi, J., Tran, H.J., Barham, D. and Douglas, M.A. (1993) Characterisation of deinking sludge and its ashed residue, in *Proceedings of 2nd Research Forum on Recycling*, CPPA Technical Section.
23. McKinney, R.W.J. (1992) Consumers lead green revolution, *Pulp and Paper Internat.* **34**(11), 41-43.
24. Anon (1992) Specification for Environmental Management Systems, BS7750 British Standards Institute.

Index

MIX
Papier aus verantwortungsvollen Quellen
Paper from responsible sources
FSC® C105338

If you have any concerns about our products,
you can contact us on
ProductSafety@springernature.com

In case Publisher is established outside the EU,
the EU authorized representative is:
Springer Nature Customer Service Center GmbH
Europaplatz 3, 69115 Heidelberg, Germany

Printed by Libri Plureos GmbH
in Hamburg, Germany